CHAPTER 5 – THE NORMAL DISTRIBUTION

$$Z = \frac{Y - \bar{Y}}{s}$$

$$Y = \bar{Y} + Z(s)$$

CHAPTER 6 – SAMPLING AND SAMPLING DISTRIBUTIONS

$$K = \frac{\textit{Population Size}}{\textit{Sample Size}}$$

CHAPTER 7 – ESTIMATION

$$\sigma_{\bar{Y}} = \frac{\sigma}{\sqrt{N}}$$

$$CI = \bar{Y} \pm Z(\sigma_{\bar{Y}})$$

$$s_{\bar{Y}} = \frac{s}{\sqrt{N}}$$

$$\sigma_p = \sqrt{\frac{(\pi)(1-\pi)}{N}}$$

$$s_p = \sqrt{\frac{(p)(1-p)}{N}}$$

$$CI = p \pm Z\,(s_p)$$

CHAPTER 8 – TESTING HYPOTHESES

$$Z = \frac{\bar{Y} - \mu_Y}{\sigma / \sqrt{N}}$$

$$t = \frac{\bar{Y} - \mu}{s / \sqrt{N}}$$

$$df = N - 1$$

$$\sigma_{\bar{Y}_1 - \bar{Y}_2} = \sqrt{\frac{\sigma_1^2}{N_1} + \frac{\sigma_2^2}{N_2}}$$

$$s_{\bar{Y}_1 - \bar{Y}_2} = \sqrt{\frac{(N_1 - 1)s_1^2 + (N_2 - 1)s_2^2}{(N_1 + N_2) - 2}}\sqrt{\frac{N_1 + N_2}{N_1 N_2}}$$

$$t = \frac{\bar{Y}_1 - \bar{Y}_2}{s_{\bar{Y}_1 - \bar{Y}_2}}$$

$$df = (N_1 + N_2) - 2$$

ESSENTIALS OF SOCIAL STATISTICS FOR A DIVERSE SOCIETY

Fourth Edition

Sara Miller McCune founded SAGE Publishing in 1965 to support the dissemination of usable knowledge and educate a global community. SAGE publishes more than 1000 journals and over 800 new books each year, spanning a wide range of subject areas. Our growing selection of library products includes archives, data, case studies and video. SAGE remains majority owned by our founder and after her lifetime will become owned by a charitable trust that secures the company's continued independence.

Los Angeles | London | New Delhi | Singapore | Washington DC | Melbourne

ESSENTIALS OF SOCIAL STATISTICS FOR A DIVERSE SOCIETY

Fourth Edition

Anna Leon-Guerrero
Pacific Lutheran University

Chava Frankfort-Nachmias
University of Wisconsin–Milwaukee

Georgiann Davis
University of Nevada, Las Vegas

Los Angeles | London | New Delhi
Singapore | Washington DC | Melbourne

FOR INFORMATION:

SAGE Publications, Inc.
2455 Teller Road
Thousand Oaks, California 91320
E-mail: order@sagepub.com

SAGE Publications Ltd.
1 Oliver's Yard
55 City Road
London EC1Y 1SP
United Kingdom

SAGE Publications India Pvt. Ltd.
B 1/I 1 Mohan Cooperative Industrial Area
Mathura Road, New Delhi 110 044
India

SAGE Publications Asia-Pacific Pte. Ltd.
18 Cross Street #10-10/11/12
China Square Central
Singapore 048423

Acquisitions Editor: Helen Salmon
Editorial Assistant: Natalie Elliott
Production Editor: Andrew Olson
Typesetter: Hurix Digital
Proofreader: Larry Baker
Indexer: Integra
Cover Designer: Candice Harman
Marketing Manager: Shari Countryman

Printed in Canada

ISBN: 978-1-5443-7250-1

This book is printed on acid-free paper.

20 21 22 23 24 10 9 8 7 6 5 4 3 2 1

MIX
Paper from responsible sources
FSC® C103567
FSC www.fsc.org

BRIEF CONTENTS

*Website to be found at: **https://edge.sagepub.com/ssdsess4e**

DETAILED CONTENTS

CHAPTER 7 • Testing Hypotheses 201

---

*Website to be found at: **https://edge.sagepub.com/ssdsess4e**

PREFACE

You may be reading this introduction on your first day of class. We know you have some questions and concerns about what your course will be like. Math, formulas, and calculations? Yes, those will be part of your learning experience. But there is more.

Throughout our text, we highlight the relevance of statistics in our daily and professional lives. Data are used to predict public opinion, consumer spending, and even a presidential election. How Americans feel about a variety of political and social topics—the Black Lives Matter movement, gun control, immigration, the economy, health care reform, or terrorism—are measured by surveys and polls and reported daily by the news media. Your recent Amazon purchase of hand sanitizer during the COVID-19 pandemic didn't go unnoticed. The study of consumer trends, specifically focusing on young adults, helps determine commercial programming, product advertising and placement, and, ultimately, consumer spending. And as we prepare this text, the world struggles to comprehend political divides around the seriousness of COVID-19, police brutality, migration patterns, and transgender rights, among other things.

Statistics are not just a part of our lives in the form of news bits or information. And it isn't just numbers either. As social scientists, we rely on statistics to help us understand our social world. We use statistical methods and techniques to track demographic trends, to assess social differences, and to better inform social policy. We encourage you to move beyond just being a consumer of statistics and determine how you can use statistics to gain insight into important social issues that affect you and others.

TEACHING AND LEARNING GOALS

Three teaching and learning goals continue to be the guiding principles of our book, as they were in previous editions.

Our first goal is to introduce you to social statistics and demonstrate its value. Although most of you will not use statistics in your own student research, you will be expected to read and interpret statistical information presented by others in professional and scholarly publications, in the workplace, and in the popular media. This book will help you understand the concepts behind the statistics so that you will be able to assess the circumstances in which certain statistics should and should not be used.

A special characteristic of this book is its integration of statistical techniques with substantive issues of particular relevance in the social sciences. Our second goal is to demonstrate that substance and statistical techniques are truly related in social science research. Your learning will not be limited to statistical calculations and formulas. Rather, you will become proficient in statistical techniques while learning about social differences and inequality through numerous substantive examples and real-world data applications. Because the world we live in is characterized by a growing diversity—where personal and social realities are increasingly shaped by race, class, gender, and other categories of experience—this book teaches you basic statistics while incorporating social science research related to the dynamic interplay of our social worlds.

Our third goal is to enhance your learning by using straightforward prose to explain statistical concepts and by emphasizing intuition, logic, and common sense over rote memorization and derivation of formulas.

DISTINCTIVE AND UPDATED FEATURES OF OUR BOOK

Our learning goals are accomplished through a variety of specific and distinctive features throughout this book.

A Close Link Between the Practice of Statistics, Important Social Issues, and Real-World Examples. This book is distinct for its integration of statistical techniques with pressing social issues of particular concern to society and social science. We emphasize how the conduct of social science is the constant interplay between social concerns and methods of inquiry. In addition, the examples throughout the book—mostly taken from news stories, government reports, public opinion polls, scholarly research, and the National Opinion Research Center's General Social Survey—are formulated to emphasize to students like you that we live in a world in which statistical arguments are common. Statistical concepts and procedures are illustrated with real data and research, providing a clear sense of how questions about important social issues can be studied with various statistical techniques.

A Focus on Diversity: The United States and International. A strong emphasis on race, class, and gender as central substantive concepts is mindful of a trend in the social sciences toward integrating issues of diversity in the curriculum. This focus on the richness of social differences within our society and our global neighbors is manifested in the application of statistical tools to examine how race, class, gender, and other categories of experience shape our social world and explain social behavior.

Chapter Reorganization and Content. Each revision presents many opportunities to polish and expand the content of our text. In this edition, we have made a number of changes in response to feedback from reviewers and fellow instructors. Measures of central tendency and variability are covered in a single chapter. We also continued to refine our discussion of the interpretation and application of descriptive statistics (variance and standard deviation) and inferential tests (t, Z, F ratio, and regression and correlation). Chapter Exercises do not require the use of computer software.

Reading the Research Literature, Statistics in Practice, A Closer Look, and Data at Work. In your student career and in the workplace, you may be expected to read and interpret statistical information presented by others in professional and scholarly publications. These statistical analyses are a good deal more complex than most class and textbook presentations. To guide you in reading and interpreting research reports written by social scientists, most of our chapters include a Reading the Research Literature and a Statistics in Practice feature, presenting excerpts of published research reports or specific SPSS calculations using the statistical concepts under discussion. Being statistically literate involves more than just completing a calculation; it also includes learning how to apply and interpret statistical information and being able to say what it means. We include an A Closer Look discussion in each chapter, advising students about the common errors and limitations in quantitative data collection and analysis.

SPSS, Excel, and GSS 2018. IBM® SPSS® Statistics[1] and Microsoft Excel[2] are used throughout this book, although the use of computers is not required to learn from the text. Real data are used to motivate and make concrete the coverage of statistical topics. As a companion to the fourth edition's SPSS and Excel demonstrations and exercises, we provide three GSS 2018 data sets on the study site at **https://edge.sagepub.com/ssdsess4e.** Two of the GSS data sets (GSS18SSDS-A and GSS18SSDS-B) are formatted for SPSS, while the third data set (GSS18SSDS-E) is ready for use in Excel. These demonstrations and exercises, available on the study site, rely on variables from these modules. There is ample opportunity for instructors to develop their own exercises using these data.

Tools to Promote Effective Study. Each chapter concludes with a list of Main Points and Key Terms discussed in that chapter. Boxed definitions of the Key Terms also appear in the body of the chapter, as do Learning Checks keyed to the most important points. Key Terms are also clearly defined and explained in the Glossary, another special feature in our book. Answers to all the Odd-Numbered Exercises and Learning Checks in the text are included at the end of the book, as well as on the study site at **https://edge.sagepub.com/ssdsess4e.** Complete step-by-step solutions are provided in the Instructor's Manual, available on the study site.

A NOTE ABOUT ROUNDING

Throughout this text and in ancillary materials, we followed these rounding rules: If the number you are rounding is followed by 5, 6, 7, 8, or 9, round the number up. If the number you are rounding is followed by 0, 1, 2, 3, or 4, do not change the number. For rounding long decimals, look only at the number in the place you are rounding to and the number that follows it.

[1]SPSS is a registered trademark of International Business Machines Corporation.
[2]Microsoft Excel is a registered trademark of Microsoft Corporation.

On the companion website at: **https://edge.sagepub.com/ssdsess4e**

Instructor site:

- LMS-ready test bank, also available in Microsoft Word
- Editable PowerPoint slides
- Instructors Manual: lecture notes, and answers to all end-of-chapter questions from the book, and the SPSS and Excel problems posted on the student study site. Answers to odd-numbered questions are in the back of the book for students to check their answers.
- Datasets and codebooks
- SPSS and Excel demonstrations and problems to accompany each chapter

Student site:

- eFlashcards of the glossary terms
- Datasets and codebooks
- SPSS and Excel walk-through videos
- SPSS and Excel demonstrations and problems to accompany each chapter
- Appendix F: Basic Math Review

ACKNOWLEDGMENTS

We are grateful to Jeff Lasser and Helen Salmon, publishers for SAGE Publications, for their commitment to our book and for their invaluable assistance through the production process.

We are grateful to Andrew Olson, Kate Russillo, and Gillian Dickens for guiding the book through the production process. We would also like to acknowledge Tiara Beatty, Tara Slagle, Shelly Gupta, and the rest of the SAGE staff for their assistance on this edition.

We are especially indebted to Torisha Khonach, a sociology PhD student at the University of Nevada, Las Vegas, for offering us a fresh pair of eyes throughout the revision process. We deeply appreciate her meticulousness.

Anna Leon-Guerrero would like to thank her Pacific Lutheran University students for inspiring her to be a better teacher. My love and thanks to my husband, Brian Sullivan.

Chava Frankfort-Nachmias would like to thank and acknowledge her friends and colleagues for their unending support; she also would like to thank her students: I am grateful to my students at the University of Wisconsin–Milwaukee, who taught me that even the most complex statistical ideas can be simplified. The ideas presented in this book are the products of many years of classroom testing. I thank my students for their patience and contributions. Finally, I thank my partner, Marlene Stern, for her love and support.

Georgiann Davis would like to thank her students, colleagues, and mentors at the various universities she has been affiliated with over the years for offering her the space to learn and enjoy statistics: I'm especially thankful for my students whose excitement, and sometimes anxiety, continues to nurture my love for quantitative data analysis. A special thank you to my partner, who sometimes shares my love for statistics despite being a hardcore ethnographer.

Anna Leon-Guerrero
Pacific Lutheran University

Chava Frankfort-Nachmias
University of Wisconsin–Milwaukee

Georgiann Davis
University of Nevada, Las Vegas

ABOUT THE AUTHORS

Anna Leon-Guerrero is Professor of Sociology at Pacific Lutheran University in Washington. She received her PhD in sociology from the University of California–Los Angeles. A recipient of the university's Faculty Excellence Award and the K. T. Tang Award for Excellence in Research, she teaches courses in statistics, social theory, and social problems. She is also the author of *Social Problems: Community, Policy, and Social Action*.

Chava Frankfort-Nachmias is an Emeritus Professor of Sociology at the University of Wisconsin–Milwaukee. She is the coauthor of *Research Methods in the Social Sciences* (with David Nachmias), coeditor of *Sappho in the Holy Land* (with Erella Shadmi), and numerous publications on ethnicity and development, urban revitalization, science and gender, and women in Israel. She was the recipient of the University of Wisconsin System teaching improvement grant on integrating race, ethnicity, and gender into the social statistics and research methods curriculum.

Georgiann Davis is Associate Professor of Sociology at the University of Nevada, Las Vegas. An award-winning instructor, researcher, and scholar-activist, she received her PhD in sociology from the University of Illinois at Chicago. She is the author of *Contesting Intersex: The Dubious Diagnosis*.

1 THE WHAT AND THE WHY OF STATISTICS

Chapter Learning Objectives

1. Describe the five stages of the research process.
2. Define independent and dependent variables.
3. Distinguish between the three levels of measurement.
4. Apply descriptive and inferential statistical procedures.

Are you taking statistics because it is required in your major—not because you find it interesting? If so, you may be feeling intimidated because you associate statistics with numbers, formulas, and abstract notations that seem inaccessible and complicated. Perhaps you feel intimidated not only because you're uncomfortable with math but also because you suspect that numbers and math don't leave room for human judgment or have any relevance to your own personal experience. In fact, you may even question the relevance of statistics to understanding people, social behavior, or society.

In this book, we will show you that statistics can be a lot more interesting and easier to understand than you may have been led to believe. In fact, as we draw on your previous knowledge and experience and relate statistics to interesting and important social issues, you'll begin to see that statistics is not just a course you have to take but a useful tool as well.

There are two reasons why learning statistics may be of value to you. First, you are constantly exposed to statistics every day of your life. Marketing surveys, voting polls, and social research findings appear daily in the news media. By learning statistics, you will become a sharper consumer of statistical material. Second, as a major in the social sciences, you may be expected to read and interpret statistical information related to your occupation or work. Even if conducting research is not a part of your work, you may still be expected to understand and learn from other people's research or to be able to write reports based on statistical analyses.

Just what is statistics, anyway? You may associate the word with numbers that indicate COVID-19 hospitalization rates, support for the Black Lives Matter movement, and so on. But the word statistics also refers to a set of procedures used by

Statistics: A set of procedures used by social scientists to organize, summarize, and communicate numerical information.

social scientists to organize, summarize, and communicate numerical information. Only information represented by numbers can be the subject of statistical analysis. Such information is called **data**; researchers use statistical procedures to analyze data to answer research questions and test theories. It is the latter usage—answering research questions and testing theories—that this textbook explores.

Data: Information represented by numbers, which can be the subject of statistical analysis.

THE RESEARCH PROCESS

To give you a better idea of the role of statistics in social research, let's start by looking at the **research process**. We can think of the research process as a set of activities in which social scientists engage so that they can answer questions, examine ideas, or test theories.

Research process: A set of activities in which social scientists engage to answer questions, examine ideas, or test theories.

As illustrated in Figure 1.1, the research process consists of five stages:

1. Asking the research question
2. Formulating the hypotheses
3. Collecting data
4. Analyzing data
5. Evaluating the hypotheses

Figure 1.1 The Research Process

Asking the research question
Formulating the hypotheses
THEORY
Evaluating the hypotheses
Analyzing data
Collecting data

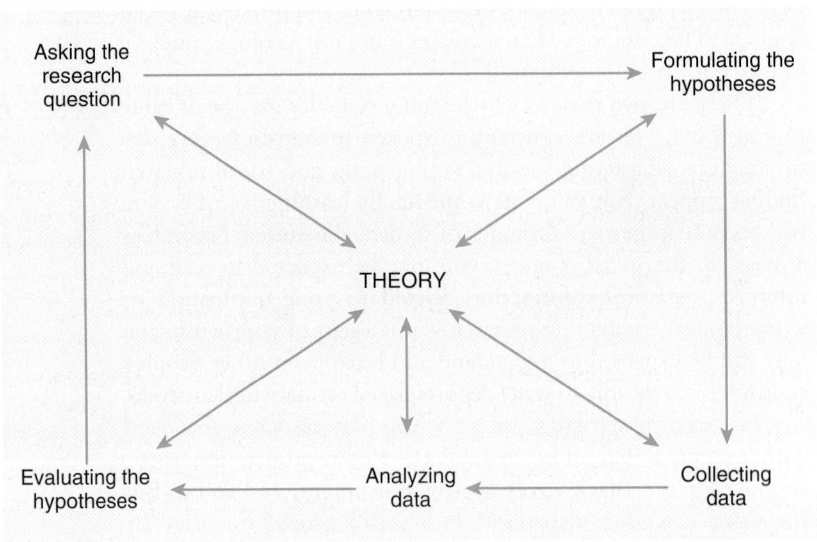

Each stage affects the theory and is affected by it as well. Statistics is most closely tied to the data analysis stage of the research process. As we will see in later chapters, statistical analysis of the data helps researchers test the validity and accuracy of their hypotheses.

ASKING RESEARCH QUESTIONS

The starting point for most research is asking a research question. Consider the following research questions taken from several social science journals:

> How does the expansion of police presence in poor urban communities affect educational outcomes?
>
> What does it mean to be a wounded warrior and how does the term impact the way wounded veterans think about themselves?
>
> How do Lebanese women use their informal social networks to engage in political activism for women's rights?
>
> What factors affect the economic mobility of female workers?

These are all questions that can be answered by conducting **empirical research**—research based on information that can be verified by using our direct experience. To answer research questions, we cannot rely on reasoning, speculation, moral judgment, or subjective preference. For example, the questions "Is racial equality good for society?" and "Is an urban lifestyle better than a rural lifestyle?" cannot be answered empirically because the terms good and better are concerned with values, beliefs, or subjective preference and, therefore, cannot be independently verified. One way to study these questions is by defining good and better in terms that can be verified empirically. For example, we can define good in terms of economic growth and better in terms of psychological well-being. These questions could then be answered by conducting empirical research.

Empirical research: A research based on evidence that can be verified by using our direct experience.

You may wonder how to come up with a research question. The first step is to pick a question that interests you. If you are not sure, look around! Ideas for research problems are all around you, from media sources to personal experience or your own intuition. Talk to other people, write down your own observations and ideas, or learn what other social scientists have written about.

Take, for instance, the relationship between gender and work. As a college student about to enter the labor force, you may wonder about the similarities and differences between women's and men's work experiences and about job opportunities when you graduate. Here are some facts and observations based on research reports: In 2018, women who were employed full-time earned about $794 (in current dollars) per week on average; men who were employed full-time earned $993 (in current dollars) per week on average.[1] Women's and men's work are also very different. Women continue to be the minority in many of the higher-ranking and higher-salaried positions in professional and managerial occupations. For example, in 2017, women made up 18.4% of software developers and 28% of chief executives. In comparison, among all those employed as secretaries and administrative assistants, 96% were women. Among all receptionists and information clerks in 2017, 93% were women.[2] These observations may prompt us to ask research questions such as the following: How much change has there been in women's work over time? Are women paid, on average, less than men for the same type of work?

LEARNING CHECK 1.1

Identify one or two social science questions amenable to empirical research. You can almost bet that you will be required to do a research project sometime in your college career.

THE ROLE OF THEORY

You may have noticed that each preceding research question was expressed in terms of a relationship. This relationship may be between two or more attributes of individuals or groups, such as gender and income or gender segregation in the workplace and income disparity. The relationship between attributes or characteristics of individuals and groups lies at the heart of social scientific inquiry.

Most of us use the term theory quite casually to explain events and experiences in our daily life. You may have a theory about why your roommate has been so nice to you lately or why you didn't do so well on your last exam. In a somewhat similar manner, social scientists attempt to explain the nature of social reality. Whereas our theories about events in our lives are commonsense explanations based on educated guesses and personal experience, to the social scientist, a theory is a more precise explanation that is frequently tested by conducting research.

Theory: A set of assumptions and propositions used to explain, predict, and understand social phenomena.

A **theory** is a set of assumptions and propositions used by social scientists to explain, predict, and understand the phenomena they study.[3] The theory attempts to establish a link between what we observe (the data) and our conceptual understanding of why certain phenomena are related to each other in a particular way.

For instance, suppose we wanted to understand the reasons for the income disparity between men and women; we may wonder whether the types of jobs men and women have and the organizations in which they work have something to do with their wages. One explanation for gender wage inequality is gender segregation in the workplace—the fact that American men and women are concentrated in different kinds of jobs and occupations. What is the significance of gender segregation in the workplace? In our society, people's occupations and jobs are closely associated with their level of prestige, authority, and income. The jobs in which women and men are segregated are not only different but also unequal. Although the proportion of women in the labor force has markedly increased, women are still concentrated in occupations with low pay, low prestige, and few opportunities for promotion. Thus, gender segregation in the workplace is associated with unequal earnings, authority, and status. In particular, women's segregation into different jobs and occupations from those of men is the most immediate cause of the pay gap. Women receive lower pay than men do even when they have the same level of education, skill, and experience as men in comparable occupations.

FORMULATING THE HYPOTHESES

So far, we have come up with several research questions about the income disparity between men and women in the workplace. We have also discussed a possible explanation—a theory—that helps us make sense of gender inequality in wages. Is that enough? Where do we go from here?

Our next step is to test some of the ideas suggested by the gender segregation theory. But this theory, even if it sounds reasonable and logical to us, is too general and does not contain enough specific information to be tested. Instead, theories suggest specific concrete predictions or **hypotheses** about the way that observable attributes of people or groups are interrelated in real life. Hypotheses are tentative because they can be verified only after they have been tested empirically.[4] For example, one hypothesis we can derive from the gender segregation theory is that wages in occupations in which the majority of workers are female are lower than the wages in occupations in which the majority of workers are male.

Hypothesis: A statement predicting the relationship between two or more observable attributes.

Not all hypotheses are derived directly from theories. We can generate hypotheses in many ways—from theories, directly from observations, or from intuition. Probably, the greatest source of hypotheses is the professional or scholarly literature. A critical review of the scholarly literature will familiarize you with the current state of knowledge and with hypotheses that others have studied.

Let's restate our hypothesis:

> Wages in occupations in which the majority of workers are female are lower than the wages in occupations in which the majority of workers are male.

Variable: A property of people or objects that takes on two or more values.

Note that this hypothesis is a statement of a relationship between two characteristics that vary: wages and gender composition of occupations. Such characteristics are called variables. A **variable** is a property of people or objects that takes on two or more values. For example, people can be classified into a number of social class categories, such as upper class, middle class, or working class. Family income is a variable; it can take on values from zero to hundreds of thousands of dollars or more. Similarly, gender composition is a variable. The percentage of females (or males) in an occupation can vary from 0 to 100. Wages is a variable, with values from zero to thousands of dollars or more. See Table 1.1 for examples of some variables and their possible values.

Unit of analysis: The object of research, such as individuals, groups, organizations, or social artifacts.

Social scientists must also select a **unit of analysis**; that is, they must select the object of their research. We often focus on individual characteristics or behavior, but we could also examine groups of people such as families, formal organizations like elementary schools or corporations, or social artifacts such as children's books or advertisements. For example, we may be interested in the relationship between an individual's educational degree and annual income. In this case, the unit of analysis is the individual. On the other hand, in a study of how corporation profits are associated with employee benefits, corporations are the unit of analysis. If we examine how often women are featured in prescription drug advertisements, the advertisements are the unit of analysis. Figure 1.2 illustrates different units of analysis frequently employed by social scientists.

Table 1.1 Variables and Value Categories

Variable	Categories
Social class	Lower Working Middle Upper
Gender	Male Female
Education	Less than high school High school Some college College graduate

Figure 1.2 Examples of Units of Analysis

Individual as unit of analysis:

How old are you?
What are your political views?
What is your occupation?

Family as unit of analysis:

How many children are in the family?
Who does the housework?
How many wage earners are there?

Organization as unit of analysis:

How many employees are there?
What is the gender composition?
Do you have a diversity office?

City as unit of analysis:

What was the crime rate last year?
What is the population density?
What type of government runs things?

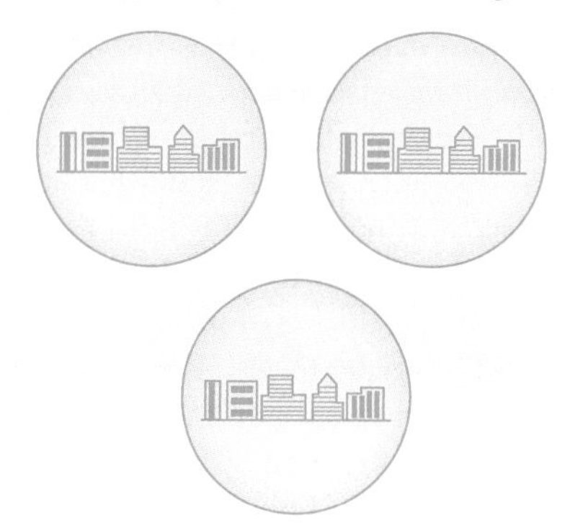

LEARNING CHECK 1.2

Remember that research question you came up with? Formulate a testable hypothesis based on your research question. Remember that your variables must take on two or more values and you must determine the unit of analysis. What is your unit of analysis?

Independent and Dependent Variables: Causality

Hypotheses are usually stated in terms of a relationship between an independent and a dependent variable. The distinction between an independent and a dependent variable is important in the language of research. Social theories often intend to provide an explanation for social patterns or causal relations between variables. For example, according to the gender segregation theory, gender segregation in the workplace is the primary explanation (although certainly not the only one) of the male-female earning gap. Why should jobs where the majority of workers are women pay less than jobs that employ mostly men? One explanation is that

> societies undervalue the work women do, regardless of what those tasks are, because women do them. . . . For example, our culture tends to devalue caring or nurturant work at least partly because women do it. This tendency accounts for childcare workers' low rank in the pay hierarchy.[5]

Dependent variable: The variable to be explained (the "effect").

Independent variable: The variable expected to account for (the "cause" of) the dependent variable.

In the language of research, the variable the researcher wants to explain ("the effect") is called the **dependent variable**. The variable that is expected to "cause" or account for the dependent variable is called the **independent variable**. Therefore, in our example, *gender composition of occupations* is the independent variable, and *wages* is the dependent variable.

Cause-and-effect relationships between variables are not easy to infer in the social sciences. To establish that two variables are causally related, your analysis must meet three conditions: (1) The cause has to precede the effect in time, (2) there has to be an empirical relationship between the cause and the effect, and (3) this relationship cannot be explained by other factors.

Let's consider the decades-old debate about controlling crime through the use of prevention versus punishment. Some people argue that special counseling for youths at the first sign of trouble and strict controls on access to firearms would help reduce crime. Others argue that overhauling federal and state sentencing laws to stop early prison releases is the solution. In the early 1990s, Washington and California adopted "three strikes and you're out" legislation, imposing life prison terms on three-time felony offenders. Such laws are also referred to as habitual or persistent offender laws. Twenty-six other states and the federal government adopted similar measures, all advocating a "get tough" policy on crime; the most recent legislation was in 2012 in the state of Massachusetts. In 2012, California voters supported a revision to the original law, imposing a life sentence only when the new felony conviction is serious or violent. Let's suppose that years after the measure was introduced, the crime rate declined in some of these states (in fact, advocates of the measure have identified declining crime rates as evidence of its success). Does the observation that the incidence of crime declined mean that the new measure caused this reduction? Not necessarily! Perhaps the rate of crime had been going down for other reasons, such as improvement in the economy, and the new measure had nothing to do with it. To demonstrate a cause-and-effect relationship, we would need to show three things: (1) The reduction of crime actually occurred after the enactment of this measure, (2) the enactment of the "three strikes and you're out" measure was empirically associated with a decrease in crime, and (3) the relationship between the reduction in crime and the "three strikes and you're out" policy is not due to the influence of another variable (e.g., the improvement of overall economic conditions).

Independent and Dependent Variables: Guidelines

Because it is difficult to infer cause-and-effect relationships in the social sciences, be cautious about using the terms cause and effect when examining relationships between variables. However, using the terms independent variable and dependent variable is still appropriate even when this relationship is not articulated in terms of direct cause and effect. Here are a few guidelines that may help you identify the independent and dependent variables:

1. The dependent variable is always the property that you are trying to explain; it is always the object of the research.
2. The independent variable usually occurs earlier in time than the dependent variable.
3. The independent variable is often seen as influencing, directly or indirectly, the dependent variable.

The purpose of the research should help determine which is the independent variable and which is the dependent variable. In the real world, variables are neither dependent nor independent; they can be switched around depending on the research problem. A variable defined as independent in one research investigation may be a dependent variable in another.[6] For instance, *educational attainment* may be an independent variable in a study attempting to explain how education influences political attitudes. However, in an investigation of whether a person's level of education is influenced by the social status of his or her family of origin, *educational attainment* is the dependent variable. Some variables, such as race, age, and ethnicity, because they are primordial characteristics that cannot be explained by social scientists, are never considered dependent variables in a social science analysis.

LEARNING CHECK 1.3

Identify the independent and dependent variables in the following hypotheses:

- Older Americans are more likely to support stricter immigration laws than younger Americans.

(Continued)

(Continued)

- People who attend church regularly are more likely to oppose abortion than people who do not attend church regularly.
- Elderly women are more likely to live alone than elderly men.
- Individuals with postgraduate education are likely to have fewer children than those with less education.

What are the independent and dependent variables in your hypothesis?

COLLECTING DATA

Once we have decided on the research question, the hypothesis, and the variables to be included in the study, we proceed to the next stage in the research cycle. This step includes measuring our variables and collecting the data. As researchers, we must decide how to measure the variables of interest to us, how to select the cases for our research, and what kind of data collection techniques we will be using. A wide variety of data collection techniques are available to us, from direct observations to survey research, experiments, or secondary sources. Similarly, we can construct numerous measuring instruments. These instruments can be as simple as a single question included in a questionnaire or as complex as a composite measure constructed through the combination of two or more questionnaire items. The choice of a particular data collection method or instrument to measure our variables depends on the study objective. For instance, suppose we decide to study how one's social class is related to attitudes about women in the labor force. Since attitudes about working women are not directly observable, we need to collect data by asking a group of people questions about their attitudes and opinions. A suitable method of data collection for this project would be a survey that uses a questionnaire or interview guide to elicit verbal reports from respondents. The questionnaire could include numerous questions designed to measure attitudes toward working women, social class, and other variables relevant to the study.

How would we go about collecting data to test the hypothesis relating the gender composition of occupations to wages? We want to gather information on the proportion of men and women in different occupations and the average earnings for these occupations. This kind of information is routinely collected and disseminated by the U.S. Department of Labor, the Bureau of Labor Statistics, and the U.S. Census Bureau. We could use these data to test our hypothesis.

Levels of Measurement

The statistical analysis of data involves many mathematical operations, from simple counting to addition and multiplication. However, not every operation can be used with every variable. The type of statistical operation we employ depends on how our variables are measured. For example, for the variable *gender*, we can use the number 1 to represent females and the number 2 to represent males. Similarly, 1 can also be used as a numerical code for the category "one child" in the variable *number of children*. Clearly, in the first example, the number is an arbitrary symbol that does not correspond to the property "female," whereas in the second example, the number 1 has a distinct numerical meaning that does correspond to the property "one child." The correspondence between the properties we measure and the numbers representing these properties determines the type of statistical operations we can use. The degree of correspondence also leads to different ways of measuring—that is, to distinct levels of measurement. In this section, we will discuss three levels of measurement: (1) nominal, (2) ordinal, and (3) interval-ratio.

Nominal Level of Measurement

With a **nominal level of measurement**, numbers or other symbols are assigned a set of categories for the purpose of naming, labeling, or classifying the observations. *Gender* is an example of a nominal-level variable (Table 1.2). Using the numbers 1 and 2, for instance, we can classify our observations into the categories "females" and "males," with 1 representing females and 2 representing males. We could use any of a variety of symbols to represent the different categories of a nominal variable; however, when numbers are used to represent the different categories, we do not imply anything about the magnitude or quantitative difference between the categories. Nominal categories cannot be rank-ordered. Because the different categories (e.g., males vs. females) vary in the quality inherent in each but not in quantity, nominal variables are often

Nominal level of measurement: Numbers or other symbols are assigned to a set of categories for the purpose of naming, labeling, or classifying the observations. Nominal categories cannot be rank-ordered.

Table 1.2 Nominal Variables and Value Categories

Variable	Categories
Gender	Male Female
Religion	Protestant Christian Jewish Muslim
Marital status	Married Single Widowed Other

called qualitative. Other examples of nominal-level variables are political party, religion, and race.

Nominal variables should include categories that are both exhaustive and mutually exclusive. Exhaustiveness means that there should be enough categories composing the variables to classify every observation. For example, the common classification of the variable *marital status* into the categories "married," "single," and "widowed" violates the requirement of exhaustiveness. As defined, it does not allow us to classify same-sex couples or heterosexual couples who are not legally married. We can make every variable exhaustive by adding the category "other" to the list of categories. However, this practice is not recommended if it leads to the exclusion of categories that have theoretical significance or a substantial number of observations.

Mutual exclusiveness means that there is only one category suitable for each observation. For example, we need to define religion in such a way that no one would be classified into more than one category. For instance, the categories Protestant and Methodist are not mutually exclusive because Methodists are also considered Protestant and, therefore, could be classified into both categories.

LEARNING CHECK 1.4

Review the definitions of exhaustive and mutually exclusive. Now look at Table 1.2. What other categories could be added to each variable to be exhaustive and mutually exclusive?

Ordinal Level of Measurement

Ordinal level of measurement: Numbers are assigned to rank-ordered categories ranging from low to high or high to low.

Whenever we assign numbers to rank-ordered categories ranging from low to high or high to low, we have an **ordinal level of measurement**. *Social class* is an example of an ordinal variable. We might classify individuals with respect to their social class status as "upper class," "middle class," or "working class." We can say that a person in the category "upper class" has a higher class position than a person in a "middle-class" category (or that a "middle-class" position is higher than a "working-class" position), but we do not know the magnitude of the differences between the categories—that is, we don't know how much higher "upper class" is compared with the "middle class."

Many attitudes that we measure in the social sciences are ordinal-level variables. Take, for instance, the following statement used to measure attitudes toward working women: "Women should return to their traditional role in society." Respondents are asked to identify the number representing their degree of agreement or disagreement with this statement. One form in which a number might be made to correspond with the answers can be seen in

Table 1.3 Ordinal Ranking Scale	
Rank	**Value**
1	Strongly agree
2	Agree
3	Neither agree nor disagree
4	Disagree
5	Strongly disagree

Table 1.3. Although the differences between these numbers represent higher or lower degrees of agreement with the statement, the distance between any two of those numbers does not have a precise numerical meaning.

Like nominal variables, ordinal variables should include categories that are mutually exhaustive and exclusive.

Interval-Ratio Level of Measurement

If the categories (or values) of a variable can be rank-ordered and if the measurements for all the cases are expressed in the same units and equally spaced, then an **interval-ratio level of measurement** has been achieved. Examples of variables measured at the interval-ratio level are *age*, *income*, and *SAT scores*. With all these variables, we can compare values not only in terms of which is larger or smaller but also in terms of how much larger or smaller one is compared with another. In some discussions of levels of measurement, you will see a distinction made between interval-ratio variables that have a natural zero point (where zero means the absence of the property) and those variables that have zero as an arbitrary point. For example, weight and length have a natural zero point, whereas temperature has an arbitrary zero point. Variables with a natural zero point are also called *ratio variables*. In statistical practice, however, ratio variables are subjected to operations that treat them as interval and ignore their ratio properties. Therefore, we make no distinction between these two types in this text.

Interval-ratio level of measurement: Measurements for all cases are expressed in the same units and equally spaced. Interval-ratio values can be rank-ordered.

Cumulative Property of Levels of Measurement

Variables that can be measured at the interval-ratio level of measurement can also be measured at the ordinal and nominal levels. As a rule, properties that can be measured at a higher level (interval-ratio is the highest) can also be measured at lower levels, but not vice versa. Let's take, for example, *gender composition of occupations*, the independent variable in our research example. Table 1.4 shows the percentage of women in five major occupational groups.

Table 1.4 Gender Composition of Five Major Occupational Groups, 2018	
Occupational Group	**Women in Occupation (%)**
Management, professional, and related occupations	51.5
Service occupations	57.5
Production, transportation, and materials occupations	23.1
Sales and office occupations	61.1
Natural resources, construction, and maintenance occupations	5.1

Source: U.S. Department of Labor, 2018, Labor Force Statistics from the Current Population Survey 2018, Table 11.

The variable *gender composition* (measured as the percentage of women in the occupational group) is an interval-ratio variable and, therefore, has the properties of nominal, ordinal, and interval-ratio measures. For example, we can say that the management group differs from the natural resources group (a nominal comparison), that service occupations have more women than the other occupational categories (an ordinal comparison), and that service occupations have 34.4 percentage points more women (57.5–23.1) than production occupations (an interval-ratio comparison).

The types of comparisons possible at each level of measurement are summarized in Table 1.5 and Figure 1.3. Note that differences can be established at each of the three levels, but only at the interval-ratio level can we establish the magnitude of the difference.

Levels of Measurement of Dichotomous Variables

Dichotomous variable: A variable that has only two values.

A variable that has only two values is called a **dichotomous variable**. Several key social factors, such as gender, employment status, and marital status, are dichotomies—that is, you are male or female, employed or unemployed, married or not married. Such variables may seem to be measured at the nominal level: You fit in either one category or the other. No category is naturally higher or lower than the other, so they can't be ordered.

Table 1.5 Levels of Measurement and Possible Comparisons			
Level	**Different or Equivalent**	**Higher or Lower**	**How Much Higher**
Nominal	Yes	No	No
Ordinal	Yes	Yes	No
Interval-ratio	Yes	Yes	Yes

Figure 1.3 Levels of Measurement and Possible Comparisons: Education Measured on Nominal, Ordinal, and Interval-Ratio Levels

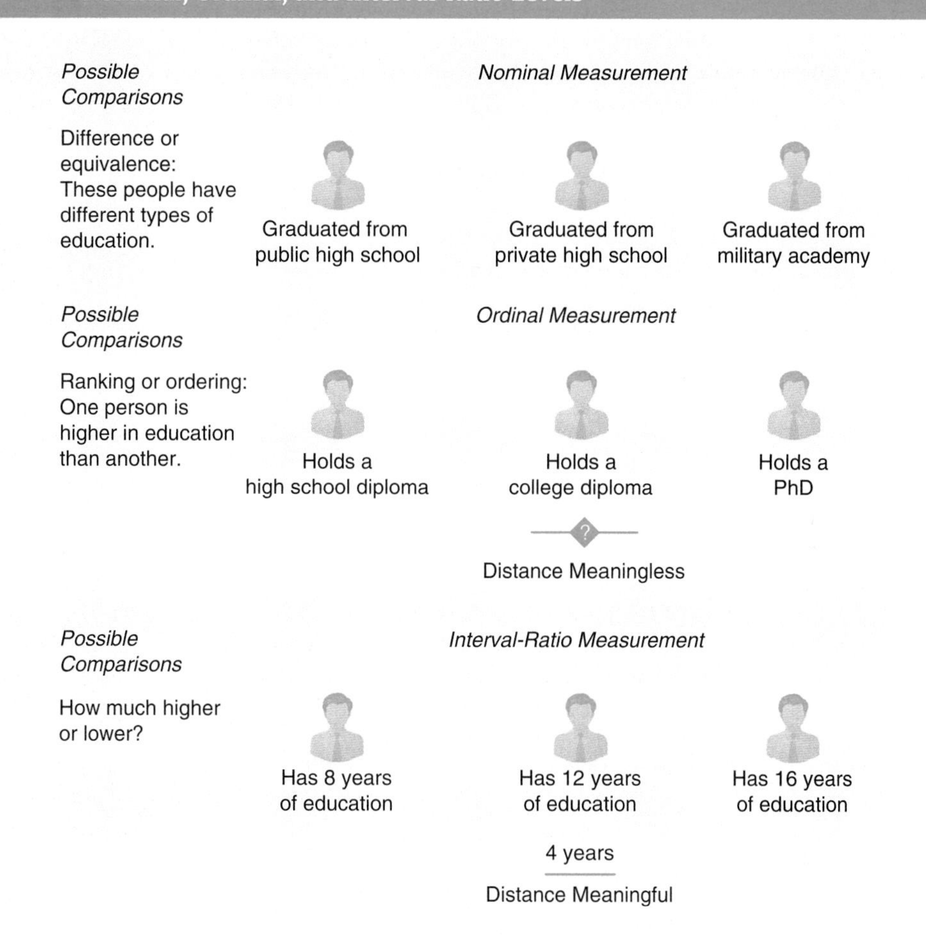

However, because there are only two possible values for a dichotomy, we can measure it at the ordinal or the interval-ratio level. For example, we can think of "femaleness" as the ordering principle for gender, so that "female" is higher and "male" is lower. Using "maleness" as the ordering principle, "female" is lower and "male" is higher. In either case, with only two classes, there is no way to get them out of order; therefore, gender could be considered at the ordinal level.

Dichotomous variables can also be considered to be interval-ratio level. Why is this? In measuring interval-ratio data, the size of the interval between the categories is meaningful: The distance between 4 and 7, for example, is the same as the distance between 11 and 14. But with a dichotomy, there is only one interval. Therefore, there is really no other distance to which we can compare it.

Mathematically, this gives the dichotomy more power than other nominal-level variables (as you will notice later in the text).

For this reason, researchers often dichotomize some of their variables, turning a multicategory nominal variable into a dichotomy. For example, you may see race dichotomized into "white" and "nonwhite." Though we would lose the ability to examine each unique racial category and we may collapse categories that are not similar, it may be the most logical statistical step to take. When you dichotomize a variable, be sure that the two categories capture a distinction that is important to your research question (e.g., a comparison of the number of white vs. nonwhite U.S. senators).

LEARNING CHECK 1.5

Make sure you understand these levels of measurement. As the course progresses, your instructor is likely to ask you what statistical procedure you would use to describe or analyze a set of data. To make the proper choice, you must know the level of measurement of the data.

Discrete and Continuous Variables

The statistical operations we can perform are also determined by whether the variables are continuous or discrete. Discrete variables have a minimum-sized unit of measurement, which cannot be subdivided. The number of children per family is an example of a discrete variable because the minimum unit is one child. A family may have two or three children, but not 2.5 children. The variable *wages* in our research example is a discrete variable because currency has a minimum unit (1 cent), which cannot be subdivided. One can have $101.21 or $101.22 but not $101.21843. Wages cannot differ by less than 1 cent—the minimum-sized unit.

Unlike discrete variables, continuous variables do not have a minimum-sized unit of measurement; their range of values can be subdivided into increasingly smaller fractional values. *Length* is an example of a continuous variable because there is no minimum unit of length. A particular object may be 12 in. long, it may be 12.5 in. long, or it may be 12.532011 in. long. Although we cannot always measure all possible length values with absolute accuracy, it is possible for objects to exist at an infinite number of lengths.[7] In principle, we can speak of a tenth of an inch, a ten thousandth of an inch, or a ten trillionth of an inch. The variable *gender composition of occupations* is a continuous variable because it is measured in proportions or percentages (e.g., the percentage of women civil engineers), which can be subdivided into smaller and smaller fractions.

This attribute of variables—whether they are continuous or discrete—affects subsequent research operations, particularly measurement procedures, data analysis, and methods of inference and generalization. However, keep in mind that, in practice, some discrete variables can be treated as if they were continuous, and vice versa.

LEARNING CHECK 1.6

Name three continuous and three discrete variables. Determine whether each of the variables in your hypothesis is continuous or discrete.

A CLOSER LOOK 1.1

A Cautionary Note: Measurement Error

Social scientists attempt to ensure that the research process is as error free as possible, beginning with how we construct our measurements. We pay attention to two characteristics of measurement: (1) reliability and (2) validity.

Reliability means that the measurement yields consistent results each time it is used. For example, asking a sample of individuals, "Do you approve or disapprove of President Donald Trump's job performance?" is more reliable than asking "What do you think of President Donald Trump's job performance?" While responses to the second question are meaningful, the answers might be vague and could be subject to different interpretations. Researchers look for the consistency of measurement over time, in relationship with other related measures, or in measurements or observations made by two or more researchers. Reliability is a prerequisite for validity: We cannot measure a phenomenon if the measure we are using gives us inconsistent results.

Validity refers to the extent to which measures indicate what they are intended to measure. While standardized IQ tests are reliable, it is still debated whether such tests measure intelligence or one's test-taking ability. A measure may not be valid due to individual error (individuals may want to provide socially desirable responses) or method error (questions may be unclear or poorly written).

Specific techniques and practices for determining and improving measurement reliability and validity are the subject of research methods courses.

ANALYZING DATA AND EVALUATING THE HYPOTHESES

Following the data collection stage, researchers analyze their data and evaluate the hypotheses of the study. The data consist of codes and numbers used to represent their observations. In our example, two scores would represent each occupational group: (1) the percentage of women and (2) the average wage. If we had collected information on 100 occupations, we would end up with 200 scores, 2 per occupational group. However, the typical research project includes more variables; therefore, the amount of data the researcher confronts is considerably larger. We now must find a systematic way to organize these data, analyze them, and use some set of procedures to decide what they mean. These last steps make up the statistical analysis stage, which is the main topic of this textbook. It is also at this point in the research cycle that statistical procedures will help us evaluate our research hypothesis and assess the theory from which the hypothesis was derived.

Descriptive and Inferential Statistics

Population: The total set of individuals, objects, groups, or events in which the researcher is interested.

Statistical procedures can be divided into two major categories: (1) descriptive statistics and (2) inferential statistics. Before we can discuss the difference between these two types of statistics, we need to understand the terms population and sample. A **population** is the total set of individuals, objects, groups, or events in which the researcher is interested. For example, if we were interested in looking at voting behavior in the last presidential election, we would probably define our population as all citizens who voted in the election. If we wanted to understand the employment patterns of Latinas in our state, we would include in our population all Latinas in our state who are in the labor force.

Sample: A subset of cases selected from a population.

Sampling: The process of identifying and selecting the subset of the population for study.

Descriptive statistics: Procedures that help us organize and describe data collected from either a sample or a population.

Inferential statistics: The logic and procedures concerned with making predictions or inferences about a population from observations and analyses of a sample.

Although we are usually interested in a population, quite often, because of limited time and resources, it is impossible to study the entire population. Imagine interviewing all the citizens of the United States who voted in the last election or even all the Latinas who are in the labor force in our state. Not only would that be very expensive and time-consuming, but we would also probably have a very hard time locating everyone! Fortunately, we can learn a lot about a population if we carefully select a subset from that population. A subset of cases selected from a population is called a **sample**. The process of identifying and selecting this subset is referred to as **sampling**. Researchers usually collect their data from a sample and then generalize their observations to the population. The ultimate goal of sampling is to have a subset that closely resembles the characteristics of the population. Because the sample is intended to represent the population that we are interested in, social scientists take sampling seriously. We'll explore different sampling methods in Chapter 5.

Descriptive statistics includes procedures that help us organize and describe data collected from either a sample or a population. Occasionally, data are collected on an entire population, as in a census. **Inferential statistics**, on the other hand, make predictions or inferences about a population based on observations and analyses of a sample. For instance, the General Social Survey (GSS), from which numerous examples presented in this book are drawn, is conducted every other year by the National Opinion Research Center (NORC) on a representative sample of several thousands of respondents. The survey, which includes several hundred questions (the data collection interview takes approximately 90 minutes), is designed to provide social science researchers with a readily accessible database of socially relevant attitudes, behaviors, and attributes of a cross section of the U.S. adult (18 years of age or older) population. Since 2006, the survey has been administered in English and Spanish. NORC has verified that the composition of the GSS samples closely resembles census data. But because the data are based on a sample rather than on the entire population, the average of the sample does not equal the average of the population as a whole.

Evaluating the Hypotheses

At the completion of these descriptive and inferential procedures, we can move to the next stage of the research process: the assessment and evaluation of our hypotheses and theories in light of the analyzed data. At this next stage, new questions might be raised about unexpected trends in the data and about other variables that may have to be considered in addition to our original variables. For example, we may have found that the relationship between gender composition of occupations and earnings can be observed with respect to some groups of occupations but not others. Similarly, the relationship between these variables may apply for some racial/ethnic groups but not for others.

These findings provide evidence to help us decide how our data relate to the theoretical framework that guided our research. We may decide to revise our theory and hypothesis to take account of these later findings. Recent studies are modifying what we know about gender segregation in the workplace. These studies suggest that race as well as gender shape the occupational structure in the United States and help explain disparities in income. This reformulation of the theory calls for a modified hypothesis and new research, which starts the circular process of research all over again.

Statistics provides an important link between theory and research. As our example on gender segregation demonstrates, the application of statistical techniques is an indispensable part of the research process. The results of statistical analyses help us evaluate our hypotheses and theories, discover unanticipated patterns and trends, and provide the impetus for shaping and reformulating our theories. Nevertheless, the importance of statistics should not diminish the significance of the preceding phases of the research process. Nor does the use of statistics lessen the importance of our own judgment in the entire process. Statistical analysis is a relatively small part of the research process, and even the most rigorous statistical procedures cannot speak for themselves. If our research questions are poorly conceived or our data are flawed due to errors in our design and measurement procedures, our results will be useless.

EXAMINING A DIVERSE SOCIETY

The increasing diversity of American society is relevant to social science. By the middle of this century, if current trends continue unchanged, the United States will no longer be comprised predominantly of European immigrants and their descendants. Due mostly to renewed immigration and higher birthrates, in time, nearly half the U.S. population will be of African, Asian, Latinx, or Native American ancestry.

Less partial and distorted explanations of social relations tend to result when researchers, research participants, and the research process itself reflect that diversity. A consciousness of social differences shapes the research questions we ask, how we observe and interpret our findings, and the conclusions we draw. Although diversity has been traditionally defined by race, class, and gender, other social characteristics such as sexual identity, physical ability, religion, and age have been identified as important dimensions of diversity. Statistical procedures and quantitative methodologies can be used to describe our diverse society, and we will begin to look at some applications in the next chapter. For now, we will preview some of these statistical procedures.

In Chapter 2, we will learn how to organize information using descriptive statistics, frequency distributions, bivariate tables, and graphic techniques. These statistical tools can also be employed to learn about the characteristics and experiences of groups in our society that have not been as visible as other groups. For example, in a series of special reports published by the U.S. Census Bureau over the past few years, these descriptive statistical techniques have been used to describe the characteristics and experiences of ethnic minorities and those who are foreign born. Using data published by the U.S. Census Bureau, we discuss various graphic devices that can be used to explore the differences and similarities among the many social groups coexisting within the American society. These devices are also used to emphasize the changing age composition of the U.S. population.

In Chapter 3, we describe how to calculate and describe the similarities and commonalities in social experiences (measures of central tendency) and the differences and diversity within social groups (measures of variability). We examine a variety of social demographic variables, including the ethnic composition of the 50 U.S. states.

We will learn about inferential statistics and bivariate analyses in Chapters 4 through 10. First, we review the bases of inferential statistics—the normal distribution, sampling and probability, and estimation—in Chapters 4 to 6. In Chapters 7 to 10, we examine the ways in which class, sex, and ethnicity influence various social behaviors and attitudes. Inferential statistics, such as the t test, chi-square, and the F statistic, help us determine the error involved in using our samples to answer questions about the population from which they are drawn. In addition, we review several methods of bivariate analysis, which are especially suited for examining the association between different social behaviors and attitudes and variables such as race, class, ethnicity, gender, and religion. We use these methods of analysis to show not only how each of these variables operates independently in shaping behavior but also how they interlock to shape our experience as individuals in society.[8]

Whichever model of social research you use—whether you follow a traditional one or integrate your analysis with qualitative data, whether you focus on social differences or any other aspect of social behavior—remember that any application of statistical procedures requires a basic understanding of the statistical concepts and techniques. This introductory text is intended to familiarize you with the range of descriptive and inferential statistics widely applied in the social sciences. Our emphasis on statistical techniques should not diminish the importance of human judgment and your awareness of the person-made quality of statistics. Only with this awareness can statistics become a useful tool for understanding diversity and social life.

DATA AT WORK

At the end of each chapter, the Data at Work feature will introduce you to people who use quantitative data and research methods in their professional lives. They represent a wide range of career fields—education, clinical psychology, international studies, public policy, publishing, politics, and research. Some may have been led to their current positions because of the explicit integration of quantitative data and research, while others are accidental data analysts—quantitative data became part of their work portfolio. Although "data" or "statistics" are not included in their job titles, these individuals are collecting, disseminating, and/or analyzing data.

We encourage you to review each profile and imagine how you could use quantitative data and methods at work.

MAIN POINTS

- Social scientists use statistics to organize, summarize, and communicate information. Only information represented by numbers can be the subject of statistical analysis.
- The research process is a set of activities in which social scientists engage to answer questions, examine ideas, or test theories. It consists of the following stages: asking the research question, formulating the hypotheses, collecting data, analyzing data, and evaluating the hypotheses.
- A theory is a set of assumptions and propositions used for explanation, prediction, and understanding of social phenomena. Theories offer specific concrete predictions about the way observable attributes of people or groups would be interrelated in real

life. These predictions, called hypotheses, are tentative answers to research problems.

- A variable is a property of people or objects that takes on two or more values. The variable that the researcher wants to explain (the "effect") is called the dependent variable. The variable that is expected to "cause" or account for the dependent variable is called the independent variable.
- Three conditions are required to establish causal relations: (1) The cause has to precede the effect in time, (2) there has to be an empirical relationship between the cause and the effect, and (3) this relationship cannot be explained by other factors.
- At the nominal level of measurement, numbers or other symbols are assigned to a set of categories to name, label, or classify the observations. At the ordinal level of measurement, categories can be rank-ordered from low to high (or vice versa). At the interval-ratio level of measurement, measurements for all cases are expressed in the same unit.
- A population is the total set of individuals, objects, groups, or events in which the researcher is interested. A sample is a relatively small subset selected from a population. Sampling is the process of identifying and selecting the subset.
- Descriptive statistics includes procedures that help us organize and describe data collected from either a sample or a population. Inferential statistics is concerned with making predictions or inferences about a population from observations and analyses of a sample.

KEY TERMS

data 2
dependent variable 8
descriptive statistics 18
dichotomous variable 14
empirical research 3
hypothesis 5
independent variable 8
inferential statistics 18
interval-ratio level of measurement 13
nominal level of measurement 11
ordinal level of measurement 12
population 18
research process 2
sample 18
sampling 18
statistics 1
theory 4
unit of analysis 6
variable 6

DIGITAL RESOURCES

Access key study tools at https://edge.sagepub.com/ssdsess4e

- eFlashcards of the glossary terms
- Datasets and codebooks
- SPSS and Excel walk-through videos
- SPSS and Excel demonstrations and problems to accompany each chapter
- Appendix F: Basic Math Review

CHAPTER EXERCISES

1. In your own words, explain the relationship of data (collecting and analyzing) to the research process. (Refer to Figure 1.1.)
2. Construct potential hypotheses or research questions to relate the variables in each of the following examples. Also, write a brief statement explaining why you believe there is a relationship between the variables as specified in your hypotheses.
    a. Political party and support of a U.S.–Mexico border wall
    b. Income and race/ethnicity
    c. The crime rate and the number of police in a city
    d. Life satisfaction and marital status
    e. Age and support for marijuana legalization
    f. Care of elderly parents and ethnicity
3. Determine the level of measurement for each of the following variables:
    a. The number of people in your statistics class
    b. The percentage of students who are first-generation college students at your school
    c. The name of each academic major offered in your college
    d. The level of support for the Black Lives Matter movement, on a scale from "strong support" to "no support."
    e. The type of transportation a person takes to school (e.g., bus, walk, car)
    f. The percentage of community members who tested positive for COVID-19
    g. The rating of the overall quality of your campus coffee shop, on a scale from "excellent" to "poor"

4. For each of the variables in Exercise 3 that you classified as interval-ratio, identify whether it is discrete or continuous.

5. Why do you think men and women, on average, do not earn the same amount of money? Develop your own theory to explain the difference. Use three independent variables in your theory, with annual income as your dependent variable. Construct hypotheses to link each independent variable with your dependent variable.

6. For each of the following examples, indicate whether it involves the use of descriptive or inferential statistics. Justify your answer.

   a. The number of unemployed people in the United States
   b. Determining students' opinion about the quality of food at the cafeteria based on a sample of 100 students
   c. The national incidence of breast cancer among Asian women
   d. Conducting a study to determine the rating of the quality of a new smartphone, gathered from 1,000 new buyers
   e. The average GPA of various majors (e.g., sociology, psychology, English) at your university
   f. The change in the number of immigrants coming to the United States from Southeast Asian countries between 2010 and 2015

7. Adela García-Aracil (2007)[9] identified how several factors affected the earnings of young European higher-education graduates. Based on data from several EU (European Union) countries, her statistical models included the following variables: annual income (actual dollars), gender (male or female), the number of hours worked per week (actual hours), and years of education (actual years) for each graduate. She also identified each graduate by current job title (senior officials and managers, professionals, technicians, clerks, or service workers).

   a. What is García-Aracil's dependent variable?
   b. Identify two independent variables in her research. Identify the level of measurement for each.
   c. Based on her research, García-Aracil can predict the annual income for other young graduates with similar work experiences and characteristics like the graduates in her sample. Is this an application of descriptive or inferential statistics? Explain.

8. Construct measures of political participation at the nominal, ordinal, and interval-ratio levels. (*Hint:* You can use behaviors such as voting frequency or political party membership.) Discuss the advantages and disadvantages of each.

9. Variables can be measured according to more than one level of measurement. For the following variables, identify at least two levels of measurement. Is one level of measurement better than another? Explain.

    a. Individual age
    b. Annual income
    c. Religiosity
    d. Student performance
    e. Social class
    f. Number of children

2

THE ORGANIZATION AND GRAPHIC PRESENTATION OF DATA

Demographers examine the size, composition, and distribution of human populations. Changes in the birth, death, and migration rates of a population affect its composition and social characteristics.[1] To examine a large population, researchers often have to deal with very large amounts of data. For example, imagine the amount of data it takes to describe the immigrant or elderly population in the United States. To make sense out of these data, a researcher must organize and summarize the data in some systematic fashion. In this chapter, we review three such methods used by social scientists: (1) the creation of frequency distributions, (2) the construction of bivariate tables and (3) the use of graphic presentation.

Chapter Learning Objectives

1. Construct and analyze frequency, percentage, and cumulative distributions.
2. Calculate proportions and percentages.
3. Compare and contrast frequency and percentage distributions for nominal, ordinal, and interval-ratio variables.
4. Create a bivariate table
5. Construct and interpret a pie chart, bar graph, histogram, the statistical map, line graph, and time-series chart.

FREQUENCY DISTRIBUTIONS

The most basic way to organize data is to classify the observations into a frequency distribution. A **frequency distribution** is a table that reports the number of observations that fall into each category of the variable we are analyzing. Constructing a frequency distribution is usually the first step in the statistical analysis of data.

Frequency distribution: A table reporting the number of observations falling into each category of the variable.

Immigration has been described as "remaking America with political, economic, and cultural ramifications."[2] Globalization has fueled migration, particularly since the beginning of the 21st century. Workers migrate because of the promise of employment and higher standards of living than what is attainable in their home countries. Data reveal that many migrants seek specifically to move to the United States.[3] The U.S. Census Bureau uses the term foreign born to refer to those who are not U.S. citizens at birth. The U.S. Census estimates that 13.5% of the U.S. population, or approximately 44 million people, are foreign born.[4] Immigrants are not one homogeneous group but are many diverse groups. Table 2.1 shows the frequency distribution of the world region of birth for the foreign-born population.

The frequency distribution is organized in a table, which has a number (2.1) and a descriptive title. The title indicates the kind of data presented: "Frequency Distribution for Categories of Region of Birth for Foreign-Born Population." The table consists of two columns. The first column identifies the variable (*world region of birth*) and its categories. The second column, with the heading "Frequency (f)," tells the number of cases in each category as well as the total number of cases (N = 43,681,654). Note also that the source of the table is clearly identified. It tells us that the data are from a 2018 report by Jynnah Radford and Abby Budiman (although the information is based on 2016 American Community Survey data from the U.S. Census). The source of the data can be reported as a source note or in the title of the table. This table is also referred to as a **univariate frequency table**, as it presents frequency information for a single variable.

Univariate frequency table: A table that displays the distribution of one variable

What can you learn from the information presented in Table 2.1? The table shows that as of 2016, approximately 44 million people were classified as foreign born. Out of this group, most—about 11.7 million people—were from South and East Asia, just under 11.6 million were from Mexico, followed by about 5.8 million from Europe or Canada.

Table 2.1 Frequency Distribution for Categories of Region of Birth for Foreign-Born Population, 2016

Region of Birth	Frequency (f)
South and East Asia	11,731,584
Mexico	11,568,060
Europe/Canada	5,785,135
Caribbean	4,300,022
Central America	3,463,389
South America	2,927,145
Middle East	1,875,264
Sub-Saharan Africa	1,769,778
All other	261,277
Total	43,681,654

Source: "2016, Foreign-Born Population in the United States Statistical Portrait", Pew Research Center, Washington, D.C (September 14, 2018), https://www.pewhispanic.org/2018/09/14/2016-statistical-information-on-foreign-born-in-united-states/.

PROPORTIONS AND PERCENTAGES

Frequency distributions are helpful in presenting information in a compact form. However, when the number of cases is large, the frequencies may be difficult to grasp. To standardize these raw frequencies, we can translate them into relative frequencies—that is, proportions or percentages.

A **proportion** is a relative frequency obtained by dividing the frequency in each category by the total number of cases. To find a proportion (p), divide the frequency (f) in each category by the total number of cases (N):

Proportion: A relative frequency obtained by dividing the frequency in each category by the total number of cases.

$$p = \frac{f}{N} \tag{2.1}$$

where

f = frequency

N = total number of cases

We've calculated the proportion for the three largest groups of foreign born. First, the proportion of foreign born originally from South and East Asia is

$$\frac{11{,}731{,}584}{43{,}681{,}654} = .269$$

The proportion of foreign born who were originally from Mexico is

$$\frac{11{,}568{,}060}{43{,}681{,}654} = .265$$

The proportion of foreign born who were originally from Europe or Canada is

$$\frac{5{,}785{,}135}{43{,}681{,}654} = .132$$

The proportion of foreign born who were originally from all other reported areas (combining the category "All other" with those from the Caribbean, Central and South America, Middle East, and sub-Saharan Africa) is

$$\frac{14{,}596{,}875}{43{,}681{,}654} = .334$$

Proportions should always sum to 1.00 (allowing for some rounding errors). Thus, in our example, the sum of the six proportions is

$$0.27 + 0.27 + 0.13 + 0.33 = 1.0$$

To determine a frequency from a proportion, we simply multiply the proportion by the total N:

$$f = p(N) \tag{2.2}$$

Thus, the frequency of foreign born from South and East Asia can be calculated as

$$0.27\ (43{,}681{,}654) = 11{,}794{,}047$$

The obtained frequency differs somewhat from the actual frequency of 11,731,584. This difference is due to rounding off of the proportion. If we use the actual proportion instead of the rounded proportion, we obtain the correct frequency:

$$0.268570050026036\ (43{,}681{,}654) = 11{,}731{,}584$$

Percentage: A relative frequency obtained by dividing the frequency in each category by the total number of cases and multiplying by 100.

We can also express frequencies as percentages. A **percentage** is a relative frequency obtained by dividing the frequency in each category by the total number of cases and multiplying by 100. In most statistical reports, frequencies are presented as percentages rather than proportions. Percentages express the size of the frequencies as if there were a total of 100 cases.

To calculate a percentage, multiply the proportion by 100:

$$\text{Percentage (\%)} = \frac{f}{N}(100) \tag{2.3}$$

or

$$\text{Percentage (\%)} = p(100) \tag{2.4}$$

Thus, the percentage of respondents who were originally from Mexico is

$$0.27\ (100) = 27\%$$

LEARNING CHECK 2.1

Calculate the proportion and percentage of males and females in your statistics class. What proportion is female?

Percentage distribution: A table showing the percentage of observations falling into each category of the variable.

PERCENTAGE DISTRIBUTIONS

Percentages are usually displayed as percentage distributions. A **percentage distribution** is a table showing the percentage of observations falling into each

Table 2.2 Frequency Distribution for Categories of Region of Birth for Foreign-Born Population, 2016

Region of Birth	Frequency (f)	Percentage (%)
South and East Asia	11,731,584	27
Mexico	11,568,060	27
Europe/Canada	5,785,135	13
Caribbean	4,300,022	10
Central America	3,463,389	8
South America	2,927,145	7
Middle East	1,875,264	4
Sub-Saharan Africa	1,769,778	4
All other	261,277	1
Total	43,681,654	100* (rounded)

Source: "2016, Foreign-Born Population in the United States Statistical Portrait", Pew Research Center, Washington, DC (September 14, 2018), https://www.pewhispanic.org/2018/09/14/2016-statistical-information-on-foreign-born-in-united-states/.

category of the variable. For example, Table 2.2 presents the frequency distribution of categories of places of origin (Table 2.1) along with the corresponding percentage distribution. Percentage distributions (or proportions) should always show the base (N) on which they were computed. Thus, in Table 2.2, the base on which the percentages were computed is $N = 43,681,654$.

THE CONSTRUCTION OF FREQUENCY DISTRIBUTIONS

In this section, you will learn how to construct frequency distributions. Most often, we can use statistical software to accomplish this, but it is important to go through the process to understand how frequency distributions are actually constructed.

For nominal and ordinal variables, constructing a frequency distribution is quite simple. To do so, count and report the number of cases that fall into each category of the variable along with the total number of cases (N). For the purpose of illustration, let's take a small random sample of 40 cases from a General Social Survey (GSS) sample and record their scores on the following variables: gender, a nominal-level variable; degree, an ordinal measurement of education; and age and number of children, both interval-ratio variables. The use of "male" and "female" in parts of this book is in keeping with the GSS categories for the variable *sex* (respondent's sex).

The interviewer recorded the gender of each respondent at the beginning of the interview. To measure degree, researchers asked everyone to indicate the highest degree completed: less than high school, high school, some college, bachelor's degree, and graduate degree. The first category represented the lowest level of education. Researchers calculated respondents' age based on the respondent's birth year. The number of children was determined by the question, "How many children have you ever had?" The answers given by our subsample of 40 respondents are displayed in Table 2.3. Note that each row in the table represents a respondent, whereas each column represents a variable. This format is conventional in the social sciences.

You can see that it is going to be difficult to make sense of these data just by eyeballing Table 2.3. How many of these 40 respondents are males? How many said that they had a graduate degree? How many were older than 50 years of age? To answer these questions, we construct a frequency distribution for each variable.

Table 2.3 A GSS Subsample of 40 Respondents

Gender of Respondent	Degree	Number of Children	Age
M	Bachelor	1	43
F	High school	2	71
F	High school	0	71
M	High school	0	37
M	High school	0	28
F	High school	6	34
F	High school	4	69
F	Graduate	0	51
F	Bachelor	0	76
M	Graduate	2	48
M	Graduate	0	49
M	Less than high school	3	62
F	Less than high school	8	71
F	High school	1	32

Gender of Respondent	Degree	Number of Children	Age
F	High school	1	59
F	High school	1	71
M	High school	0	34
M	Bachelor	0	39
F	Bachelor	2	50
M	High school	3	82
F	High school	1	45
M	High school	0	22
M	High school	2	40
F	High school	2	46
M	High school	0	29
F	High school	1	75
F	High school	0	23
M	Bachelor	2	35
M	Bachelor	3	44
F	High school	3	47
M	High school	1	84
F	Graduate	1	45
F	Less than high school	3	24
F	Graduate	0	47
F	Less than high school	5	67
F	High school	1	21
F	High school	0	24
F	High school	3	49
F	High school	3	45
F	Graduate	3	37

Note: M = male; F = female.

Table 2.4 Frequency Distribution of the Variable *Gender*: GSS Subsample

Gender	Tallies	Frequency (*f*)	Percentage (%)
Male	卌 卌 卌	15	37.5
Female	卌 卌 卌 卌 卌	25	62.5
Total (*N*)		40	100.0

Frequency Distributions for Nominal Variables

Let's begin with the nominal variable, *gender*. First, we tally the number of males, then the number of females (the column of tallies has been included in Table 2.4 for the purpose of illustration). The tally results are then used to construct the frequency distribution presented in Table 2.4. The table has a title describing its content ("Frequency Distribution of the Variable Gender: GSS Subsample"). Its categories (male and female) and their associated frequencies are clearly listed; in addition, the total number of cases (*N*) is also reported. The Percentage column is the percentage distribution for this variable. To convert the Frequency column to percentages, simply divide each frequency by the total number of cases and multiply by 100. Percentage distributions are routinely added to almost any frequency table and are especially important if comparisons with other groups are to be considered. Immediately, we can see that it is easier to read the information. There are 25 females and 15 males in this sample. Based on this frequency distribution, we can also conclude that the majority of sample respondents are female.

LEARNING CHECK 2.2

Construct a frequency and percentage distribution for males and females in your statistics class.

Frequency Distributions for Ordinal Variables

To construct a frequency distribution for ordinal-level variables, follow the same procedures outlined for nominal-level variables. Table 2.5 presents the frequency distribution for the variable *degree*. The table shows that 60.0%, a majority, indicated that their highest degree was a high school degree.

The major difference between frequency distributions for nominal and ordinal variables is the order in which the categories are listed. The categories

Table 2.5 Frequency Distribution of the Variable *Degree*: GSS Subsample

Degree	Tallies	Frequency (*f*)	Percentage (%)
Less than high school	𝍷𝍷𝍷𝍷	4	10.0
High school	𝍸 𝍸 𝍸 𝍸 𝍷𝍷𝍷𝍷	24	60.0
Bachelor	𝍸 𝍷	6	15.0
Graduate	𝍸 𝍷	6	15.0
Total (*N*)		40	100.0

for nominal-level variables do not have to be listed in any particular order. For example, we could list females first and males second without changing the nature of the distribution. Because the categories or values of ordinal variables are rank-ordered, however, they must be listed in a way that reflects their rank—from the lowest to the highest or from the highest to the lowest. Thus, the data on degree in Table 2.5 are presented in declining order from "less than high school" (the lowest educational category) to "graduate" (the highest educational category).

Frequency Distributions for Interval-Ratio Variables

We hope that you agree by now that constructing frequency distributions for nominal- and ordinal-level variables is rather straightforward. Simply list the categories and count the number of observations that fall into each category. Building a frequency distribution for interval-ratio variables with relatively few values is also easy. For example, when constructing a frequency distribution for number of children, simply list the number of children and report the corresponding frequency, as shown in Table 2.6.

Very often interval-ratio variables have a wide range of values, which makes simple frequency distributions very difficult to read. For example, look at the frequency distribution for the variable *age* in Table 2.7. The distribution contains age values ranging from 21 to 84 years. For a more concise picture, the large number of different scores could be reduced into a smaller number of groups, each containing a range of scores. Table 2.8 displays such a grouped frequency distribution of the data in Table 2.7. Each group, known as a class interval, now contains 10 possible scores instead of 1. Thus, the ages of 21, 22, 23, 24, 28, and 29 all fall into a single class interval of 20–29. The second column of Table 2.8, Frequency, tells us the number of respondents

Table 2.6 Frequency Distribution of Variable *Number of Children*: GSS Subsample

Number of Children	Frequency (*f*)	Percentage (%)
0	13	32.5
1	9	22.5
2	6	15.0
3	8	20.0
4	1	2.5
5	1	2.5
6	1	2.5
7+	1	2.5
Total (*N*)	40	100.0

Table 2.7 Frequency Distribution of the Variable *Age*: GSS Subsample

Age of Respondent	Frequency (*f*)	Age of Respondent	Frequency (*f*)
21	1	40	1
22	1	43	1
23	1	44	1
24	2	45	3
28	1	46	1
29	1	47	2
32	1	48	1
34	2	49	2
35	1	50	1
37	2	51	1
39	1	59	1

Age of Respondent	Frequency (*f*)	Age of Respondent	Frequency (*f*)
62	1		
67	1		
69	1		
71	4		
75	1		
76	1		
82	1		
84	1		

Table 2.8 Grouped Frequency Distribution of the Variable *Age*: GSS Subsample

Age Category	Frequency (*f*)	Percentage (%)
20–29	7	17.5
30–39	7	17.5
40–49	12	30.0
50–59	3	7.5
60–69	3	7.5
70–79	6	15.0
80–89	2	5.0
Total (*N*)	40	100.0

who fall into each of the intervals—for example, that seven respondents fall into the class interval of 20–29. Having grouped the scores, we can clearly see that the biggest single age group is between 40 and 49 years (12 out of 40, or 30% of sample). The percentage distribution that we have added to Table 2.8 displays the relative frequency of each interval and emphasizes this pattern as well.

The decision as to how many groups to use and, therefore, how wide the intervals should be is usually up to the researcher and depends on what makes sense in terms of the purpose of the research. The rule of thumb is that an interval width should be large enough to avoid too many categories but not so large that significant differences between observations are concealed. Obviously, the number of intervals depends on the width of each. For instance, if you are working with scores ranging from 10 to 60 and you establish an interval width of 10, you will have five intervals.

LEARNING CHECK 2.3

Can you verify that Table 2.8 was constructed correctly? Use Table 2.7 to determine the frequency of cases that fall into the categories of Table 2.8.

LEARNING CHECK 2.4

If you are having trouble distinguishing between nominal, ordinal, and interval-ratio variables, review the section on levels of measurement in Chapter 1. The distinction between these levels of measurement will be important throughout the book.

CUMULATIVE DISTRIBUTIONS

Sometimes, we may be interested in locating the relative position of a given score in a distribution. For example, we may be interested in finding out how many or what percentage of our sample was younger than 40 or older than 60. Frequency distributions can be presented in a cumulative fashion to answer such questions. A **cumulative frequency distribution** shows the frequencies at or below each category of the variable.

Cumulative frequency distribution: A distribution showing the frequency at or below each category (class interval or score) of the variable.

Cumulative frequencies are appropriate only for variables that are measured at an ordinal level or higher. They are obtained by adding to the frequency in each category the frequencies of all the categories below it.

Let's look at Table 2.9. It shows the cumulative frequencies based on the frequency distribution from Table 2.8. The cumulative frequency column, denoted by *Cf*, shows the number of persons at or below each interval. For example, you can see that 14 of the 40 respondents were 39 years old or younger, and 29 respondents were 59 years old or younger.

To construct a cumulative frequency distribution, start with the frequency in the lowest class interval (or with the lowest score, if the data are

Table 2.9 Grouped Frequency Distribution and Cumulative Frequency for the Variable *Age*: GSS Subsample

Age Category	Frequency (*f*)	Cumulative Frequency (*Cf*)
20–29	7	7
30–39	7	14
40–49	12	26
50–59	3	29
60–69	3	32
70–79	6	38
80–89	2	40
Total (*N*)	40	

ungrouped), and add to it the frequencies in the next highest class interval. Continue adding the frequencies until you reach the last class interval. The cumulative frequency in the last class interval will be equal to the total number of cases (*N*). In Table 2.9, the frequency associated with the first class interval (20–29) is 7. The cumulative frequency associated with this interval is also 7, since there are no cases below this class interval. The frequency for the second class interval is 7. The cumulative frequency for this interval is $7 + 7 = 14$. To obtain the cumulative frequency of 26 for the third interval, we add its frequency (12) to the cumulative frequency associated with the second class interval (14). Continue this process until you reach the last class interval. Therefore, the cumulative frequency for the last interval is equal to 40, the total number of cases (*N*).

We can also construct a cumulative percentage distribution (*C*%), which has wider applications than the cumulative frequency distribution (*Cf*). A **cumulative percentage distribution** shows the percentage at or below each category (class interval or score) of the variable. A cumulative percentage distribution is constructed using the same procedure as for a cumulative frequency distribution except that the percentages—rather than the raw frequencies—for each category are added to the total percentages for all the previous categories.

Cumulative percentage distribution: A distribution showing the percentage at or below each category (class interval or score) of the variable.

In Table 2.10, we have added the cumulative percentage distribution to the frequency and percentage distributions shown in Table 2.8. The cumulative percentage distribution shows, for example, that 35% of the sample was 39 years or younger.

Table 2.10 Grouped Frequency Distribution and Cumulative Percentages for the Variable *Age*: GSS Subsample

Age Category	Frequency (f)	Percentage (%)	Cumulative Percentage ($C\%$)
20–29	7	17.5	17.5
30–39	7	17.5	35.0
40–49	12	30.0	65.0
50–59	3	7.5	72.5
60–69	3	7.5	80.0
70–79	6	15.0	95.0
80–89	2	5.0	100.0
Total (N)	40	100.0	

RATES

Terms such as birthrate, unemployment rate, and marriage rate are often used by social scientists and demographers and then quoted in the popular media to describe population trends. But what exactly are rates, and how are they constructed? A **rate** is obtained by dividing the number of actual occurrences in a given time period by the number of possible occurrences.

Rate: A number obtained by dividing the number of actual occurrences in a given time period by the number of possible occurrences.

$$\text{Rate} = \frac{f}{\text{Population}} \tag{2.5}$$

For example, we can use data from the American Community Survey to determine the 2017 poverty rate by dividing (actual occurrences) by the total population in 2017 (possible occurrences). The 2017 rate can be expressed as

$$\text{Poverty rate, 2017} = \frac{\text{Number of people in poverty in 2017}}{\text{Total population in 2017}}$$

Since 42,583,651 people were poor in 2017 and the number for the total population was 317,741,588, the poverty rate for 2017 is

$$\text{Poverty rate, 2017} = \frac{42{,}583{,}651}{317{,}741{,}588} = 0.13$$

We can thus conclude that the poverty rate in 2017 was 13% (.13 × 100). This means that for every 1,000 people, 130 were poor according to the American

Community Survey definition. Rates are often expressed as rates per thousand or hundred thousand to eliminate decimal points and make the number easier to interpret.

The preceding poverty rate can be referred to as a crude rate because it is based on the total population. Rates can be calculated on the general population or on a more narrowly defined select group. For instance, poverty rates are often given for the number of people who are 18 years or younger—highlighting how our young are vulnerable to poverty. The poverty rate for those 18 years or younger is as follows:

$$\text{Poverty rate for those 18 years or younger, 2017} = \frac{13{,}353{,}202}{72{,}452{,}925} = 0.18$$

We could even take a look at the poverty rate for older Americans:

$$\text{Poverty rate for those 65 years of age or older, 2017} = \frac{4{,}581{,}772}{49{,}500{,}479} = 0.09$$

LEARNING CHECK 2.5

Law enforcement agencies routinely record crime rates (the number of crimes committed relative to the size of a population), arrest rates (the number of arrests made relative to the number of crimes reported), and conviction rates (the number of convictions relative to the number of cases tried). What other variables can be expressed as rates?

BIVARIATE TABLES

Cross-tabulation: A technique for analyzing the relationship between two variables that have been organized in a table.

Bivariate analysis: A statistical method designed to detect and describe the relationship between two variables.

Bivariate table: A table that displays the distribution of one variable across the categories of another variable.

Cross-tabulation is a technique for analyzing the relationship between two variables (an independent and a dependent variable) that have been organized in a table. A cross-tabulation is a type of **bivariate analysis**, a method designed to detect and describe the relationship between two nominal or ordinal variables. We demonstrate not only how to detect whether two variables are associated but also how to determine the strength of the association and, when appropriate, its direction in Chapter 8 The Chi-Square Test and Measures of Association.[5]

HOW TO CONSTRUCT A BIVARIATE TABLE

A **bivariate table** displays the distribution of one variable across the categories of another variable. It is obtained by classifying cases based on their joint scores on two nominal or ordinary variables. It can be thought of as a series of frequency distributions joined to make one table. The data in Table 2.11 represent a sample of General Social Survey (GSS) respondents by race and whether they own or rent their home (in this case, both variables are nominal-level measurements).

Table 2.11 Race and Home Ownership for 20 GSS Respondents

Respondent	Race	Home Ownership
1	Black	Own
2	Black	Own
3	White	Rent
4	White	Rent
5	White	Own
6	White	Own
7	White	Own
8	Black	Rent
9	Black	Rent
10	Black	Rent
11	White	Own
12	White	Own
13	White	Rent
14	White	Own
15	Black	Rent
16	White	Own
17	Black	Rent
18	White	Rent
19	Black	Own
20	Black	Rent

Table 2.12 Home Ownership by Race (Absolute Frequencies), GSS

	Race			
	Black	White		
Own	3	7	10	(Row total)
Rent	6	4	10	
	9	11	20	Total cases (*N*)
	(Column total)			

To make sense out of these data, we must first construct the table in which these individual scores will be classified. In Table 2.12, the 20 respondents have been classified according to joint scores on race and home ownership.

The table has the following features typical of most bivariate tables:

1. The table's title is descriptive, identifying its content in terms of the two variables.

2. It has two dimensions, one for race and one for home ownership. The variable *home ownership* is represented in the rows of the table, with one row for owners and another for renters. The variable *race* makes up the columns of the table, with one column for each racial group. A table may have more columns and more rows, depending on how many categories the variables represent. For example, had we included a group of Latinx people, there would have been three columns (not including the row total column). Usually, the independent variable is the column variable and the dependent variable is the row variable.

3. The intersection of a row and a column is called a cell. For example, the two individuals represented in the upper left cell are blacks who are also home owners.

4. The column and row totals are the frequency distribution for each variable, respectively. The column total is the frequency distribution for *race*, and the row total is for *home ownership*. Row and column totals are sometimes called marginals. The total number of cases (*N*) is the number reported at the intersection of the row and column totals. (These elements are all labeled in the table.)

Column variable: A variable whose categories are the columns of a bivariate table.

Row variable: A variable whose categories are the rows of a bivariate table.

Cell: The intersection of a row and a column in a bivariate table.

Marginals: The row and column totals in a bivariate table.

5. The table is a 2 × 2 table because it has two rows and two columns (not counting the marginals). We usually refer to this as an $r \times c$ table, in which r represents the number of rows and c the number of columns. Thus, a table in which the row variable has three categories and the column variable has two categories would be designated as a 3 × 2 table.

6. The source of the data should also be clearly noted in a source note to the table.

LEARNING CHECK 2.6

Examine Table 2.12. Make sure you can identify all the parts just described and that you understand how the numbers were obtained. Can you identify the independent and dependent variables in the table? You will need to know this to convert the frequencies to percentages.

HOW TO COMPUTE PERCENTAGES IN A BIVARIATE TABLE

To compare home ownership status for blacks and whites, we need to convert the raw frequencies to percentages because the column totals are not equal. Percentages are especially useful for comparing two or more groups that differ in size. There are two basic rules for computing and analyzing percentages in a bivariate table:

1. Calculate percentages within each category of the independent variable.

2. Interpret the table by comparing the percentage point difference for different categories of the independent variable.

Calculating Percentages Within Each Category of the Independent Variable

The first rule means that we have to calculate percentages within each category of the variable that the investigator defines as the independent variable. When the independent variable is arrayed in the columns, we compute percentages within each column separately. The frequencies within each cell and the row marginals are divided by the total of the column in which they are located, and the column totals should sum to 100%. When the

Table 2.13 Home Ownership by Race (in Percentages)			
Home Ownership	Race		Total
	Black	White	
Own	33%	64%	50%
Rent	67%	36%	50%
Total	100%	100%	100%
(*N*)	(9)	(11)	(20)

independent variable is arrayed in the rows, we compute percentages within each row separately. The frequencies within each cell and the column marginals are divided by the total of the row in which they are located, and the row totals should sum to 100%.

In our example, we are interested in *race* as the independent variable and in its relationship with *home ownership*. Therefore, we are going to calculate percentages by using the column total of each racial group as the base of the percentage. For example, the percentage of black respondents who own their homes is obtained by dividing the number of black home owners by the total number of blacks in the sample.

Table 2.13 presents percentages based on the data in Table 2.12. Notice that the percentages in each column add up to 100%, including the total column percentages. Always show the *N*s that are used to compute the percentages—in this case, the column totals.

Comparing the Percentages Across Different Categories of the Independent Variable

The second rule tells us to compare how home ownership varies between blacks and whites. Comparisons are made by examining differences between percentage points across different categories of the independent variable. Some researchers limit their comparisons to categories with at least a 10–percentage point difference. In our comparison, we can see that there is a 31–percentage point difference between the percentage of white home owners (64%) and black home owners (33%). In other words, in this group, whites are more likely to be home owners than blacks.[5] Therefore, we can conclude that one's race appears to be associated with the likelihood of being a home owner.

Note that the same conclusion would be drawn had we compared the percentage of black and white renters. However, since the percentages of home owners and renters within each racial group sum to 100%, we need to make only one comparison. In fact, for any 2 × 2 table, only one comparison needs to be made to interpret the table. For a larger table, more than one comparison can be made and used in interpretation.

LEARNING CHECK 2.7

Practice constructing a bivariate table. Use Table 2.11 to create a percentage bivariate table. Compare your table with Table 2.12. Did you remember all the parts? Are your calculations correct? If not, go back and review this section.

GRAPHIC PRESENTATION OF DATA

You have probably heard that "a picture is worth a thousand words." The same can be said about statistical graphs because they summarize hundreds or thousands of numbers. Graphs communicate information visually, rather than in words or numbers, and are often used in news stories, research reports, and government documents. Information that is presented graphically may seem more accessible than the same information when presented in frequency distributions or in other tabular forms.

In this section, you will learn about some of the most commonly used graphical techniques. We concentrate less on the technical details of how to create graphs and more on how to choose the appropriate graphs to make statistical information coherent. We also focus on how to interpret graphically presented information.

The particular story we tell here is that of the elderly in the United States and throughout the world. Demographers predict that over the next several decades, the U.S. overall population growth will be among middle-aged and older Americans, what demographers have referred to as the graying of America. "Population aging is a long-range trend that will characterize our society as we continue into the 21st century. It is a force we all will cope with for the rest of our lives," warns gerontologist Harry Moody.[7]

The different types of graphs demonstrate the many facets and challenges of our aging society. People have tended to talk about seniors as if they were a homogeneous group, but the different graphical techniques illustrate the wide variation in economic characteristics, living arrangements, and family status among people aged 65 years and older.

Here we focus on those graphical techniques most widely used in the social sciences. The first two, (1) the pie chart and (2) bar graph, are appropriate for nominal and ordinal variables. The next two, (3) histograms and (4) line graphs, are used with interval-ratio variables. We also discuss time-series charts. Time-series charts are used to show how some variables change over time.

THE PIE CHART

The elderly population of the United States is racially heterogeneous. As the data in Table 2.14 show, of the total 47,732,389 elderly (defined as persons 65 years and older) in 2013–2017, the two largest racial groups were whites (83.5%) and blacks (8.9%).

A **pie chart** shows the differences in frequencies or percentages among the categories of a nominal or an ordinal variable. The categories are displayed as segments of a circle whose pieces add up to 100% of the total frequencies. The pie chart shown in Figure 2.1 displays the same information that Table 2.14 presents (notice that due to rounding, the percentages in Table 2.14 do not add up to 100%). Although you can inspect these data in Table 2.14, you can interpret the information more easily by seeing it presented in the pie chart in Figure 2.1.

Pie chart: A graph showing the differences in frequencies or percentages among categories of a nominal or an ordinal variable. The categories are displayed as segments of a circle whose pieces add up to 100% of the total frequencies.

Did you notice that the percentages for several of the racial groups are 4.2% or less? It might be better to combine categories—American Indian or Alaska Native, Asian, Native Hawaiian or Pacific Islander, and some other race—into an "other races" category. This will leave us with four distinct categories: (1) white, (2) black, (3) two or more races, and (4) other. The revised pie chart is presented in Figure 2.2. We can highlight the diversity of the elderly population by "exploding" the pie chart, moving the nonwhite segments representing these groups slightly outward to draw them to the viewer's attention. This also highlights the largest slice of the pie chart—white elderly comprised 83.5% of the U.S. elderly population in 2013–2017.

Table 2.14 Five-Year Estimates of the U.S. Population 65 Years and Over by Race, 2013–2017

Race	Percentage (%)
White alone	83.5
Black alone	8.9
American Indian or Alaska Native	0.5
Asian alone	4.2
Native Hawaiian or Pacific Islander alone	0.1
Some other race alone	1.7
Two or more races combined	1.0
Total	99.9

Source: U.S. Census Bureau, *American Fact Finder*, Table S0103, 2017.

Figure 2.1 Five-Year Estimates of the U.S. Population 65 Years and Over by Race, 2013–2017

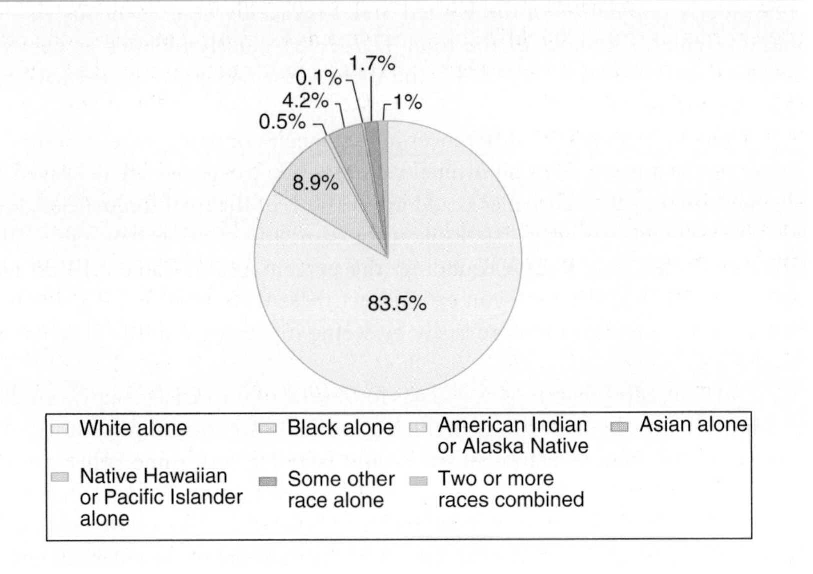

Source: U.S. Census Bureau, *American Fact Finder*, Table S0103, 2017.

Figure 2.2 Five-Year Estimates of the U.S. Population 65 Years and Over by Race, 2013–2017

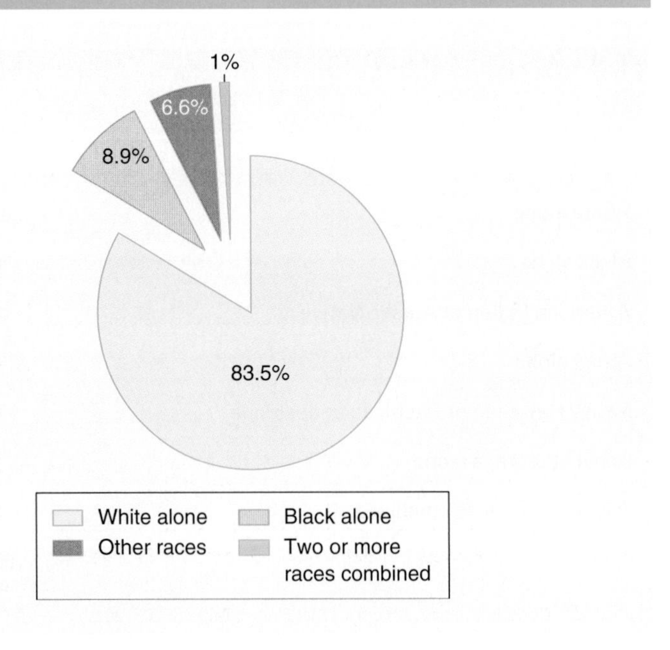

Source: U.S. Census Bureau, *American Fact Finder*, Table S0103, 2017.

THE BAR GRAPH

The bar graph provides an alternative way to graphically present nominal or ordinal data. It shows the differences in frequencies or percentages among categories of a nominal or an ordinal variable. The categories are displayed as rectangles of equal width with their height proportional to the frequency or percentage of the category.

Bar graph: A graph showing the differences in frequencies or percentages among categories of a nominal or an ordinal variable. The categories are displayed as rectangles of equal width with their height proportional to the frequency or percentage of the category.

Let's illustrate the bar graph with an overview of the marital status of the elderly. Figure 2.3 is a bar graph displaying the percentage distribution of persons 65 years old and over by marital status in 2017. This chart is interpreted similar to a pie chart except that the categories of the variable are arrayed along the horizontal axis (sometimes referred to as the *X*-axis) and the percentages along the vertical axis (sometimes referred to as the *Y*-axis). This bar graph is easily interpreted: It shows that in 2017, the majority of the elderly population were married. Specifically, 55.8% were married, 23.4% were widowed, 13.9% divorced, 1.2% separated, and 5.7% never married.

Construct a bar graph by first labeling the categories of the variables along the horizontal axis. For these categories, construct rectangles of equal width, with the height of each proportional to the frequency or percentage of the category. Note that a space separates each of the categories to make clear that they are nominal categories.

Bar graphs are often used to compare one or more categories of a variable among different groups. Suppose we want to show how the patterns in marital status differ between men and women. The longevity of women is a major factor

Figure 2.3 Marital Status of U.S. Elderly (65 Years and Older), Percentages, 2017

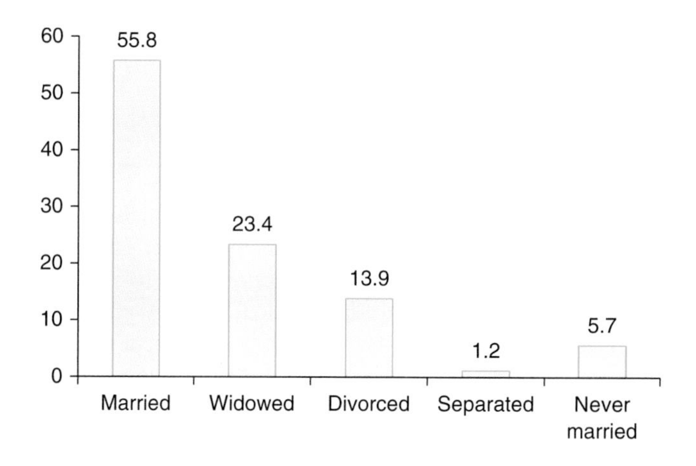

Source: U.S. Census Bureau, *American Fact Finder*, Table S0103, 2017.

Figure 2.4 Marital Status of U.S. Elderly (65 Years and Older) by Gender (Percentages), 2017

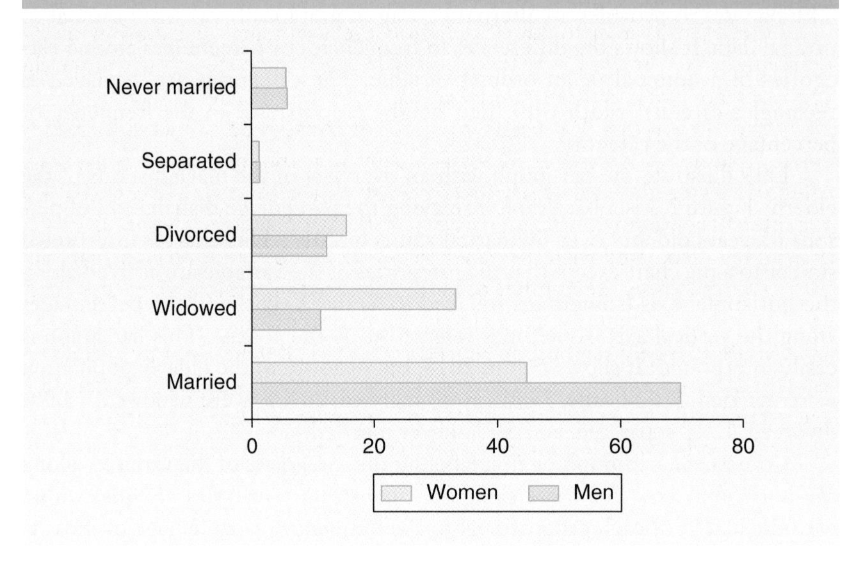

Source: U.S. Census Bureau, *American Fact Finder*, Table S1201, 2017.

in the gender differences in marital and living arrangements.[8] Additionally, elderly widowed men are more likely to remarry than elderly widowed women.

Figure 2.4 compares the marital status for women and men 65 years and older in 2017. We can also construct bar graphs horizontally, with the categories of the variable arrayed along the vertical axis and the percentages or frequencies displayed on the horizontal axis, as displayed in Figure 2.4. This presentation allows for a side-by-side visual comparison. It shows that elderly women are more likely than elderly men to be widowed (33% vs. 11%), and elderly men are more likely to be married than elderly women (70% vs. 45%).

Histogram: A graph showing the differences in frequencies or percentages among categories of an interval-ratio variable. The categories are displayed as contiguous bars, with width proportional to the width of the category and height proportional to the frequency or percentage of that category.

THE HISTOGRAM

The **histogram** is used to show the differences in frequencies or percentages among categories of an interval-ratio or ordinal variable. The categories are displayed as contiguous bars, with width proportional to the width of the category and height proportional to the frequency or percentage of that category. A histogram looks very similar to a bar graph except that the bars are contiguous to each other (touching) and may not be of equal width. In a bar graph, the spaces between the bars visually indicate that the categories are separate. Examples of variables with separate categories are marital status (married, single), gender (male, female), and employment status (employed, unemployed). In a histogram, the touching bars indicate that the categories or intervals are ordered from low to high in a meaningful way. For example, the categories of the variables hours spent studying, age, and years of school completed are contiguous, ordered intervals.

Figure 2.5 is a histogram displaying the frequency distribution of the population 65 years and over by age. To construct the histogram, arrange the age intervals along the horizontal axis and the frequencies (or percentages) along the vertical axis. For each age category, construct a bar with the height corresponding to the frequency of the elderly in the population in that age category. The width of each bar corresponds to the number of years that the age interval represents. And, in histograms, the bar for each category is touching the bar associated with the category above and below. The area that each bar occupies tells us the number of individuals that falls into a given age interval. Note that the figure title includes the notation "numbers in thousands." You should multiply each reported frequency by 1,000. For example, the largest age category is 65–69 years with 16,927,000 (16,927 × 1,000). The smallest age group is 80–84 years with 5,970,000. The total number of elderly 65 years and over can be found by summing all the reported frequencies.

Figure 2.5 Age Distribution of U.S. Elderly (65 Years and Older), 2017 (Numbers in Thousands)

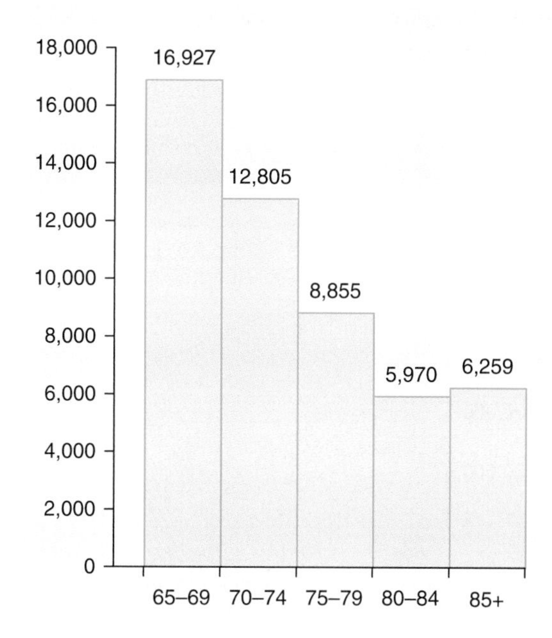

Source: U.S. Census Bureau, *American Fact Finder*, Table S0101, 2017.

Note: Ages were collapsed into categories in this example for visual purposes only. In general, histograms should be displayed with interval-ratio data that haven't been collapsed.

THE LINE GRAPH

The elderly population is growing worldwide in both developed and developing countries. In 1994, 30 nations had elderly populations of at least 2 million; demographic projections indicate that there will be 55 such nations by 2020. Japan is one of the nations that is experiencing dramatic growth of its elderly population. Figure 2.6 is a line graph displaying the elderly population of Japan for 2010 and 2014.

Line graph: A graph showing the differences in frequencies or percentages among categories of an interval-ratio variable. Points representing the frequencies of each category are placed above the midpoint of the category and are joined by a straight line.

The line graph is another way to display interval-ratio distributions; it shows the differences in frequencies or percentages among categories of an interval-ratio variable. Compared with histograms, line graphs are better suited for comparing how a variable is distributed across two or more groups or across two or more time periods, as we've done in Figure 2.6. Points representing the frequencies of each category are placed above the midpoint of the category and are joined by a straight line. Notice that in Figure 2.6, the age intervals are arranged on the horizontal axis and the frequencies along the vertical axis. Instead of using bars to represent the frequencies, however, points representing the frequencies of each interval are placed above the midpoint of the intervals. Adjacent points are then joined by straight lines.

Figure 2.6 shows how Japan's population of age 65 and over increased from 2010 to 2014. According to projections, Japan's oldest-old population—those 80 years or older—is projected to grow rapidly, from about 4.8 million (less than 4% of the total population) in 2014 to 10.8 million (8.9%) by 2020 (not depicted in the figure). This projected rise has already led to a reduction in

Figure 2.6 Population of Japan, Age 65 and Above, 2010 and 2014

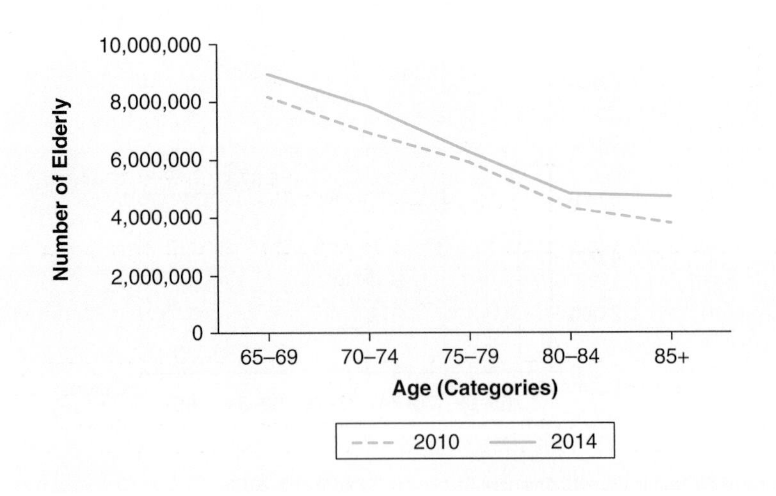

Source: United Nations, Statistics Division, *Population by Age, Sex, and Urban/Rural Residence*, 2015. Retrieved from http://data.un.org/Data.aspx?d=POP&f=tableCode%3A22

retirement benefits and other adjustments to prepare for the economic and social impact of a rapidly aging society.[9]

THE TIME-SERIES CHART

We are often interested in examining how some variables change over time. For example, we may be interested in showing changes in the labor force participation of Latinas over the past decade, changes in the public's attitude toward same-sex marriage, or changes in divorce and marriage rates. A time-series chart displays changes in a variable at different points in time. It involves two variables: (1) *time*, which is labeled across the horizontal axis, and (2) another variable of interest whose values (frequencies, percentages, or rates) are labeled along the vertical axis. To construct a time-series chart, use a series of dots to mark the value of the variable at each time interval and then join the dots by a series of straight lines.

Time-series chart: A graph displaying changes in a variable at different points in time. It shows time (measured in units such as years or months) on the horizontal axis and the frequencies (percentages or rates) of another variable on the vertical axis.

Figure 2.7 shows a time series from 2010 to 2050 of the percentage of the total population that is 65 years or older (the percentages for 2030 and 2050 are

Figure 2.7 Percentage of Total Population 65 Years and Above for Selected World Regions, 2010, 2030, and 2050

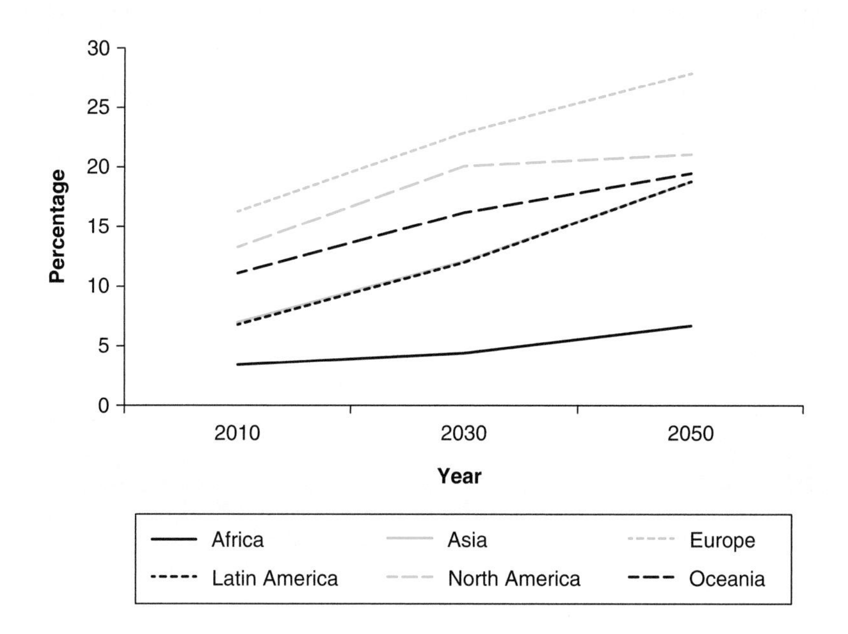

Source: Loraine West, Samantha Cole, Daniel Goodkind, and Wan He, *65+ in the United States: 2010*, Current Population Report, P23–212, 2014.

projections, as reported by the U.S. Census Bureau) for selected world regions. This time series enables us to see clearly the increase in the elderly population worldwide. As we have already mentioned, these demographic changes will have significant social, political, and economic implications, capturing the attention of policy makers and social scientists.

LEARNING CHECK 2.8

How does the time-series chart differ from a line graph? The difference is that line graphs display frequency distributions of a single variable, whereas time-series charts display two variables. In addition, time is always one of the variables displayed in a time-series chart.

STATISTICS IN PRACTICE: FOREIGN-BORN POPULATION 65 YEARS AND OVER

In their 2014 report *65+ in America*, U.S. Census Bureau researchers Loraine West, Samantha Cole, Daniel Goodkind, and Wan He describe the foreign-born population aged 65 and over in a series of tables and graphs. We present several for your review.

Frequencies and percentages presented in Table 2.15 summarize three characteristics of the 5,000,000 foreign-born elderly. The majority of these older men and women entered the U.S. prior to 1990. Almost 73% were naturalized citizens in 2010. In the same year, the largest percentage lived in the West (36%), followed by the South (29%).

Figure 2.8 is a pie chart, presenting one variable—*world region of birth*. We learn that the majority of the foreign-born elderly originally came from Latin America (37%), Asia (29%), and Europe (28%). The bar graph (Figure 2.9) presents the percentage of foreign-born elderly from each world region by their period of entry. Prior to 1990 and during 2000–2010, the largest percentage of foreign-born elderly came from Latin America and the Caribbean. However, from 1990 to 1999, the largest percentage of foreign-born elderly emigrated from Asia.

Table 2.15 Foreign-Born Population Aged 65 and Over by Period of Entry, Citizenship Status, and Region, 2010

Characteristic	Population (in Thousands)	Percentage (%)
Total	4,963	100
Period of entry		
Prior to 1990	3,769	76

Characteristic	Population (in Thousands)	Percentage (%)
1990 to 1999	644	13.0
2000 to 2010	550	11.1
Citizenship status		
Naturalized citizen	3,582	72.2
Not a U.S. citizen	1,381	27.8
Region		
Northwest	1,232	24.8
Midwest	504	10.1
South	1,442	29.1
West	1,784	36.0

Source: Loraine West, Samantha Cole, Daniel Goodkind, and Wan He, 65+ *in the United States: 2010*, Current Population Report, P23–212, 2014.

Figure 2.8 Foreign-Born Population Aged 65 Years and Over by World Region of Birth, 2010

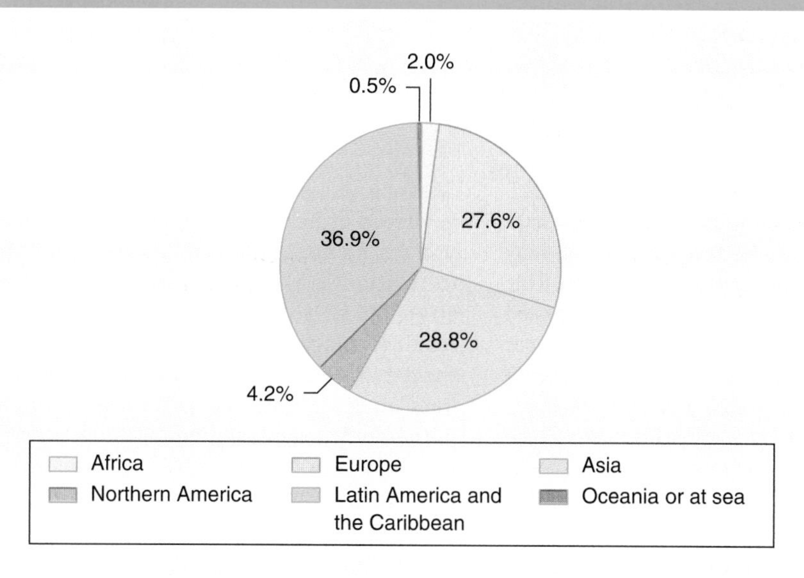

Source: Loraine West, Samantha Cole, Daniel Goodkind, and Wan He, 65+ *in the United States: 2010*, Current Population Report, P23–212, 2014.

Figure 2.9 Foreign-Born Population Aged 65 and Over by World Region of Birth and Period of Entry, 2010

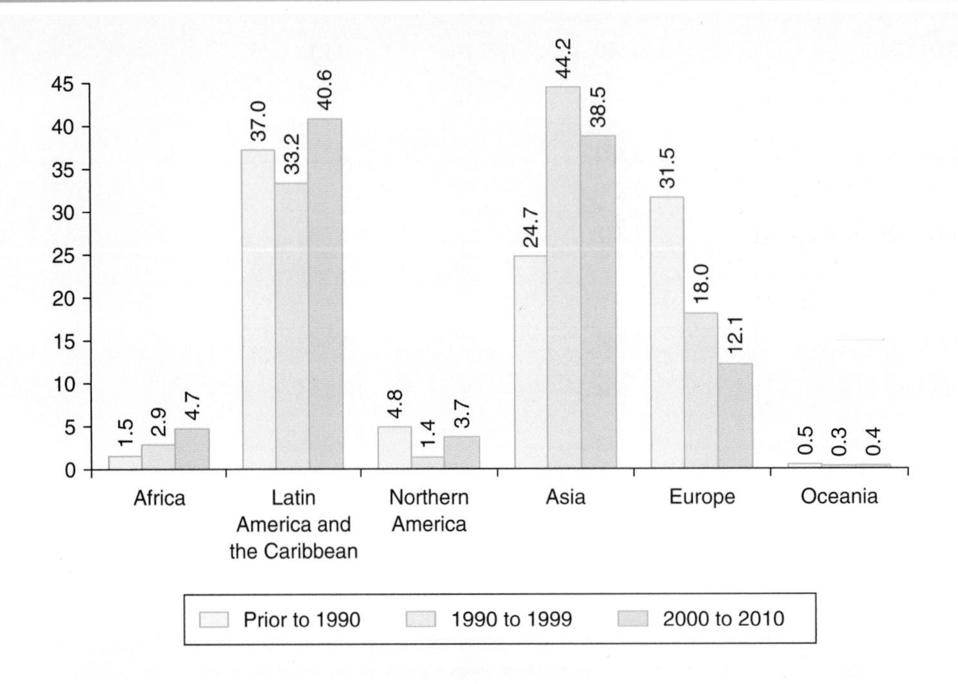

Source: Loraine West, Samantha Cole, Daniel Goodkind, and Wan He, 65+ *in the United States: 2010*, Current Population Report, P23–212, 2014.

A CLOSER LOOK 2.1

A Cautionary Note: Distortions in Graphs

In this chapter, we have seen that statistical graphs can give us a quick sense of the main patterns in the data. However, graphs not only can quickly inform us but also can also quickly deceive us. Because we are often more interested in general impressions than in detailed analyses of the numbers, we are more vulnerable to being swayed by distorted graphs. Edward Tufte in his 1983 book *The Visual Display of Quantitative Information* not only demonstrates the advantages of working with graphs but also offers a detailed discussion of some of the pitfalls in the application and interpretation of graphics.[10]

Probably the most common distortions in graphical representations occur when the distance along the vertical or horizontal axis is altered either by not using 0 as the baseline (as demonstrated in Figure 2.10a,b) or in relation to the other axis. Axes may be stretched or shrunk to create any desired result to exaggerate or disguise a pattern in the data. In Figure 2.10a,b, 2015 international data on female representation in national parliaments are presented. Without altering the data in any way, notice how the difference between the countries is exaggerated by using 30 as a baseline (as in Figure 2.10b).

Remember to interpret the graph in the context of the numerical information the graph represents.

Figure 2.10 Female Representation in National Parliaments, 2015: (a) Using 0 as the Baseline and (b) Using 30 as the Baseline

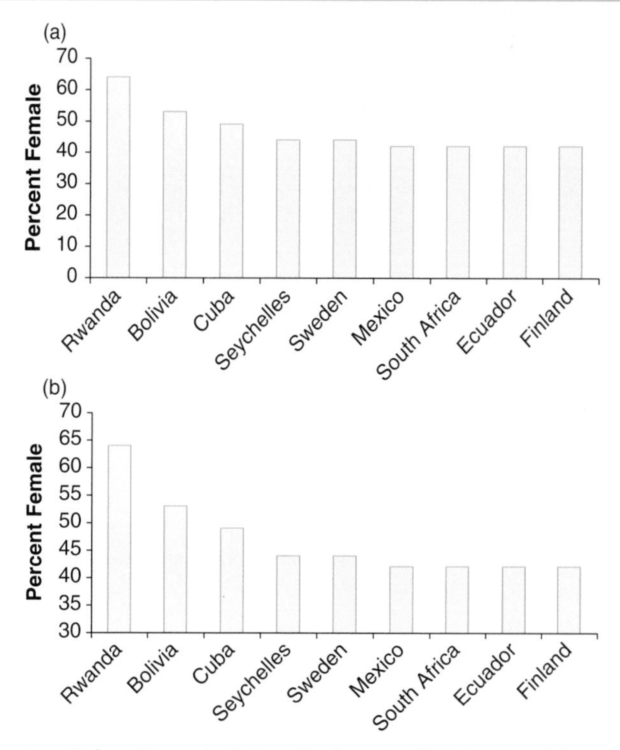

Source: Inter-Parliamentary Union, *Women in National Parliaments*, 2016. Retrieved from www.ipu.org/wmn-e/classif.htm

DATA AT WORK

Spencer Westby: Senior Editorial Analyst

Photo courtesy of Kurt Taylor Gaubatz

As a senior editorial analyst for an academic publishing company, Spencer uses data to examine the driving factors behind a successful journal publication. He tracks journal usage article submissions and citations to determine ways to improve reader outreach and revenue.

"I use bivariate tables and pie charts more than anything else as they are the quickest way to display relationships between variables. It is a simple way to display complex data for the publishing editors I am working with. Having worked at this position for a while now, I find it is just as important to

(Continued)

(Continued)

make sure the data and analysis are understandable by someone who does not know statistics well. The work needs to be beneficial to the whole business."

Spencer was introduced to this work through an internship. He hoped to gain career experience for college, but during his internship, he discovered an interest in publishing. "It was especially interesting to see how this industry worked from the inside compared to how simple it seemed on the outside as a student."

"If you are interested in a career using quantitative research, my biggest advice would be to make sure that it is something that you at least find interesting," says Spencer. "Working with numbers all day can be fairly demanding mental work if you are not ready for it. However, an inquisitive mind can turn any analytical project into a puzzle waiting to be solved. A career utilizing statistics is very satisfying for those who pursue it with creative minds."

MAIN POINTS

- The most basic way to organizing data is to classify the observations into a frequency distribution—a table that reports the number of observations that fall into each category of the variable being analyzed. A frequency distribution for a single variable is referred to as univariate table.
- Constructing a frequency distribution is usually the first step in the statistical analysis of data. To obtain a frequency distribution for nominal and ordinal variables, count and report the number of cases that fall into each category of the variable along with the total number of cases (*N*). To construct a frequency distribution for interval-ratio variables that have a wide range of values, first combine the scores into a smaller number of groups—known as class intervals—each containing a number of scores.
- Proportions and percentages are relative frequencies. To construct a proportion, divide the frequency (*f*) in each category by the total number of cases (*N*). To obtain a percentage, divide the frequency (*f*) in each category by the total number of cases (*N*) and multiply by 100.
- Percentage distributions are tables that show the percentage of observations that fall into each category of the variable. Percentage distributions are routinely added to almost any frequency table and are especially important if comparisons between groups are to be considered.
- Cumulative frequency distributions allow us to locate the relative position of a given score in

a distribution. They are obtained by adding to the frequency in each category the frequencies of all the categories below it.

- Cumulative percentage distributions have wider applications than cumulative frequency distributions. A cumulative percentage distribution is constructed by adding to the percentages in each category the percentages of all the categories below it.
- A rate is a number that expresses raw frequencies in relative terms. A rate can be calculated as the number of actual occurrences in a given time period divided by the number of possible occurrences for that period. Rates are often multiplied by some power of 10 to eliminate decimal points and make the number easier to interpret.
- The bivariate table displays the distribution of one variable across the categories of another variable. It is obtained by classifying cases based on their joint scores for two variables.
- Percentaging bivariate tables are used to examine the relationship between two variables that have been organized in a bivariate table. The percentages are always calculated within each category of the independent variable.
- A pie chart shows the differences in frequencies or percentages among categories of a nominal or an ordinal variable. The categories of the variable are segments of a circle whose pieces add up to 100% of the total frequencies.
- A bar graph shows the differences in frequencies or percentages among categories of a nominal or an ordinal variable. The categories are displayed as rectangles of equal width with their height proportional to the frequency or percentage of the category.
- Histograms display the differences in frequencies or percentages among categories of interval-ratio variables. The categories are displayed as contiguous bars with their width proportional to the width of the category and height proportional to the frequency or percentage of that category.
- A line graph shows the differences in frequencies or percentages among categories of an interval-ratio variable. Points representing the frequencies of each category are placed above the midpoint of the category (interval). Adjacent points are then joined by a straight line.
- A time-series chart displays changes in a variable at different points in time. It displays two variables: (1) *time*, which is labeled across the horizontal axis, and (2) another variable of interest whose values (e.g., frequencies, percentages, or rates) are labeled along the vertical axis.

KEY TERMS

bar graph 49
bivariate analysis 41
bivariate table 41
cell 43
column variable 43
cross-tabulation 41
cumulative frequency distribution 38
cumulative percentage distribution 39
frequency distribution 27
histogram 50
line graph 52
marginals 43
percentage 30
percentage distribution 30
pie chart 47
proportion 29
rate 40
row variable 43
time-series chart 53
univariate frequency table 28

DIGITAL RESOURCES

Access key study tools at https://edge.sagepub.com/ssdsess4e

- eFlashcards of the glossary terms
- Datasets and codebooks
- SPSS and Excel walk-through videos
- SPSS and Excel demonstrations and problems to accompany each chapter
- Appendix F: Basic Math Review

CHAPTER EXERCISES

1. Suppose you surveyed 30 people and asked them whether they are white (W) or nonwhite (N) and how many traumas (serious accidents, rapes, or crimes) they have experienced in the past year. You also asked them to tell you whether they perceive themselves as being in the upper, middle, working, or lower class. Your survey resulted in the raw data presented in the following table:
   a. Identify the level of measurement for each variable.
   b. Construct raw frequency tables for race.
   c. What proportion of the 30 individuals is nonwhite? What percentage is white?

Race	Class	Trauma	Race	Class	Trauma
W	L	1	W	W	0
W	M	0	W	M	2

Race	Class	Trauma	Race	Class	Trauma
W	M	1	W	W	1
N	M	1	W	W	1
N	L	2	N	W	0
W	W	0	N	M	2
N	W	0	W	M	1
W	M	0	W	M	0
W	M	1	N	W	1
N	W	1	W	W	0
N	W	2	W	W	0
N	M	0	N	M	0
N	L	0	N	W	0
W	U	0	N	W	1
W	W	1	W	W	0

Note: Race: W = white; N = nonwhite; Class: L = lower class; M = middle class; U = upper class; W = working class.

2. Using the data from Exercise 1, construct a frequency and percentage distribution for class.
   a. Which is the smallest perceived class group?
   b. Which two classes include the largest percentages of people?
3. Using the data from Exercise 1, construct a frequency distribution for trauma.
   a. What level of measurement is used for the trauma variable?
   b. Are people more likely to have experienced no traumas or only one trauma in the past year?
   c. What proportion has experienced one or more traumas in the past year?
4. Using the data from Exercise 1, construct appropriate graphs showing percentage distributions for race, class, and trauma.
5. GSS 2018 respondents were asked to describe how much confidence they had in the press. Results are provided in the following table for the percentage in each category by whom the respondent voted for in the 2016 U.S. presidential election. Do these data support the statement

that those who voted for Donald Trump have lower levels of confidence in the press than those who didn't vote for Trump? Why or why not?

	Trump (%)	Clinton (%)	Other (%)
A great deal	2.9	21.8	11.8
Only some	17.9	58.9	29.4
Hardly any	79.2	19.3	58.8
Total	100.0	100.0	100.0

6. How many hours per week do you spend on e-mail? Data are presented here for a GSS sample of 99 men and women, who each reported the number of hours they spent per week on e-mail.

   a. Compute the cumulative frequency and cumulative percentage distribution for the data.

   b. What proportion of the sample spent 3 hours or less per week on e-mail?

   c. What proportion of the sample spent 6 or more hours per week on e-mail?

   d. Construct a graph that best displays these data. Explain why the graph you selected is appropriate for these data.

E-mail Hours per Week	Frequency
0	19
1	20
2	13
3	5
4	2
5	6
6	5
7	2
8	3
9	1
10 or more	23

7. The time-series chart shown below displays trends for presidential election voting rates by race and Hispanic origin for 1980–2016. U.S. Census Senior Sociologist Thom File noted how for the first time in the 2012 presidential

election, black voting rates exceeded the rates for non-Hispanic whites. However, in the 2016 presidential election, there was a remarkable shift in voting rates by race and Hispanic origin. Describe the variation in voting rates in 2016 for the four racial and Hispanic origin groups and then compare that variation with that presented for the four groups from 1980 onward.

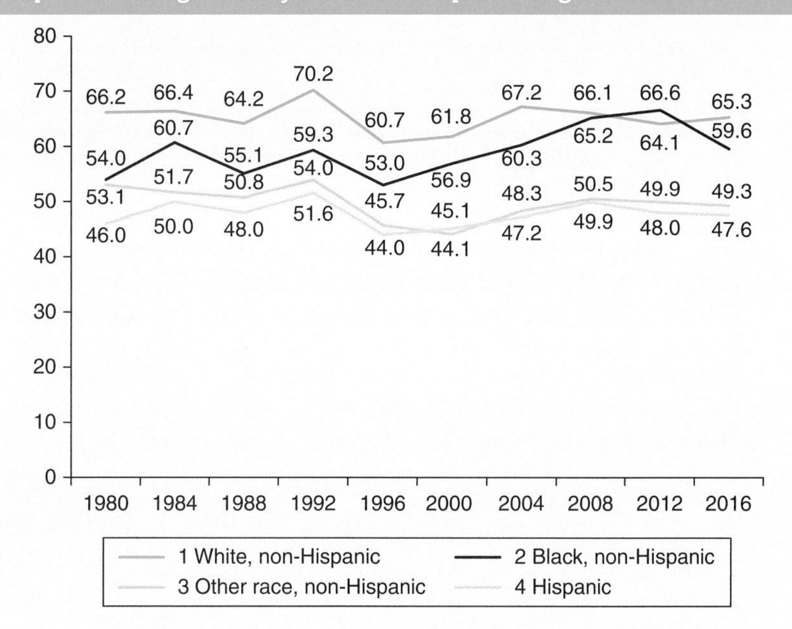

Source: Thom File, *Voting in America: A Look at the 2016 Presidential Election*, 2017. Retrieved from https://www.census.gov/newsroom/blogs/random-samplings/2017/05/voting_in_america.html

8. According to the Pew Research Center (2015), recent immigrants are better educated than earlier immigrants to the United States. The change was attributed to the availability of better education in each region or country of origin. The percentage of immigrants 25 years of age and older who completed at least high school is reported in this table for 1970 to 2013. Write a statement describing the change over time in the percentage who completed at least a high school degree.

	1970	1980	1990	2000	2013
Mexico	14	17	26	30	48
Other Central/ South America	52	57	53	60	66

(*Continued*)

(Continued)

	1970	1980	1990	2000	2013
Asia	75	72	75	82	84
Europe	48	68	81	87	95
Caribbean	36	48	52	58	72
Africa	81	91	88	85	85

Source: Pew Research Center, *Modern Immigration Wave Brings 59 Million to U.S., Driving Population Growth and Change Through 2065*, 2015. Retrieved from https://www.pewresearch.org/hispanic/2015/09/28/modern-immigration-wave-brings-59-million-to-u-s-driving-population-growth-and-change-through-2065/

9. Older Americans are often described as more politically engaged than younger Americans. One measure of political engagement is election voting. In a 2017 report, U.S. Census senior sociologist Thom File presented the following time-series chart of voting rates by age in presidential elections from 1980 to 2016. Based on this time-series chart, are older Americans more likely to vote in the presidential election than younger Americans? (Before you answer, define which age groups are younger vs. older.)

Reported Voting Rates by Age: 1980–2016

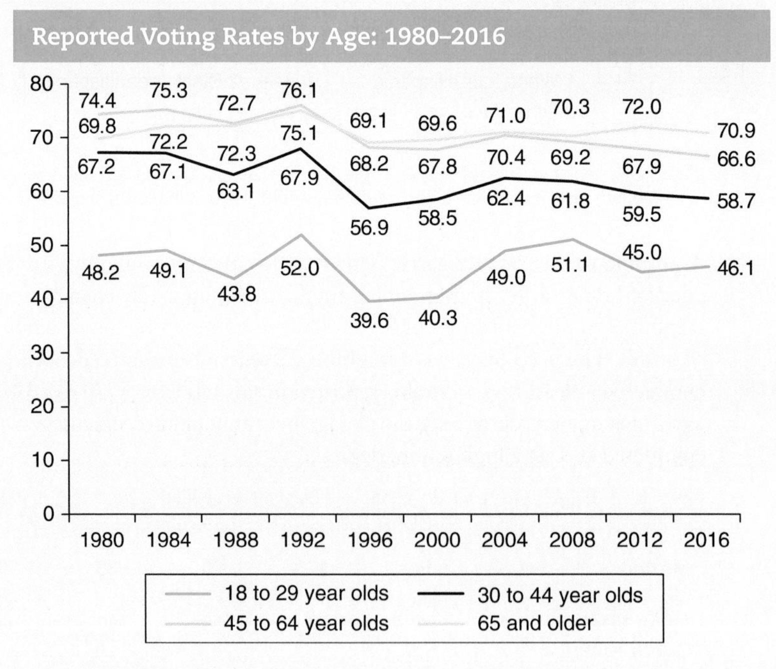

Source: Thom File, *Voting in America: A Look at the 2016 Presidential Election*, 2017. Retrieved from https://www.census.gov/newsroom/blogs/random-samplings/2017/05/voting_in_america.html

10. The following cross-tabulation, based on GSS18SSDS-B data, examines the relationship between age (the independent variable) and presidential candidate one voted for in the 2016 U.S. presidential election (dependent variable). Notice we have recoded age into four categories: 18–29, 30–39, 40–49, and 50–59. Respondents of all other ages have been excluded from the analysis.

Vote Clinton or Trump * RE_AGE Crosstabulation

			RE_AGE				
			18–29	30–39	40–49	50–59	Total
Vote Clinton or Trump	Clinton	Count	65	96	77	78	316
		% within RE_AGE	65.7%	61.1%	53.8%	45.6%	55.4%
	Trump	Count	19	42	55	86	202
		% within RE_AGE	19.2%	26.8%	38.5%	50.3%	35.4%
	Other candidate (specify)	Count	12	17	10	5	44
		% within RE_AGE	12.1%	10.8%	7.0%	2.9%	7.7%
	Didn't vote for president	Count	3	2	1	2	8
		% within RE_AGE	3.0%	1.3%	0.7%	1.2%	1.4%
Total		Count	99	157	143	171	570
		% within RE_AGE	100.0%	100.0%	100.0%	100.0%	100.0%

a. Which is the independent variable?

b. What is the level of measurement for the independent variable and the dependent variable?

c. How would you describe the relationship between the two variables? Use percentages in your answer.

11. One of your classmates hypothesizes that people of color are far less likely to own their own home than white people. Use the following data, which draws from GSS18SSDS-B, to test your classmate's hypothesis.

Home Ownership	Race		
	White	Black	Total
Own or is buying	475	64	539
	67.7%	41.3%	63%

(Continued)

(Continued)

Home Ownership	Race		Total
	White	Black	
Pays rent	227	91	318
	32.3%	58.7%	37%
Total	702	155	857
	100%	100%	100%

a. Based on your classmate's argument, what is the dependent variable? The independent variable?

b. What percentage of those surveyed own their own home?

c. Using the percentages in the table, describe the relationship between race and home ownership.

12. Youth were asked in the Monitoring the Future (MTF) 2017 survey to report how often they were drunk in the past 12 months. Responses for 3,176 twelfth graders are reported by race.

Drunk in the past 12 Months	Race			Total
	Black	White	Hispanic	
None	359	1,081	480	1,920
1–2 times	63	342	125	530
3–5 times	24	168	50	242
6 or more times	31	389	64	484
Total	477	1,980	719	3,176

Calculate the percentages using *race* as the independent variable. Is there a relationship between race and frequency of drunkenness?

13. Drawing on International Social Survey Programme data, Tsui-o Tai and Janeen Baxter (2018) examined variation in household division of labor. The following table has been adapted from their analysis to show the relationship between perceived unfairness and frequency of housework disagreement for women in their sample.

a. What is the dependent variable?

b. How many people are included in this table?

c. What percentage of the sample felt housework divisions were unfair to women?

d. Is there a relationship between perceived unfairness and frequency of housework disagreement for women?

Housework Disagreement	Women's Perceptions of Housework Divisions, % (*N*)		
	Fair to Women	Unfair to Women	Total
Never	43.65% (1,393)	25.62% (1,013)	33.68% (2,406)
Rarely	32.09% (1,024)	30.16% (1,192)	31.02% (2,216)
Several times a year	13.19% (421)	17.56% (694)	15.61% (1,115)
Several times a month	7.80% (249)	17.28% (683)	13.05% (932)
Several times a week	3.27% (104)	9.39% (371)	6.65% (475)
Total	100% (3,191)	100% (3,953)	100% (7,144)

Source: Adapted from Tsui-o Tai and Janeen Baxter, "Perceptions of Fairness and Housework Disagreement: A Comparative Analysis," *Journal of Family Issues* 39, no. 8 (2018): 2461–2485 at p. 2471.

MEASURES OF CENTRAL TENDENCY AND VARIABILITY

Frequency distributions and graphical techniques are useful tools for describing data. The main advantage of using frequency distributions or graphs is to summarize quantitative information in ways that can be easily understood even by a lay audience. Often, however, we need to describe a large set of multivariate data for which graphs and tables may not be the most efficient tools. For instance, let's say that we want to present information on the income, education, and political party affiliation of both men and women. Presenting this information might require up to six frequency distributions or graphs. The more variables we add, the more difficult it becomes to present the information clearly.

We may also describe a distribution by calculating numbers that describe what is typical about a distribution or how much variation and diversity there is in the distribution. Numbers that describe what is average or typical of the distribution are called **measures of central tendency**. Numbers that describe variation are called **measures of variability**. You are probably somewhat familiar with these measures.

We will learn about both types of measures in this chapter, beginning with three measures of central tendency.

Chapter Learning Objectives

1. Explain the importance of measures of central tendency.
2. Calculate and interpret the mode, the median, and the mean.
3. Identify the relative strengths and weaknesses of the three measures.
4. Determine and explain the shape of the distribution.
5. Explain the importance of measuring variability.
6. Calculate and interpret range, interquartile range, the variance, and the standard deviation.
7. Identify the relative strengths and weaknesses of the measures.

MEASURES OF CENTRAL TENDENCY

We will learn about three measures of central tendency: (1) the mode, (2) the median, and (3) the mean. Each describes what is most typical, central, or representative of the distribution. We will also learn about how these measures differ from one another. We will see that the choice of an appropriate measure of central tendency for representing a distribution depends on three factors: the way the variables are measured (their level of measurement), the shape of the distribution, and the purpose of the research.

Measures of central tendency: Numbers that describe what is average or typical of the distribution.

Measures of variability: Numbers that describe diversity or variability in the distribution.

Mode: A measure of central tendency. The category or score with the highest frequency (or percentage) in the distribution of main points.

The Mode

The mode is the category or score with the largest frequency or percentage in the distribution. Of all the measures of central tendency discussed in this chapter, the mode is the easiest one to identify. Simply locate the category represented by the highest frequency in the distribution.

We can use the mode to determine, for example, the most common foreign language spoken in the United States today. English is clearly the language of choice in public communication in the United States, but you may be surprised by the U.S. Census Bureau's finding that at least 350 languages are spoken in U.S. homes. In the New York metro area alone, at least 192 languages are spoken; 38% of the metro population age 5 years and over speak a language other than English.[1]

What is the most common foreign language spoken in the United States today? To answer this question, look at Table 3.1, which lists the 10 most commonly spoken foreign languages in the United States during 2009–2013 and the number of people who speak each language. The table shows that Spanish is the most common; more than 37 million people speak Spanish. In this example, we refer to "Spanish" as the mode—the category with the largest frequency in the distribution.

The mode is always a category or score, *not* a frequency. Do not confuse the two. That is, the mode in the previous example is "Spanish," not its frequency of 37,458,624. The mode is not necessarily the category with the majority (i.e., more than 50%) of cases, as it is in Table 3.1; it is simply the category in which the largest number (or proportion) of cases fall.

Table 3.1 Ten Most Common Foreign Languages Spoken in the United States, 2009–2013

Language	Number of Speakers
Spanish	37,458,624
Chinese	2,896,766
Tagalog	1,613,346
Vietnamese	1,399,936
French	1,307,742
Korean	1,117,343
German	1,063,773
Arabic	924,374
Russian	879,434
French Creole	739,725

Source: U.S. Census Bureau, *Census Bureau Reports at Least 350 Languages Spoken in U.S. Homes*, 2015.

The mode is the only measure of central tendency that can be used with nominal-level variables. Recall that with nominal variables—such as foreign languages spoken in the United States, race/ethnicity, or religious affiliation—we are only able to classify respondents based on a qualitative and not on a quantitative property. However, the mode can also be used to describe the most commonly occurring category in any distribution. For example, the variable HEALTH presented in Figure 3.1 is an ordinal variable, measuring responses to the question, "Would you say your own health, in general, is excellent, good, fair, or poor?" Which is the modal category?

In some distributions, there are two scores or categories with the highest frequency. For instance, Figure 3.2 is a bar graph showing the response of GSS 2018 respondents to the following question: "If you were asked to use one of the four names for your social class, which would you say you belong to: the lower class, the working class, the middle class, or the upper class?" The same percentage of respondents (44%) identified themselves as "working" or "middle" class. Both response categories have the highest frequency, and therefore, both are the modes. We can describe this distribution as bimodal. When two scores or categories with the highest frequencies are quite close (but not identical) in frequency, the distribution is still "essentially" bimodal. In these situations, you should not rely on merely reporting the (true) mode but instead report the two highest frequency categories.

Figure 3.1 Respondent's Health (HEALTH) (Percentages), GSS 2018

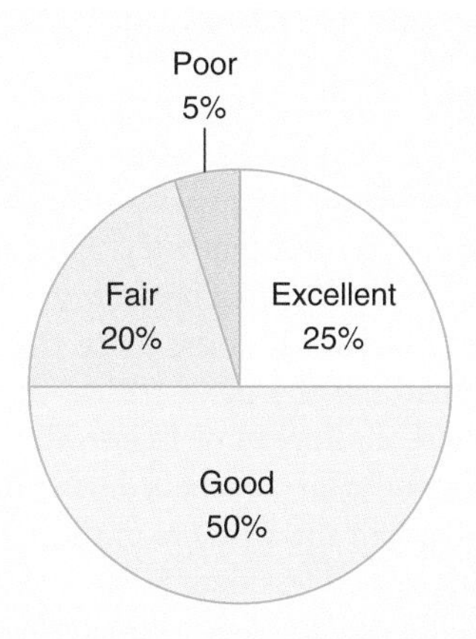

LEARNING CHECK 3.1

Listed below are the political party affiliations of 15 individuals. Find the mode.

Democrat	Republican	Democrat	Republican	Republican
Independent	Democrat	Democrat	Democrat	Republican
Independent	Democrat	Independent	Republican	Democrat

Figure 3.2 Respondent's Social Class (CLASS) (Percentages), GSS 2018

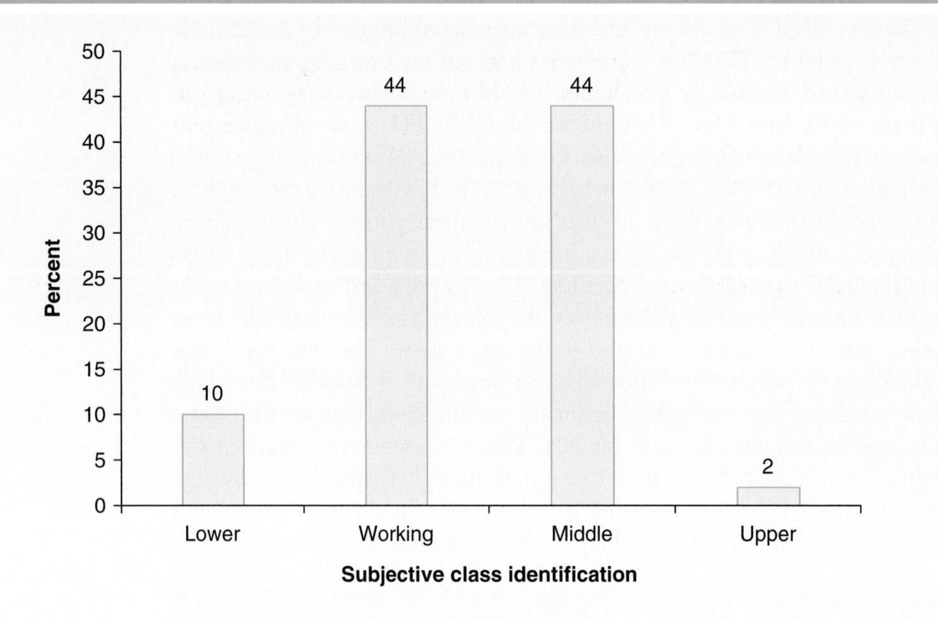

The Median

Median: A measure of central tendency. The score that divides the distribution into two equal parts so that half the cases are above and half below.

The median is a measure of central tendency that can be calculated for variables that are at least at an ordinal level of measurement. The median represents the exact middle of a distribution; it is the score that divides the distribution into two equal parts so that half the cases are above it and half below it. For example, according to the U.S. Bureau of Labor Statistics, the median weekly earnings of full-time wage and salary workers during the fourth quarter of 2018 was $900.[2] This means that half the workers in the United States earned more than $900 a week and half earned less than $900.

Because many variables used in social research are ordinal, the median is an important measure of central tendency. The median is a suitable measure for those variables whose categories or scores can be arranged in order of magnitude from the lowest to the highest. Therefore, the median can be used with ordinal or interval-ratio variables, for which scores can be at least rank-ordered but cannot be calculated for variables measured at the nominal level.

Finding the Median in Sorted Data

It is very easy to find the median. In most cases, a simple inspection of the sorted data can do it. The location of the median score differs somewhat, depending on whether the number of observations is odd or even. Let's first consider two examples with an odd number of cases.

An Odd Number of Cases Suppose we are looking at five individual responses to the question, "Thinking about the economy, how would you rate economic conditions in this country today?" Following are the responses of these five hypothetical persons:

Poor
Good
Only fair
Poor
Excellent
Total (N) = 5

To locate the median, first arrange the responses in order from the lowest to the highest (or the highest to the lowest):

Poor
Poor
Only fair
Good
Excellent
Total (N) = 5

The median is the response associated with the middle case. Find the middle case when N is odd by adding 1 to N and dividing by 2: $(N + 1)/2$. Because N is 5, you calculate $(5 + 1)/2 = 3$. The middle case is thus the third case, and the median is "only fair," the response associated with the third case. Notice that the median divides the distribution exactly into half so that there are two respondents who are more satisfied and two respondents who are less satisfied.

Now let's look at another example. The following is a list of the number of hate crimes reported in the nine most populous U.S. states in 2017.[3]

Number of Hate Crimes	States
190	Texas
1,094	California
145	Florida
552	New York
78	Pennsylvania
82	Illinois
456	Michigan
495	New Jersey
166	North Carolina
Total (N) = 9	

To locate the median, first arrange the number of hate crimes in order from the lowest to the highest (as illustrated in Figure 3.3). The middle case is $(9 + 1)/2 = 5$, the fifth state, Texas. The median is 190, the number of hate crimes associated with Texas. It divides the distribution exactly into half, so that there are four states with fewer hate crimes and four with more (this is illustrated in Figure 3.3a).

An Even Number of Cases Now let's delete the last score to make the number of states even (Figure 3.3b). The scores have already been arranged in ascending order. To locate the median, first arrange the number of hate crimes in order form the lowest to the highest:

Number of Hate Crimes	States
78	Pennsylvania
82	Illinois
145	Florida
166	North Carolina
190	Texas
456	Michigan
495	New Jersey
552	New York
Total (N) = 8	

Figure 3.3 Finding the Median Number of Hate Crimes for (a) Nine States and (b) Eight States

(a) Odd Number of Cases

1. Order the cases from the lowest to the highest:

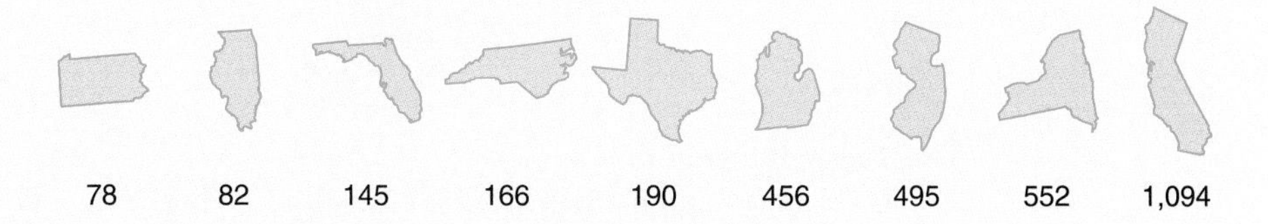

2. In this situation, we need the 5th case: (9 + 1) ÷ 2 = 5.

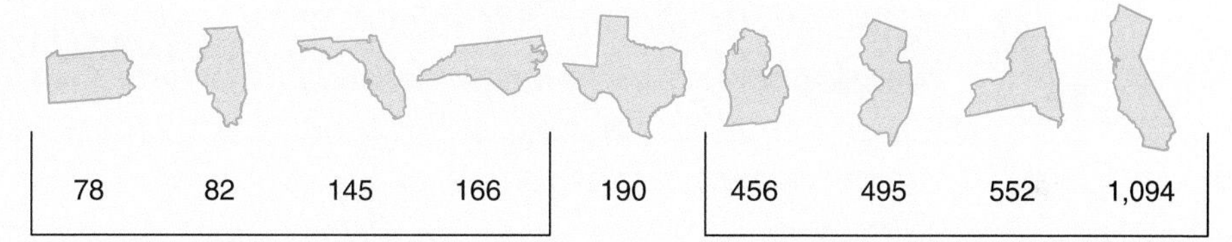

4 cases on this side of the median

4 cases on this side of the median

(b) Even Number of Cases

1. Order the cases from the lowest to the highest:

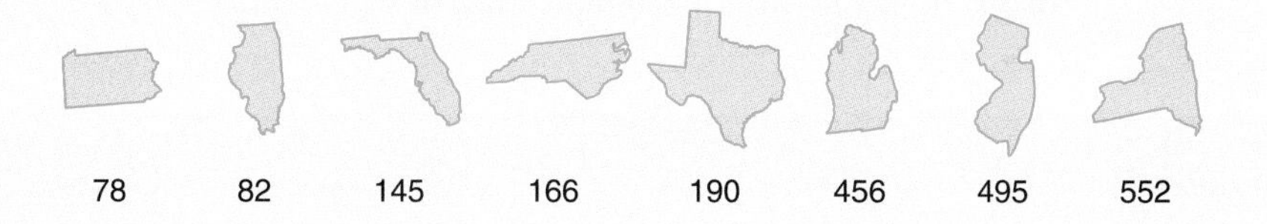

2. In this situation, we take the average of the two cases nearest the 4.5th case: (8 + 1) ÷ 2 = 4.5

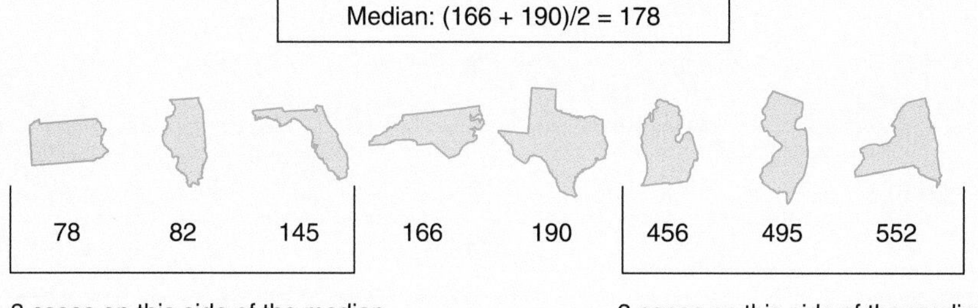

3 cases on this side of the median

3 cases on this side of the median

When N is even (eight states), we no longer have a single middle case. The median is therefore located halfway between the two middle cases. Find the two middle cases by using the previous formula: $(N + 1)/2$, or $(8 + 1)/2 = 4.5$. In our example, this means that you average the scores for the fourth and fifth states, North Carolina and Texas. The numbers of hate crimes associated with these states are 166 and 190. To find the median for this interval-ratio variable, simply average the two middle numbers:

$$\frac{166+190}{2}=178$$

The median is therefore 178.

As a note of caution, when data are ordinal, averaging the middle two scores is no longer appropriate. The median simply falls between two middle values.

LEARNING CHECK 3.2

Find the median of the following distribution of an interval-ratio variable: 22, 15, 18, 33, 17, 5, 11, 28, 40, 19, 8, 20.

Finding the Median in Frequency Distributions

Often our data are arranged in frequency distributions. Take, for instance, the frequency distribution displayed in Table 3.2. It shows the political views of GSS 2018 respondents.

To find the median, we need to identify the category associated with the observation located at the middle of the distribution. We begin by specifying N,

Table 3.2 Respondents' Political Views, GSS 2018

Political View	Frequency (*f*)	Cumulative Frequency (*Cf*)	Percentage (%)	Cumulative Percentage (*C*%)
Extremely liberal	61	61	5.5	5.5
Liberal	133	194	12	17.5
Slightly liberal	130	324	11.7	29.2
Moderate	411	735	37	66.2
Slightly conservative	151	886	13.6	79.8
Conservative	178	1,064	16	95.8
Extremely conservative	47	1,111	4.2	100
Total (*N*)	1,111		100	

the total number of respondents. In this particular example, $N = 1,111$. We then use the formula $(N + 1)/2$, or $(1,111 + 1)/2 = 556$. The median is the value of the category associated with the 556th case. The cumulative frequency (*Cf*) of the 556th case falls in the category "moderate"; thus, the median is "moderate." This may seem odd; however, the median is always the value of the response category, not the frequency.

We can also locate the median in a frequency distribution by using the cumulative percentages column, as shown in the last column of Table 3.2. In this example, the percentages are cumulated from "extremely liberal" to "extremely conservative." We could also cumulate the other way, from "extremely conservative" to "extremely liberal." To find the median, we identify the response category that contains a cumulative percentage value equal to 50%. The median is the value of the category associated with this observation.[4] Looking at Table 3.2, the percentage value equal to 50% falls within the category "moderate." The median for this distribution is therefore "moderate." If you are not sure why the middle of the distribution—the 50% point—is associated with the category "moderate," look again at the cumulative percentage column (*C*%). Notice that 33.8% (100 – 66.2) of the observations are accumulated below the category "moderate" and that 66.2% are accumulated up to and including the category "moderate." We know, then, that the percentage value equal to 50% is located somewhere within the "moderate" category.

LEARNING CHECK 3.3

For a review of cumulative distributions, refer to Chapter 2.

Locating Percentiles in a Frequency Distribution

The median is a special case of a more general set of measures of location called percentiles. A **percentile** is a score at or below which a specific percentage of the distribution falls. The *n*th percentile is a score below which *n*% of the distribution falls. For example, the 75th percentile is a score that divides the distribution so that 75% of the cases are below it. The median is the 50th percentile. It is a score that divides the distribution so that 50% of the cases fall below it. Like the median, percentiles require that data be ordinal or higher in level of measurement. Percentiles are easy to identify when the data are arranged in frequency distributions.

Percentile: A score below which a specific percentage of the distribution falls.

To help illustrate how to locate percentiles in a frequency distribution, we display in Table 3.3 the frequency distribution, the percentage distribution, and the cumulative percentage distribution of how often teens have dinner with their parents during an average week. The data were collected in the 2014 Monitoring the Future survey, a national survey measuring the behaviors and attitudes of 12th-grade students. As you can see from the cumulative percentage column, the 50th percentile (the median) falls somewhere in the fifth category, "4–5 days per week." Fifty percent of all teens have dinner with their parents 4–5 days a week or less, and 50% of all teens have dinner with their parents 4–5 days a week or more.

Table 3.3 Days per Week Having Dinner With Parent(s), Monitoring the Future 2014

Days per Week Having Dinner With Parent	Frequency (f)	Percentage (%)	Cumulative Percentage ($C\%$)
Less than 1 day	1,445	16	16
1 day per week	492	6	22
2 days per week	711	8	30
3 days per week	1,024	11	41
4–5 days per week	1,909	21	62
6–7 days per week	3,326	37	99
Total (N)	8,907	99[a]	

[a]Due to rounding, the percentage total does not add up to 100.

Percentiles are widely used to evaluate relative performance on standardized achievement tests, such as the SAT or ACT. Let's suppose that your ACT score was 29. To evaluate your performance for the college admissions officer, the testing service translated your score into a percentile rank. Your percentile rank was determined by comparing your score with the scores of all other students who took the test at the same time. Suppose for a moment that 90% of all students received a lower ACT score than you (and 10% scored above you). Your percentile rank would have been 90. If, however, there were more students who scored better than you—let's say that 15% scored above you and 85% scored lower than you—your percentile rank would have been 85.

Another widely used measure of location is the quartile. The lower quartile is equal to the 25th percentile and the upper quartile is equal to the 75th percentile. (Can you locate the upper quartile in Table 3.3?) A college admissions office interested in accepting the top 25% of its applicants based on their SAT scores could calculate the upper quartile (the 75th percentile) and accept everyone whose score is equivalent to the 75th percentile or higher. (Note that they would be calculating percentiles based on the scores of their applicants, not of all students in the nation who took the SAT.)

Mean: A measure typically used to describe central tendency in interval-ratio variables. The arithmetic average obtained by adding up all the scores and dividing by the total number of scores.

The Mean

The arithmetic **mean** is by far the best-known and most widely used measure of central tendency. The mean is what most people call the "average." The mean is typically used to describe central tendency in interval-ratio variables such as income, age, and education. You are probably already familiar with how to calculate the mean. Simply add up all the scores and divide by the total number of scores.

This calculation can be reflected in a mathematical formula. Beginning with this section, we introduce several formulas that will help you calculate some of the statistical concepts we review in each chapter. A formula is a shorthand way to explain what operations we need to follow to obtain a certain result. So instead of saying "add all the scores together and then divide by the number of scores," we can define the mean by the following formula:

$$\bar{Y} = \frac{\Sigma Y}{N} \tag{3.1}$$

Let's take a moment to consider these new symbols because we continue to use them in later chapters. We use Y to represent the raw scores in the distribution of the variable of interest; $\bar{Y}$ is pronounced "Y-bar" and is the mean of the variable of interest. The symbol represented by the Greek letter Σ is pronounced "sigma," and it is used often from now on. It is a summation sign (just like the Σ sign) and directs us to sum whatever comes after it. Therefore, ΣY means "add up all the raw Y scores." Finally, the letter N, as you know by now, represents the number of cases (or observations) in the distribution.

Let's summarize the symbols as follows:

Y = the raw scores of the variable Y

$\bar{Y}$ = the mean of Y

ΣY = the sum of all the Y scores

N = the number of observations or cases

Now that we know what the symbols mean, let's work through an example. The following are the ages of the 10 students in a graduate statistics class:

21, 32, 23, 41, 20, 30, 36, 22, 25, 27

What is the mean age of the students?

For these data, the ages included in this group are represented by Y, $N = 10$ is the number of students in the class, and ΣY is the sum of all the ages:

$$\Sigma Y = 21 + 32 + 23 + 41 + 20 + 30 + 36 + 22 + 25 + 27 = 277$$

Thus, the mean age is

$$\bar{Y} = \frac{\Sigma Y}{N} = \frac{277}{10} = 27.7$$

Let's take a look at one more example. Table 3.4 shows the 2016 incarceration rates (per 100,000 population) for 10 of the most populous U.S. states. We want to summarize the information presented in this table by calculating some measure of central tendency. Because the variable *incarceration rate* is an interval-ratio variable, we will select the arithmetic mean as our measure of central tendency.

To find the mean incarceration rate (number of people in federal or state prison per 100,000 population) for the data presented in Table 3.4, add up the incarceration rates for all states and divide the sum by the number of states:

$$\frac{(430+761+601+325+442+484+578+676+438+530)}{10}=526.5$$

The mean incarceration rate for 10 of the most populous states is 526.5.[5] For these 10 states, the average number of prisoners is 526.5.

The mean can also be calculated when the data are arranged in a frequency distribution. We have presented an example involving a frequency distribution in A Closer Look 3.1.

Table 3.4 2016 Incarceration Rates per 100,000 People for 10 of the Most Populous States

State	Incarceration per 100,000
California	430
Texas	761
Florida	601
New York	325
Illinois	442
Pennsylvania	484
Ohio	578
Georgia	676
North Carolina	438
Michigan	530

Source: E. Ann Carson, *Prisoners in 2016*. Bureau of Justice Statistics, NCJ 251149, 2018.

A CLOSER LOOK 3.1

Finding the Mean in a Frequency Distribution

When data are arranged in a frequency distribution, we must give each score its proper weight by multiplying it by its frequency. We can use the following modified formula to calculate the mean:

$$\bar{Y} = \frac{\sum(fY)}{N}$$

where

Y = the raw scores of the variable Y

$\bar{Y}$ = the mean of Y

$\sum(fY)$ = the sum of all the fYs

N = the number of observations or cases

We now illustrate how to calculate the mean from a frequency distribution using the preceding formula. In the 2018 GSS, respondents were asked about what they think is the ideal number of children for a family. Their responses are presented in the following table.

Ideal Number of Children, GSS 2018

Number of Children (Y)	Frequency (f)	Frequency $\sum(fY)$
0	11	0
1	17	17
2	367	734
3	203	609
4	78	312
5	9	45
6	6	36
Total	$N = 691$	$\sum(fY) = 1{,}753$

Notice that to calculate the value of $\sum(fY)$ (column 3), each score (column 1) is multiplied by its frequency (column 2), and the products are then added together. When we apply the formula,

$$\bar{Y} = \frac{\sum(fY)}{N} = \frac{1753}{691} = 2.54$$

we find that the mean for the ideal number of children is 2.54.

LEARNING CHECK 3.4

If you are having difficulty understanding how to find the mean in a frequency distribution, examine this illustrated table. It presents the process without using any notation.

Finding the Mean in a Frequency Distribution

Number of people per house	Number of houses like this	Number of people such houses contribute
1	3	3
2	5	10
3	1	3
4	1	4

Total number of people: 20
Total number of houses: 10
Mean number of people per house: 20/10 = 2

LEARNING CHECK 3.5

The following distribution is the same as the one you used to calculate the median in an earlier Learning Check: 22, 15, 18, 33, 17, 5, 11, 28, 40, 19, 8, 20. Calculate the mean. Is it the same as the median, or is it different?

Understanding Some Important Properties of the Arithmetic Mean

The following three mathematical properties make the mean the most important measure of central tendency. It is, in fact, a concept that is basic to numerous and more complex statistical operations.

Interval-Ratio Level of Measurement Because it requires the mathematical operations of addition and division, the mean can be calculated only for variables measured at the interval-ratio level. This is the only level of measurement that provides numbers that can be added and divided.

Center of Gravity Because the mean (unlike the mode and the median) incorporates all the scores in the distribution, we can think of it as the center of gravity of the distribution. That is, the mean is the point that perfectly balances all the scores in the distribution. If we subtract the mean from each score and add up all the differences, the sum will always be zero!

LEARNING CHECK 3.6

Why is the mean considered the center of gravity of the distribution? Think of the last time you were in a park on a seesaw (it may have been a long time ago) with a friend who was much heavier than you. You were left hanging in the air until your friend moved closer to the center. In short, to balance the seesaw, a light person far away from the center (the mean) can balance a heavier person who is closer to the center. Can you illustrate this principle with a simple income distribution?

Sensitivity to Extremes The examples we have used to show how to compute the mean demonstrate that, unlike with the mode or the median, every score enters into the calculation of the mean. This property makes the mean sensitive to extreme scores in the distribution. The mean is pulled in the direction of either very high or very low values. A glance at Figure 3.4 should convince you of that. Figure 3.4 shows the incomes of 10 individuals. In Figure 3.4b, the income of one individual has shifted from $5,000 to $35,000. Notice the effect it has on the mean; it shifts from $3,000 to $6,000! The mean is disproportionately affected by the relatively high income of $35,000 and is misleading as a measure of central tendency for this distribution. Notice that the median's value is not affected by this extreme score; it remains at $3,000. Thus, the median gives us better information on the typical income for this group. In the next section, we will see that because of the sensitivity of the mean, it is not suitable as a measure of central tendency in distributions that have a few very extreme values on one side of the distribution. (A few extreme values are no problem if they are not mostly on one side of the distribution.)

LEARNING CHECK 3.7

When asked to choose the appropriate measure of central tendency for a distribution, remember that the level of measurement is not the only consideration. When variables are measured at the interval-ratio level, the mean is usually the measure of choice, but remember that extreme scores in one direction make the mean unrepresentative and the median or mode may be the better choice.

Figure 3.4 The Value of the Mean Is Affected by Extreme Scores: (a) No Extreme Scores and (b) One Extreme Score

(a) No extreme scores: The mean is $3,000

Income (Y)	Frequency (f)	fY
1,000	1	1,000
2,000	2	4,000
3,000	4	12,000
4,000	2	8,000
5,000	1	5,000
	$N = 10$	$\Sigma(fY) = 30{,}000$

$$\text{Mean} = \frac{\Sigma(fY)}{N} = \frac{30{,}000}{10} = \$3{,}000$$

Median = $3,000

(b) One extreme score: The mean is $6,000

Income (Y)	Frequency (f)	fY
1,000	1	1,000
2,000	2	4,000
3,000	4	12,000
4,000	2	8,000
35,000	1	35,000
	$N = 10$	$\Sigma(fY) = 60{,}000$

$$\text{Mean} = \frac{\Sigma(fY)}{N} = \frac{60{,}000}{10} = \$6{,}000$$

Median = $3,000

Reading the Research Literature: The Case of Reporting Income

Although income levels have changed for American men and women in the past few decades, as we already noted in Chapter 1, gender income equality has yet to be achieved. We continue our investigation of gender income inequality by reviewing published income data.

As noted in the previous discussion (and in Figure 3.4), due to its sensitivity to extreme values, the mean may not be the best measure of central tendency for income. In fact, median earnings are routinely reported in scholarly research and by government publications such as those reported by the U.S. Department of Labor and the U.S. Census Bureau.

Figure 3.5 compares the median weekly earnings of women and men who are full-time wage and salary workers by age for 2017. We can use these medians to note the differences between subgroups of the population or changes over time.

For 2017, women's median earnings were 82% of men's—on average, women's median usual weekly earnings were $770 and men's median usual weekly earnings were $941. The data presented in Figure 3.5 reveal that for each age group, men earn more than women. The gender earnings gap is largest, $248 (1,103 – 855), for the 45- to 54-year age group. The smallest earnings gap is $48 (547 – 499) for Americans aged 16 to 24 years.

Figure 3.5 Median Usual Weekly Earnings of Women and Men, Full-Time Wage and Salary Workers, by Age, 2017 Annual Averages

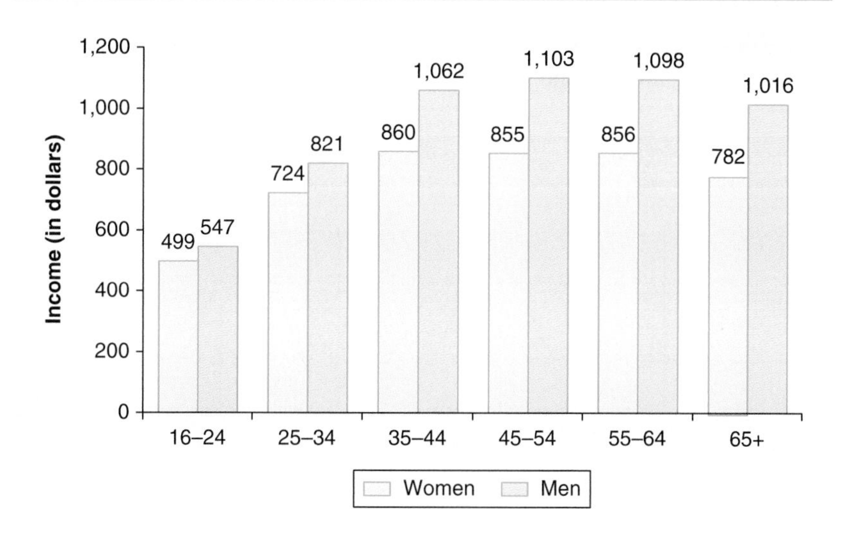

Source: U.S. Bureau of Labor Statistics. BLS Reports, "Highlights of Women's Earnings in 2017," August 2018.

When mean income scores are reported, interpret with caution as extremely low- or high-income values may skew the distribution. When a researcher reports an "average" score, assess whether the calculation was based on a median or a mean.

LEARNING CHECK 3.8

Examine Figure 3.5 and contrast the median weekly earnings of women and men. How else would you describe the relationship between age and median weekly earnings?

Statistics in Practice: The Shape of the Distribution

The distribution of interval-ratio variables can also be described by their general shape. As we learned in Chapter 2, histograms show the differences in frequencies or percentages on an interval-ratio variable. Using histograms, produced by SPSS, we will demonstrate how a distribution can be either symmetrical or skewed, depending on whether there are a few extreme values at one end of the distribution.

Symmetrical distribution: The frequencies at the right and left tails of the distribution are identical; each half of the distribution is the mirror image of the other.

The Symmetrical Distribution

A distribution is symmetrical (Figure 3.6) if the frequencies at the right and left tails of the distribution are identical, so that if it is divided into two halves, each will be the mirror image of the other. In a unimodal, symmetrical distribution, the mean, median, and mode are identical.

Figure 3.6 Types of Frequency Distributions

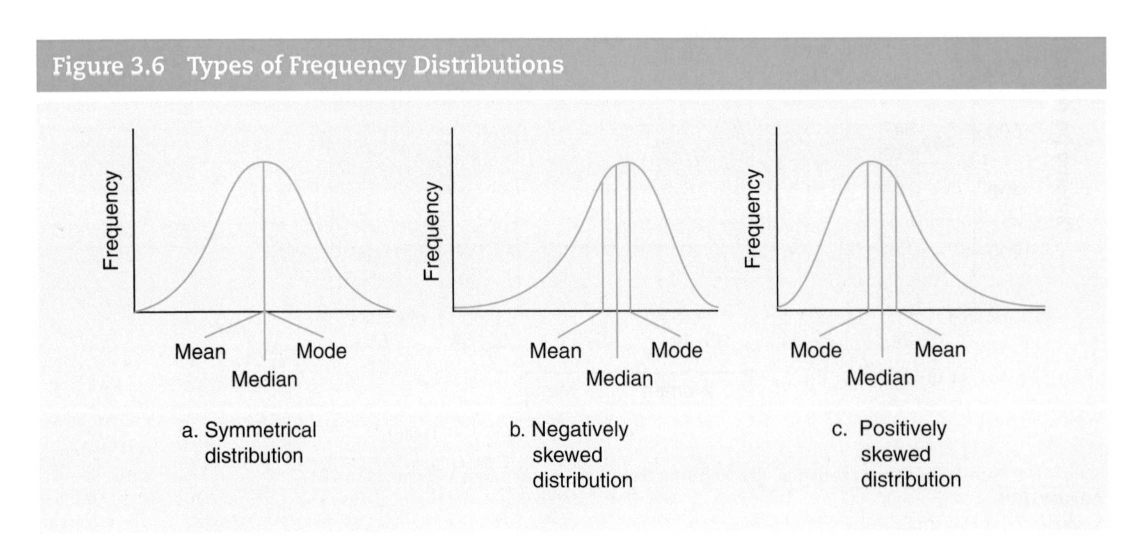

Figure 3.7 Number of Children, GSS 2014

Statistics

NChilds

N	Valid	690
	Missing	489
Mean		1.9855
Median		2.0000
Mode		2.00

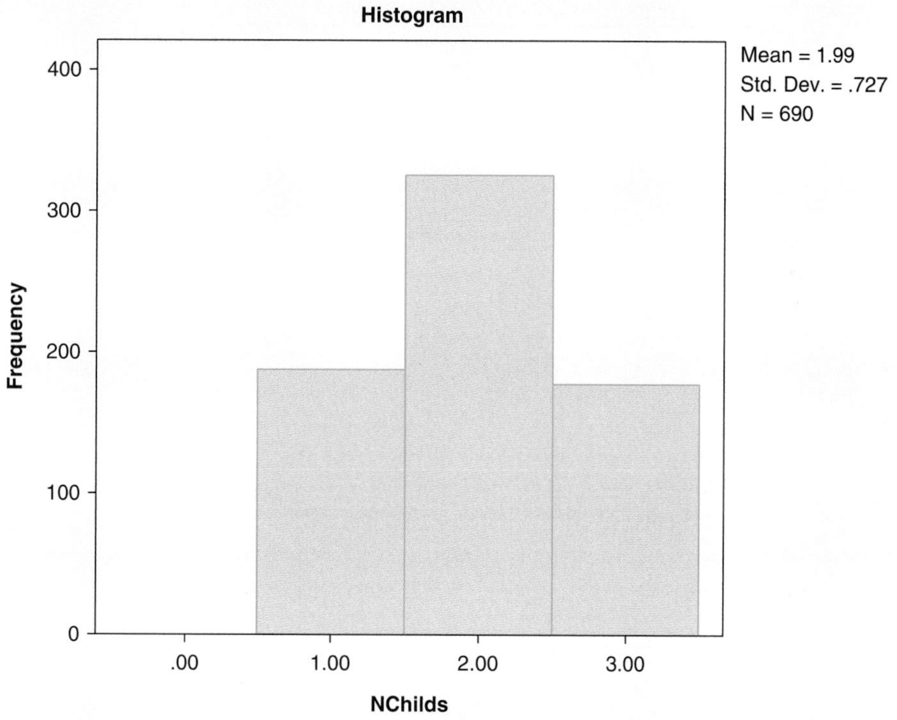

Histogram

Mean = 1.99
Std. Dev. = .727
N = 690

Frequency: 0, 100, 200, 300, 400

NChilds: .00, 1.00, 2.00, 3.00

For example, refer to Figure 3.7, displaying the distribution of the number of children. Only Categories 1–3 are shown. The distribution is symmetrical with the mean, the median, and the mode at 2.0 (the mean of 1.9 would be rounded to 2.0).

The Positively Skewed Distribution

As a general rule, for skewed distributions, the mean, the median, and the mode do not coincide. The mean, which is always pulled in the direction of extreme scores, falls closest to the tail of the distribution where a small number of extreme scores are located.

Skewed distribution: A distribution with a few extreme values on one side of the distribution.

Figure 3.8 Number of Internet Hours per Week, GSS 2014 Subsample

Statistics

wwwhr WWW HOURS PER WEEK

N	Valid	423
	Missing	756
Mean		11.91
Median		6.00
Mode		1

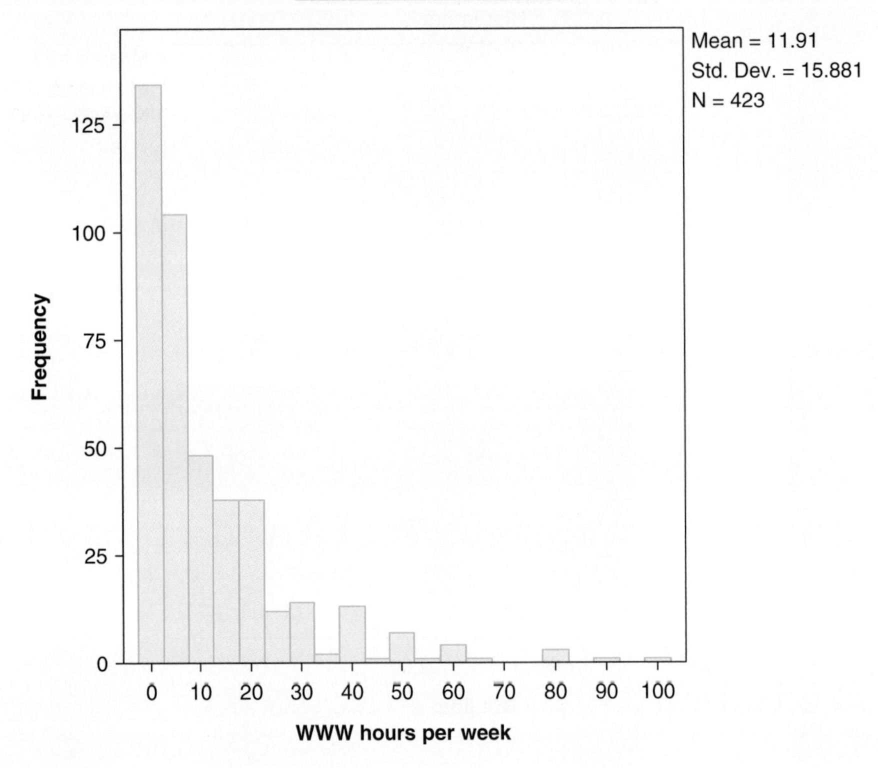

Positively skewed distribution: A distribution with a few extremely high values.

In Figure 3.8, we present a **positively skewed distribution**. The histogram reports hours per week on the Internet, with the mean (11.91) greater than the median (6.00) and the mode (1). Confirm how the shape of this distribution matches Figure 3.6.

The Negatively Skewed Distribution

Now examine Figure 3.9 for the number of years spent in school among those respondents who did not finish high school. Here you can see the opposite pattern. The distribution of the number of years spent in school for those without

Figure 3.9 Years of School Among Respondents Without a High School Degree, GSS 2014

Statistics

educ HIGHEST YEAR OF SCHOOL COMPLETED

N	Valid	169
	Missing	0
Mean		8.89
Median		10.00
Mode		11

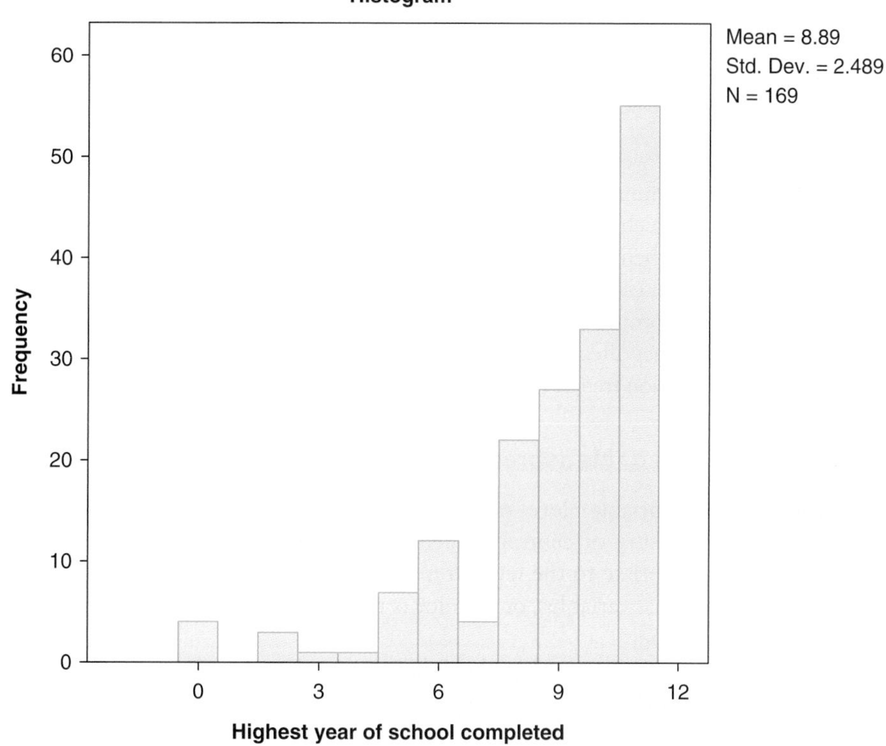

a high school diploma is a negatively skewed distribution (similar to what is presented in Figure 3.6). First, note that the largest number of years spent in school is concentrated at the high end of the scale (11 years) and that there are fewer respondents at the low end. The mean, the median, and the mode also differ in values as they did in the previous example. However, here the mode has the highest value (mode = 11.0), the median has the second highest value (median = 10.0), and the mean has the lowest value (mean = 8.89).

Negatively skewed distribution: A distribution with a few extremely low values.

Guidelines for Identifying the Shape of a Distribution

Following are some useful guidelines for identifying the shape of a distribution:

1. In unimodal distributions, when the mode, the median, and the mean coincide or are almost identical, the distribution is symmetrical.
2. When the mean is higher than the median (or is positioned to the right of the median), the distribution is positively skewed.
3. When the mean is lower than the median (or is positioned to the left of the median), the distribution is negatively skewed.

Considerations for Choosing a Measure of Central Tendency

So far, we have considered three basic kinds of measures: (1) the mode, (2) the median, and (3) the mean. Each can represent the central tendency of a distribution. But which one should we use? The mode? The median? The mean? Or, perhaps, all of them? There is no simple answer to this question. However, in general, we tend to use only one of the three measures of central tendency, and the choice of the appropriate one involves a number of considerations. These considerations and how they affect our choice of the appropriate measure are presented in the form of a decision tree in Figure 3.11.

Level of Measurement

The variable's level of measurement is the primary consideration in choosing a measure of central tendency. The measure of central tendency should be appropriate to the level of measurement. Thus, as shown in Figure 3.11, with nominal variables, our choice is restricted to the mode as a measure of central tendency.

However, with ordinal data, we have two choices: (1) the mode or (2) the median (or sometimes both). Our choice depends on what we want to know about the distribution. If we are interested in showing what is the most common or typical value in the distribution, then our choice is the mode. If, however, we want to show which value is located exactly in the middle of the distribution, then the median is our measure of choice.

When the data are measured on an interval-ratio level, the choice between the appropriate measures is a bit more complex and is restricted by the shape of the distribution.

Skewed Distribution

When the distribution is skewed, the mean may give misleading information on the central tendency because its value is affected by extreme scores in the

A CLOSER LOOK 3.2

A Cautionary Note: Representing Income

Personal income is frequently positively skewed because there are a few people with very high incomes; therefore, the mean may not be the most appropriate measure to represent average income. For example, the 2017 American Community Survey—an ongoing survey of economic and income statistics—reported the 2017 mean and median annual earnings of white, black, and Hispanic households in the United States. In Figure 3.10, we compare the mean and median income for each group.

Figure 3.10 Household Mean and Median Income by Race, 2017

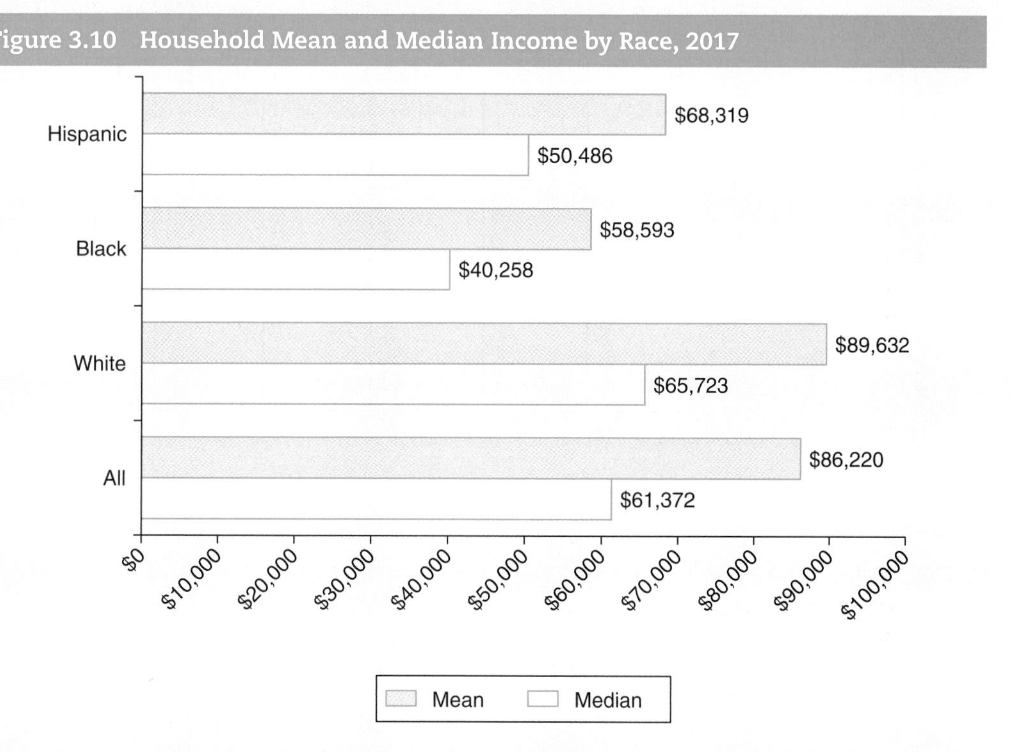

Source: U.S. Census Bureau, *Historical Income Tables: Households*, Table H-5, 2017.

As shown, for all groups in Figure 3.10, the reported mean is higher than the median. This discrepancy indicates that household income in the United States is highly skewed, with the mean overrepresenting those households in the upper-income bracket and misrepresenting the income of the average household. A preferable alternative is to use the median annual earnings of these groups.

Because the earnings of whites are the highest in comparison with all other groups, it is useful to look at each group's median earnings relative to the earnings of whites. For example, Hispanics were paid just 76 cents ($68,319/$89,632) and blacks were paid 65 cents ($58,593/$89,632) for every $1 paid to whites.

Figure 3.11 How to Choose a Measure of Central Tendency

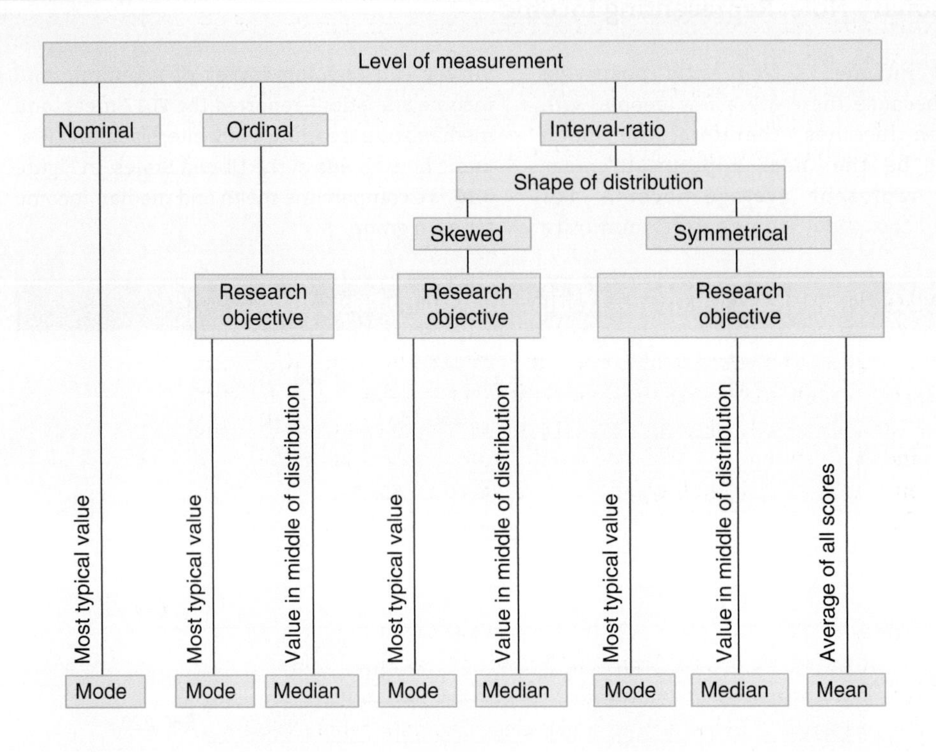

distribution. The median (see, e.g., A Closer Look 3.2) or the mode can be chosen as the preferred measure of central tendency because neither is influenced by extreme scores.

Symmetrical Distribution

When the distribution we want to analyze is symmetrical, we can use any of the three averages. Again, our choice depends on the research objective and what we want to know about the distribution. In general, however, the mean is our best choice because it contains the greatest amount of information and is easier to use in more advanced statistical analyses.

MEASURES OF VARIABILITY

Although measures of central tendency can be very helpful, they tell only part of the story. In fact, when used alone, they may mislead rather than inform. Another way of summarizing a distribution of data is by selecting a single number that describes how much variation and diversity there are in the distribution. Numbers that describe diversity or variation are called measures of variability. Researchers often use measures of central tendency along with measures of variability to describe their data.

We will discuss four measures of variability: (1) the range, (2) the interquartile range, (3) the standard deviation, and (4) the variance. Before we discuss these measures, let's explore why they are important.

The Importance of Measuring Variability

The importance of looking at variation and diversity can be illustrated by thinking about the differences in the experiences of U.S. women. Are women united by their similarities or are they divided by their differences? The answer is *both*. To address the similarities without dealing with differences is "to misunderstand and distort that which separates as well as that which binds women together."[6] Even when we focus on one particular group of women, it is important to look at the differences as well as the commonalities. Take, for example, Asian American women. As a group, they share a number of characteristics.

> Their participation in the workforce is higher than that of women in any other ethnic group. Many . . . live life supporting others, often allowing their lives to be subsumed by the needs of the extended family. . . . However, there are many circumstances when these shared experiences are not sufficient to accurately describe the condition of a particular Asian-American woman. Among Asian-American women there are those who were born in the United States . . . and . . . those who recently arrived in the United States. Asian-American women are diverse in their heritage or country of origin: China, Japan, the Philippines, Korea . . . and . . . India. . . . Although the majority of Asian-American women are working class—contrary to the stereotype of the "ever successful" Asians—there are poor, "middle-class," and even affluent Asian-American women.[7]

One form of stereotyping is treating a group as if it were totally characterized by its central value, ignoring the diversity within the group. The complete story of a particular group, like Asian American women, can best be told by examining their commonalities and their differences. We learned earlier in this chapter how measures of central tendency can be used to document what is common or average for a group of individuals, and now, we'll learn different measures of variation to understand the diversity of experiences. The concept of variability has implications not only for describing the diversity of social groups such as Asian American women but also for issues that are important in your everyday life.

The Range

Demographers track population trends, focusing on state and regional growth. It is predicted that the U.S. population, currently over 320 million, will cross the 400 million mark in 2058. Table 3.5 displays the percentage change in the population from 2010 to 2016 by region and by state as documented by the U.S. Census Bureau. We can apply several measures of variability to better understand U.S. population growth.

Range: A measure of variation in interval-ratio variables. It is the difference between the highest (maximum) and the lowest (minimum) scores in the distribution.

The simplest and most straightforward measure of variation is the **range**, which measures variation in interval-ratio variables. It is the difference between the highest (maximum) and the lowest (minimum) scores in the distribution:

$$\text{Range} = \text{Highest score} - \text{Lowest score}$$

What is the range in the percentage change in state population for the United States? To find the ranges in a distribution, simply pick out the highest and the lowest scores in the distribution and subtract. Washington, D.C. (a city, not a state), has the highest percentage change, with 13.2%, and West Virginia has the lowest change, with −1.2%. The range is 14.4 percentage points, or 13.2% to −1.2%.

Table 3.5 Percentage Change in the Population by Region and State, 2010–2016

Region and State	Percentage Change	Region and State	Percentage Change
Northeast		Iowa	2.9
Connecticut	0.1	Kansas	1.9
Delaware	6.0	Michigan	0.4
Maine	0.2	Minnesota	4.1
Massachusetts	4.0	Missouri	1.7
New Hampshire	1.4	Nebraska	4.4
New Jersey	1.7	North Dakota	12.7
New York	1.9	Ohio	0.7
Pennsylvania	0.6	South Dakota	6.3
Rhode Island	0.3	Wisconsin	1.6
Vermont	−0.2	**South**	
Midwest		Alabama	4.7
Indiana	2.3	Arkansas	2.5
Illinois	−0.2	Florida	9.6

Region and State	Percentage Change
Georgia	6.4
Kentucky	2.2
Louisiana	3.3
Maryland	4.2
Mississippi	0.7
North Carolina	6.4
Oklahoma	4.6
South Carolina	7.3
Tennessee	4.8
Texas	10.8
Virginia	5.1
Washington, D.C.	13.2
West Virginia	–1.2
West	
Alaska	4.5
Arizona	8.4
California	5.4
Colorado	10.2
Hawaii	5.0
Idaho	7.4
Montana	5.4
Nevada	8.9
New Mexico	1.1
Oregon	6.8
Utah	10.4
Washington	8.4
Wyoming	3.9

Source: ProQuest, *Statistical Abstract of the United States 2018: The National Data Book*, Table 20.

Although the range is simple and quick to calculate, it is a rather crude measure because it is based on only the lowest and the highest scores. These two scores might be extreme and rather atypical, which might make the range a misleading indicator of the variation in the distribution. For instance, note that among the 50 states and Washington, D.C., listed in Table 3.5, no other state has a percentage decrease as that of West Virginia, and no other state has the percentage increase as that of Washington, D.C. The range of 14.4 percentage points does not give us information about the variation in states between Washington, D.C., and West Virginia.

LEARNING CHECK 3.9

Why can't we use the range to describe diversity in nominal variables? The range can be used to describe diversity in ordinal variables (e.g., we can say that responses to a question ranged from "somewhat satisfied" to "very dissatisfied"), but it has no quantitative meaning. Why not?

The Interquartile Range

To remedy the limitation of the range, we can employ an alternative—the **interquartile range**. The interquartile range (IQR), a measure of variation for interval-ratio and ordinal variables, is the width of the middle 50% of the

Interquartile range (IQR): The width of the middle 50% of the distribution. It is defined as the difference between the lower and upper quartiles (Q_1 and Q_3). IQR can be calculated for interval-ratio and ordinal data.

distribution. It is defined as the difference between the lower and upper quartiles (Q_1 and Q_3).

$$\text{IQR} = Q_3 - Q_1$$

Recall that the first quartile (Q_1) is the 25th percentile, the point at which 25% of the cases fall below it and 75% above it. The third quartile (Q_3) is the 75th percentile, the point at which 75% of the cases fall below it and 25% above it. The IQR, therefore, defines variation for the middle 50% of the cases.

Like the range, the IQR is based on only two scores. However, because it is based on intermediate scores, rather than on the extreme scores in the distribution, it avoids some of the instability associated with the range.

These are the steps for calculating the IQR:

1. To find Q_1 and Q_3, order the scores in the distribution from the highest to the lowest score, or vice versa. Table 3.6 presents the data of Table 3.5 arranged in order from Washington, D.C. (13.2%), to West Virginia (−1.2%).

2. Next, we need to identify the first quartile, Q_1 or the 25th percentile. We have to identify the percentage increase in the population associated with the state that divides the distribution so that 25% of the states are below it and 75% of the states are above it. To find Q_1, we multiply N by 0.25:

$$(N)(0.25) = (51)(0.25) = 12.75$$

Table 3.6 Percentage Change in the Population 2010–2016, by State, Ordered From the Highest to the Lowest

State	Percentage Change	State	Percentage Change	State	Percentage Change
Washington, D.C.	13.2	Montana	5.4	New York	1.9
North Dakota	12.7	Virginia	5.1	Kansas	1.9
Texas	10.8	Hawaii	5	New Jersey	1.7
Utah	10.4	Tennessee	4.8	Missouri	1.7
Colorado	10.2	Alabama	4.7	Wisconsin	1.6
Florida	9.6	Oklahoma	4.6	New Hampshire	1.4
Nevada	8.9	Alaska	4.5	New Mexico	1.1
Arizona	8.4	Nebraska	4.4	Ohio	0.7

State	Percentage Change	State	Percentage Change	State	Percentage Change
Washington	8.4	Maryland	4.2	Mississippi	0.7
Idaho	7.4	Minnesota	4.1	Pennsylvania	0.6
South Carolina	7.3	Massachusetts	4	Michigan	0.4
Oregon	6.8	Wyoming	3.9	Rhode Island	0.3
Georgia	6.4	Louisiana	3.3	Maine	0.2
North Carolina	6.4	Iowa	2.9	Connecticut	0.1
South Dakota	6.3	Arkansas	2.5	Illinois	−0.2
Delaware	6	Indiana	2.3	Vermont	−0.2
California	5.4	Kentucky	2.2	West Virginia	−1.2

Source: ProQuest, *Statistical Abstract of the United States 2018: The National Data Book,* Table 20.

The first quartile falls between the 12th and the 13th states. Counting from the bottom, the 12th state is New Hampshire, and the percentage increase associated with it is 1.4. The 13th state is Wisconsin, with a percentage increase of 1.6. To find the first quartile, we take the average of 1.4 and 1.6. Therefore, $(1.4 + 1.6)/2 = 1.5$ is the first quartile (Q_1).

3. To find Q_3, we have to identify the state that divides the distribution in such a way that 75% of the states are below it and 25% of the states are above it. We multiply N this time by 0.75:

$$(N)(0.75) = (51)(0.75) = 38.25$$

The third quartile falls between the 38th and the 39th states. Counting from the bottom, the 38th state is North Carolina, and the percentage increase associated with it is 6.4. The 39th state is Georgia, with a percentage increase of 6.4. To find the third quartile, we take the average of 6.4 and 6.4. Given that the percentage increase is identical for the two states, 6.4 is the third quartile (Q_3).

4. We are now ready to find the IQR:

$$\text{IQR} = Q_3 - Q_1 = 6.4 - 1.5 = 4.9$$

It may be more useful to report the full IQR (6.4 to 1.5) rather than the single value (4.9). The IQR tells us that half the states are clustered between 6.4 and 1.5, a narrower (and more meaningful) spread than the range of 14.4 points (13.2% to −1.2%).

LEARNING CHECK 3.10

Why is the IQR better than the range as a measure of variability, especially when there are extreme scores in the distribution? To answer this question, you may want to examine Figure 3.12.

Figure 3.12 The Range Versus the Interquartile Range: Number of Children Among Two Groups of Women

		Group 1: IQR = 1–3	Group 2: IQR = 1–6	
	Number of Children	Less Variable	More Variable	
	0			
Q_1 ⇨	1			⇦ Q_1
	2			
Q_3 ⇨	3			
	4			
	5			
	6			⇦ Q_3
	7			
	8			
	9			
	10			
		Range = 10	Range = 10	
		Interquartile range = 2	Interquartile range = 5	

The Box Plot

A graphic device called the box plot can visually present the range, the IQR, the median, the lowest (minimum) score, and the highest (maximum) score. The box plot provides us with a way to visually examine the center, the variation, and the shape of distributions of interval-ratio variables.

Figure 3.13 is a box plot of the distribution of the 2010–2016 percentage increase in the U.S. population displayed in Table 3.6. To construct the box plot in Figure 3.13, we used the lowest and the highest values in the distribution, the upper and lower quartiles, and the median. We can easily draw a box plot by hand following these instructions:

1. Draw a box between the lower and upper quartiles.
2. Draw a solid line within the box to mark the median.
3. Draw vertical lines (called whiskers) outside the box, extending to the lowest and highest values.

What can we learn from creating a box plot? We can obtain a visual impression of the following properties: First, the center of the distribution is easily identified by the solid line inside the box. Second, since the box is drawn between the lower and upper quartiles, the IQR is reflected in the height of the box. Similarly, the length of the vertical lines drawn outside the box (on both ends) represents the range of the distribution.[8] Both the IQR and the range give us a visual impression of the spread in the distribution. Finally, the relative position of the box and the position of the median within the box tell us whether the distribution is symmetrical or skewed. A perfectly symmetrical distribution would have the box at the center of the range as well as the median in the center of the box. When the distribution departs from symmetry, the box and/or the median will not be centered; it will be closer to the lower quartile when there are more cases with lower scores or to the upper quartile when there are more cases with higher scores.

Box plots are particularly useful for comparing distributions. To demonstrate box plots that are shaped quite differently, in Figure 3.14, we have used

Figure 3.13 Box Plot of the Percentage Change in the Population, 2010–2016

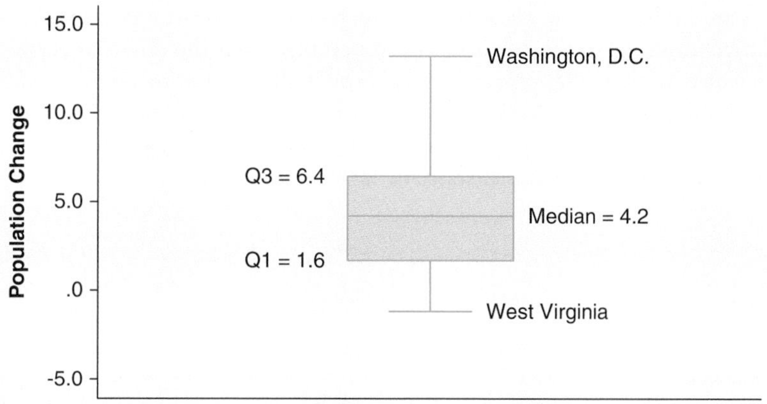

15.0
10.0
5.0
.0
-5.0
Population Change
Washington, D.C.
Q3 = 6.4
Median = 4.2
Q1 = 1.6
West Virginia

Source: ProQuest, *Statistical Abstract of the United States 2018: The National Data Book*, Table 20.

Figure 3.14 Box Plot of the Percentage Change in the Population in the Northeast and West, 2010–2016

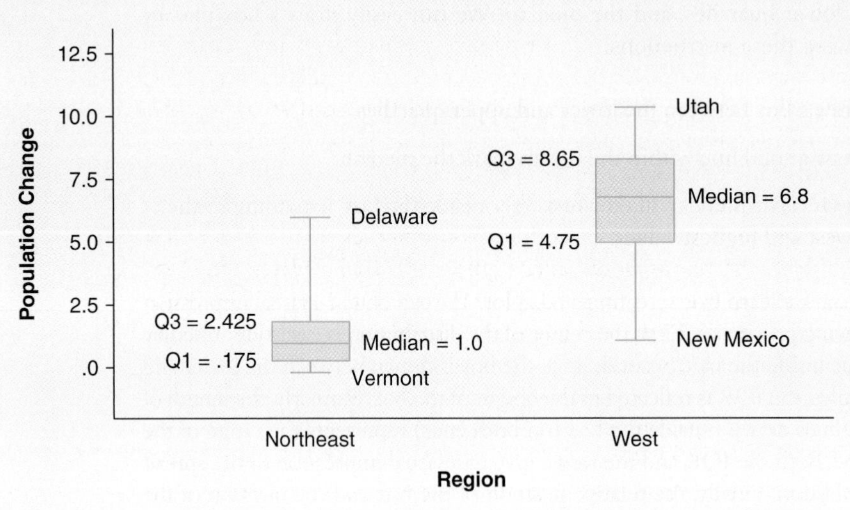

Source: ProQuest, *Statistical Abstract of the United States 2018: The National Data Book*, Table 20.

the data on the percentage increase in the population (Table 3.6) to compare the pattern of change occurring between 2010 and 2016 in the northeastern and western regions of the United States.

As you can see, the box plots differ from each other considerably. First, the positions of the medians highlight the dramatic increase in the population in the western United States. While the Northeast experienced a 1% increase (median = 1.0%) in its population, the West shows a higher percentage increase (median = 6.80%). Second, the range (illustrated by the position of the whiskers in each box plot) and the IQR (illustrated by the height of the box) are higher for the West (6.80% median and 9.30% range). Finally, the relative positions of the boxes tell us something about the different shapes of these distributions. Because its box is at the higher end of its range, the West distribution is positively skewed. In contrast, with its box off closer to the lower end of the distribution, the distribution of percentage change in the population for the Northeast states is negatively skewed.

LEARNING CHECK 3.11

Is the distribution shown in the box plot in Figure 3.13 symmetrical or skewed?

The Variance and the Standard Deviation

Table 3.7 presents the average percentage change in the population for all regions of the United States. The table shows that between 2010 and 2016, the size of the population in the United States increased by an average of 4.18%. But this average increase does not inform us about the regional variation in the population. For example, will the northeastern states show a smaller-than-average increase because of the outmigration? Is the population increase higher in the South because of the immigration of the elderly?

Although it is important to know the average percentage increase for the nation as a whole, you may also want to know whether regional increases might differ from the national average. If the regional increases are close to the national average, the figures will cluster around the mean, but if the regional increases deviate much from the national average, they will be widely dispersed around the mean.

Table 3.7 suggests that there is considerable regional variation. The percentage change ranges from 6.60% in the West to 1.60% in the Northeast, so the range is 5.00% (6.60% − 1.60% = 5.00%). Moreover, most of the regions deviate considerably from the national average of 4.18%. How large are these deviations on the average? We want a measure that will give us information about the overall variations among all regions in the United States and, unlike the range or the IQR, will not be based on only two scores.

Such a measure will reflect how much, on the average, each score in the distribution deviates from some central point, such as the mean. We use the mean as the reference point rather than other kinds of averages (the mode or the median) because the mean is based on all the scores in the distribution. Therefore, it is more useful as a basis from which to calculate average deviation. The sensitivity of the mean to extreme values carries over the

Table 3.7 Average Percentage Change in the Population by Region, 2010–2016

Region	Percentage
Northeast	1.60
South	5.29
Midwest	3.23
West	6.60
Mean ($\bar{Y}$)	4.18

Source: ProQuest, *Statistical Abstract of the United States 2018: The National Data Book*, Table 20.

calculation of the average deviation, which is based on the mean. Another reason for using the mean as a reference point is that more advanced measures of variation require the use of algebraic properties that can be assumed only by using the arithmetic mean.

The variance and the standard deviation are two closely related measures of variation that increase or decrease based on how closely the scores cluster around the mean. The **variance** is the average of the squared deviations from the center (mean) of the distribution, and the **standard deviation** is the square root of the variance. Both measure variability in interval-ratio and ordinal variables.

Variance: A measure of variation for interval-ratio and ordinal variables; it is the average of the squared deviations from the mean.

Standard deviation: A measure of variation for interval-ratio and ordinal variables; it is equal to the square root of the variance.

Calculating the Deviation From the Mean

Consider again the distribution of the percentage change in the elderly population for the four regions of the United States. Because we want to calculate the average difference of all the regions from the national average (the mean), it makes sense to first look at the difference between each region and the mean. This difference, called a deviation from the mean, is symbolized as $(Y - \bar{Y})$. The sum of these deviations can be symbolized as $\sum(Y - \bar{Y})$.

The calculations of these deviations for each region are displayed in Table 3.8 and Figure 3.15. We have also summed these deviations. Note that each region has either a positive or a negative deviation score. The deviation is positive when the percentage change in the elderly home population is above the mean. It is negative when the percentage change is below the mean. Thus, for example, the Northeast's deviation score of –2.58 means that its percentage change in the elderly population was 2.58 percentage points below the mean.

You may wonder if we could calculate the average of these deviations by simply adding up the deviations and dividing them. Unfortunately, we cannot, because the sum of the deviations of scores from the mean is always zero, or algebraically, $\sum(Y - \bar{Y})$. In other words, if we were to subtract the mean from each score and then add up all the deviations as we did in Table 3.8, the sum would be zero, which in turn would cause the average deviation (i.e., average difference) to compute to zero. This is always true because the mean is the center of gravity of the distribution.

Mathematically, we can overcome this problem either by ignoring the plus and minus signs, using instead the absolute values of the deviations, or by squaring the deviations—that is, multiplying each deviation by itself to get rid of the negative sign. Since absolute values are difficult to work with mathematically, the latter method is used to compensate for the problem.

Table 3.9 presents the same information as Table 3.8, but here we have squared the actual deviations from the mean and added together the squares.

Figure 3.15 Illustrating Deviations From the Mean

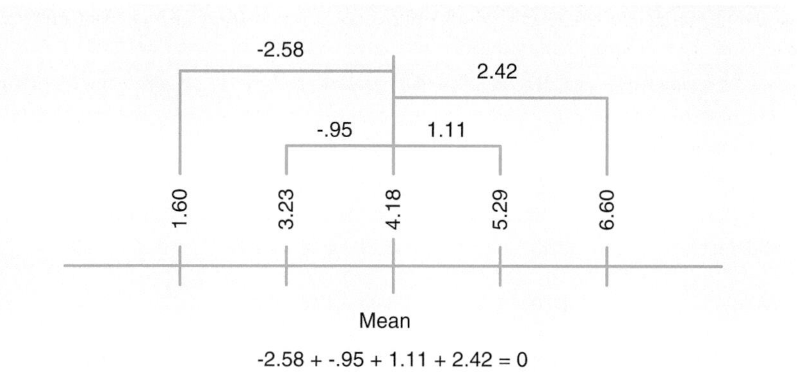

Table 3.8 Average Change in the Population, 2010–2016, by Region and Deviation From the Mean

Region	Percentage	$(Y - \bar{Y})$
Northeast	1.60	$1.60 - 4.18 = -2.58$
South	5.29	$5.29 - 4.18 = 1.11$
Midwest	3.23	$3.23 - 4.18 = -0.95$
West	6.60	$6.60 - 4.18 = 2.42$
	$\Sigma(Y) = 16.72$	$\Sigma(Y - \bar{Y}) = 0$
	$\bar{Y} = \frac{\Sigma Y}{N} = \frac{16.72}{4} = 4.18$	

Source: ProQuest, *Statistical Abstract of the United States 2018: The National Data Book*, Table 20.

The sum of the squared deviations is symbolized as $\Sigma(Y - \bar{Y})^2$. Note that by squaring the deviations, we end up with a sum representing the deviation from the mean, which is positive. (Note that this sum will equal zero if all the cases have the same value as the mean.) In our example, this sum is $\Sigma(Y - \bar{Y})^2 = 14.65$.

Table 3.9 Average Change in the Population, 2010–2016, by Region, Deviation From the Mean, and Deviation From the Mean Squared

Region	Percentage	$(Y - \bar{Y})$	$(Y - \bar{Y})^2$
Northeast	1.60	$1.60 - 4.18 = -2.58$	6.66
South	5.29	$5.29 - 4.18 = 1.11$	1.23
Midwest	3.23	$3.23 - 4.18 = -0.95$	.90
West	6.60	$6.60 - 4.18 = 2.42$	5.86
	$\Sigma(Y) = 16.72$	$\Sigma(Y - \bar{Y}) = 0$	$\Sigma(Y - \bar{Y})^2 = 14.65$

$$\text{Mean} = \bar{Y} = \frac{\Sigma Y}{N} = \frac{16.72}{4} = 4.18$$

Source: ProQuest, *Statistical Abstract of the United States 2018: The National Data Book*, Table 20.

Calculating the Variance and the Standard Deviation

The average of the squared deviations from the mean is known as the variance. The variance is symbolized as s^2. Remember that we are interested in the average of the squared deviations from the mean. Therefore, we need to divide the sum of the squared deviations by the number of scores (N) in the distribution. However, unlike the calculation of the mean, we will use $N - 1$ rather than N in the denominator.[9] The formula for the variance can be stated as

$$s^2 = \frac{\Sigma(Y - \bar{Y})^2}{N - 1} \tag{3.2}$$

where

s^2 = the variance

$(Y - \bar{Y})$ = the deviation from the mean

$\sum(Y - \bar{Y})^2$ = the sum of the squared deviations from the mean

N = the number of scores

Note that the formula incorporates all the symbols we defined earlier. This formula means that the variance is equal to the average of the squared deviations from the mean.

Follow these steps to calculate the variance:

1. Calculate the mean $\bar{Y} = \Sigma(Y)/N$
2. Subtract the mean from each score to find the deviation $Y - \bar{Y}$
3. Square each deviation $(Y - \bar{Y})^2$
4. Sum the squared deviations $\Sigma(Y - \bar{Y})^2$
5. Divide the sum by $N - 1$, $\Sigma(Y - \bar{Y})^2 / (N-1)$
6. The answer is the variance.

To assure yourself that you understand how to calculate the variance, go back to Table 3.9 and follow this step-by-step procedure for calculating the variance. Now plug the required quantities into Formula 3.2. Your result should look like this:

$$s^2 = \frac{\Sigma(Y - \bar{Y})^2}{N-1} = \frac{14.65}{3} = 4.88$$

One problem with the variance is that it is based on squared deviations and therefore is no longer expressed in the original units of measurement. For instance, it is difficult to interpret the variance of 4.88, which represents the distribution of the percentage change in the elderly population, because this figure is expressed in squared percentages. Thus, we often take the square root of the variance and interpret it instead. This gives us the standard deviation, symbolized as s, which is the square root of the variance, or

$$s = \sqrt{s^2}$$

The standard deviation for our example is

$$s = \sqrt{4.88} = 2.21$$

The formula for the standard deviation uses the same symbols as the formula for the variance:

$$s = \sqrt{\frac{\Sigma(Y - \bar{Y})^2}{N-1}} \qquad (3.3)$$

As we interpret the formula, we can say that the standard deviation is equal to the square root of the average of the squared deviations from the mean.

The advantage of the standard deviation is that unlike the variance, it is measured in the same units as the original data. For instance, the standard deviation for our example is 2.21. Because the original data were expressed in percentages, this number is expressed as a percentage as well. In other words, you could say that the standard deviation is 2.21%. But what does this mean? The actual number tells us very little by itself, but it allows us to evaluate the dispersion of the scores around the mean. We will explore this further in Chapter 4.

In a distribution where all the scores are identical, the standard deviation is zero (0). Zero is the lowest possible value for the standard deviation; in an identical distribution, all the points would be the same, with the same mean, mode, and median. There is no variation or dispersion in the scores.

The more the standard deviation departs from zero, the more variation there is in the distribution. There is no upper limit to the value of the standard deviation. In our example, we can conclude that a standard deviation of 2.21% means that the percentage change in the population for the four regions of the United States is narrowly dispersed around the mean of 4.18%.

The standard deviation can be considered a standard against which we can evaluate the positioning of scores relative to the mean and to other scores in the distribution. As we will see in more detail in Chapter 4, in most distributions, unless they are highly skewed, about 34% of all scores fall between the mean and 1 standard deviation above the mean. Another 34% of scores fall between the mean and 1 standard deviation below it. Thus, we would expect the majority of scores (68%) to fall within 1 standard deviation of the mean.

LEARNING CHECK 3.12

Take time to understand the section on standard deviation and variance. You will see these statistics in more advanced procedures. Although your instructor may require you to memorize the formulas, it is more important for you to understand how to interpret standard deviation and variance and when they can be appropriately used. Many hand calculators and all statistical software programs will calculate these measures of diversity for you, but they won't tell you what they mean. Once you understand the meaning behind these measures, the formulas will be easier to remember.

Considerations for Choosing a Measure of Variation

So far, we have considered five measures of variation: (1) the range, (2) the IQR, (3) the variance, and (4) the standard deviation. Each measure can represent the degree of variability in a distribution. But which one should we use? There is no simple answer to this question. However, in general, we tend to use only one measure of variation, and

Figure 3.16 How to Choose a Measure of Variation

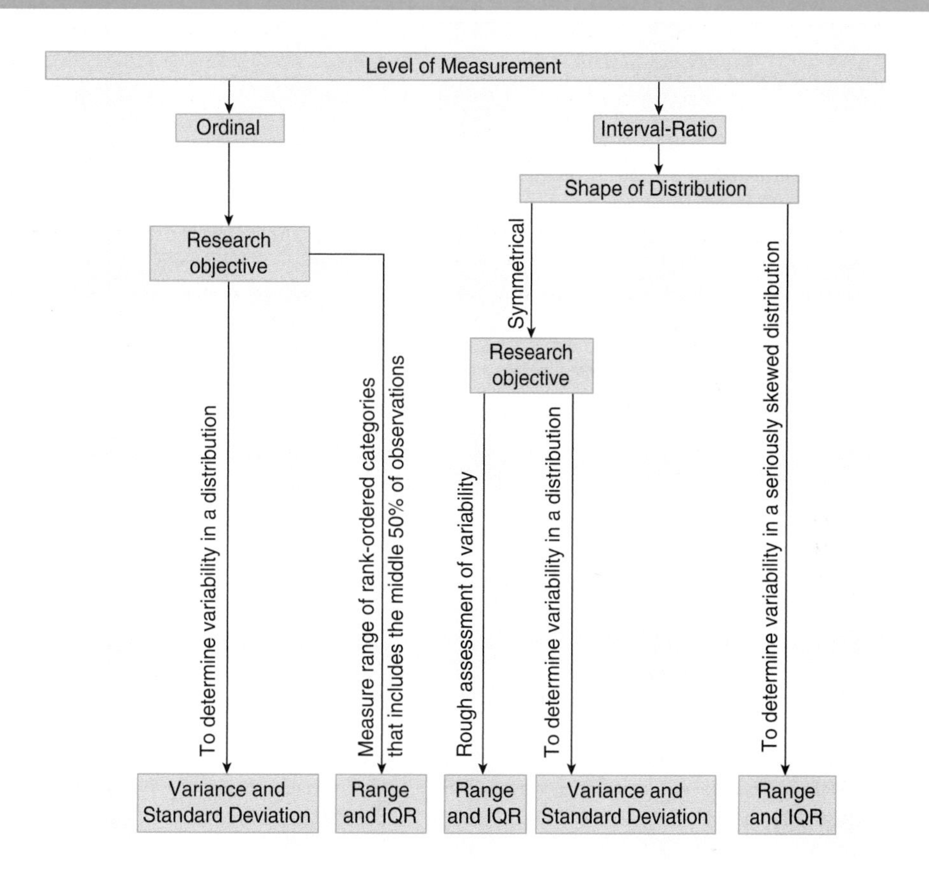

the choice of the appropriate one involves several considerations. These considerations and how they affect our choice of the appropriate measure are presented in the form of a decision tree in Figure 3.16.

As in choosing a measure of central tendency, one of the most basic considerations in choosing a measure of variability is the variable's level of measurement. Valid use of any of the measures requires that the data are measured at the level appropriate for that measure or higher, as shown in Figure 3.16.

Nominal level: We did not consider measures of variability for nominal measures.

Ordinal level: The choice of measure of variation for ordinal variables is more problematic. One possibility is to use the IQR, interpreting the IQR as the range of rank-ordered values that includes the middle 50% of the observations.[11] For example, if the IQR for income categories begins with the category $50,000 to $70,500 and ends with the category $100,000 to $120,500, the IQR can be

A CLOSER LOOK 3.3

More on Interpreting the Standard Deviation

The standard deviation can be used to describe the distribution of a specific variable. For example, in Table 3.10 we present descriptive statistics for the gross domestic product (GDP) per capita (measured in current U.S. dollars in millions) for a sample of 70 countries.

In the first column, we see the name of the variable *Gross domestic product per capita*. The next three columns tell us that there were 70 countries in our sample and that the minimum GDP per capita was $400 and the maximum was $80,700. This is quite a gap between the poorest and richest countries in our sample. The mean and standard deviation are listed in the final two columns.

Table 3.10 Descriptive Statistics for GDP

	N	Minimum (Million $)	Maximum (Million $)	Mean	Standard Deviation
Gross domestic product per capita	70	400	80,700	19,247.14	17,628.277
Valid *N* (listwise)	70				

The mean GDP per capita is $19,247.14, with a standard deviation of $17,628.28. We can expect about 68% of these countries to have GDP per capita values within a range of $1,618.86 ($19,247.14 – $17,628.28) to $36,875.42 ($19,247.14 + $17,628.28). Hence, based on the mean and the standard deviation, we have a pretty good indication of what would be considered a typical GDP per capita value for most countries in our sample. For example, we would consider a country with a GDP per capita value of $80,700 to be extremely wealthy in comparison with other countries. More than two thirds of all countries in our sample fall closer to the mean ($19,247.14) than the country with a GDP per capita value of $80,700.

Another way to interpret the standard deviation is to compare it with another distribution. For instance, Table 3.11 displays the means and standard deviations of employee age for two samples drawn from a *Fortune* 100 corporation. Samples are divided into female clerical and female technical. Note that the mean ages for both samples are about the same—approximately 39 years of age. However, the standard deviations suggest that the distribution of age is dissimilar between the two groups. Figure 3.17 loosely illustrates this dissimilarity in the two distributions.

The relatively low standard deviation for female technical indicates that this group is relatively homogeneous in age. That is to say, most of the women's ages, while not identical, are fairly similar. The average deviation from the mean age of 39.87 is 3.75 years. In contrast, the standard deviation for female clerical employees is about twice the standard deviation for female technical. This suggests a wider dispersion or greater heterogeneity in the ages of clerical workers. We can say that the average deviation from the mean age of 39.46 is 7.80 years for clerical workers. The larger standard deviation indicates a wider dispersion of points below or above the mean. On average, clerical employees are farther in age from their mean of 39.46.[10]

Table 3.11 Age Characteristics of Female Clerical and Technical Employees

Characteristics	Female Clerical ($N = 22$)	Female Technical ($N = 39$)
Mean age	39.46	39.87
Standard deviation	7.80	3.75

Source: Adapted from Marjorie Armstrong-Srassen, "The Effect of Gender and Organizational Level on How Survivors Appraise and Cope With Organizational Downsizing," *Journal of Applied Behavioral Science* 34, no. 2 (June 1998): 125–142. Reprinted with permission.

Figure 3.17 Illustrating the Means and Standard Deviations for Age Characteristics

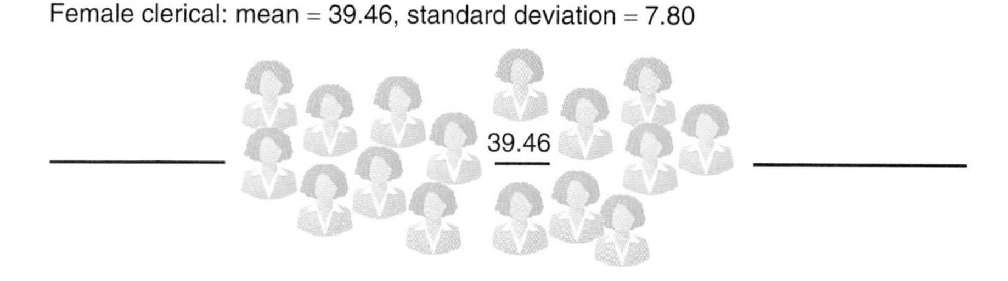

reported as between $50,000 and $120,500. However, in most instances, social science researchers treat ordinal variables as interval-ratio measures, preferring to calculate variance and standard deviation.

Interval-ratio level: For interval-ratio variables, you can choose the variance, standard deviation, the range, or the IQR. Because the range, and to a lesser extent the IQR, is based on only two scores in the distribution (and therefore tends to be sensitive if either of the two points is extreme), the variance and/or standard deviation is usually preferred. However, if a distribution is extremely skewed so that the mean is no longer representative of the central tendency in the distribution, the range and the IQR can be used. The range and the IQR will also be useful when you are reading tables or quickly scanning data to get a rough idea of the extent of dispersion in the distribution.

Reading the Research Literature: Community College Mentoring

In this Reading the Research Literature, Myron Pope (2002) explores the importance of mentoring for 250 students of color enrolled in 25 different community colleges across the country. Pope argues that students of color respond best to multiple levels of mentoring—formal and informal methods of mentoring and mentoring from different sources (faculty, staff, and other students). Pope's analysis is based on a survey measuring student perceptions about campus climate, institutional diversity, mentoring, and administrative support of diversity. For this 2002 research, Pope focused specifically on statements regarding mentoring, measuring student agreement on an ordinal 5-point scale: 1 = *no agreement* to 5 = *strong agreement*.[10] Note how mean scores and standard deviations for selected statements are reported in Table 3.12.

In Table 3.12, we present statements that measure mentoring from three different sources: (1) staff, (2) peers, and (3) students themselves. The mean agreement scores and standard deviations vary by each group—that is to say, perceptions about mentoring differ by racial or ethnic identity. The standard deviations indicate the variability of responses: A smaller standard deviation reveals more consistency of responses clustered around the mean score, whereas a larger standard deviation indicates more variation, more spread from the mean. For example, for the statement, "There are persons of color in administrative roles from whom I would seek mentoring at this institution," the larger average

Table 3.12 Minority Student Perception of Mentoring at Their Institution by Race (N = 254)

	African American (*N* = 178), *M* (*SD*)	Asian (*N* = 12), *M* (*SD*)	Hispanic (*N* = 28), *M* (*SD*)	Native American (*N* = 22), *M* (*SD*)	Multiethnic (*N* = 14), *M* (*SD*)
There are persons of color in administrative roles from whom I would seek mentoring at this institution.	3.76 (1.10)	3.50 (1.31)	3.14 (1.08)	4.09 (0.68)	4.14 (0.66)
There are peer mentors who can advise me.	3.48 (1.11)	2.17 (1.40)	3.14 (1.20)	3.91 (0.68)	3.29 (1.54)
I mentor other students.	3.30 (1.25)	2.00 (1.21)	3.00 (1.09)	3.46 (1.10)	3.29 (1.07)

Source: Adapted from Myron Pope, "Community College Mentoring Minority Student Perception," *Community College Review* 30, no. 3 (2002): 37.

Note: N = total; SD = standard deviation; M = mean.

scores for Native American and multiethnic students also correspond to smaller standard deviations. Native American and multiethnic students were more likely to agree with the statement, and there was more consistency (less variation) in their responses. For the statement, "There are peer mentors who can advise me," Native American students had the highest mean score (3.91) with the smallest standard deviation (0.68), indicating less variation of responses. Finally, for the statement, "I mentor other students," there is a lower level of agreement among all the student groups. The highest mean is 3.46 (based on a 5-point scale) for the Native American student group; the lowest mean is 2.0 for the Asian student group. The standard deviations are all above 1.0, with the most variation in responses for the group of African American students.

Pope (2002) advocates the value of multilevel mentoring:

> By way of multilevel mentoring, minority students are exposed to a variety of individuals who are committed to ensuring that they adjust to life as a college student. These individuals will be able to assist these students in overcoming some of the precollege characteristics, such as their academic preparation and first-generation college student status, along with dealing with some of the conflicts that arise as a result of their enrollment in college, such as balancing their responsibilities. . . . Mentoring programs also cannot be one dimensional, in that the mentor must provide guidance to the student in academic, personal and professional areas. The setting must incorporate an opportunity for these mentors to learn about students from various ethnic backgrounds.[11]

Pope suggests that to improve these mentoring opportunities, community college administrators "must begin more aggressively to recruit future faculty and administrators with these ethnic backgrounds."[12]

We will revisit this research example in Chapter 9.

DATA AT WORK

Sruthi Chandrasekaran: Senior Research Associate

Photo courtesy of Sruthi Chandrasekaran

Trained in economics and public policy, Sruthi is currently employed in a development economics-based research organiztion that specializes in using randomized controlled trials for impact evaluation. She emphasizes the importance of quality data as she describes her work. "As a field-based researcher, I take the lead in ensuring that the intervention follows the study design to the dot, the data collection tools elicit quality responses in an unbiased manner, the survey data is of the highest quality, the cleaning of the data is coherent and methodical, and the analysis is rigorous. The results of the study are published in leading academic journals and the policy lessons are disseminated to

(Continued)

(Continued)

stakeholders and hence it is crucial that the research is well designed and the quality of the data is impeccable."

Her research has broad implications on community stakeholders and policies. "On field visits as part of the study, there is so much that I learn that furthers my understanding, helping to piece together the results from the data analysis. Meeting communities and listening to their perspectives is intellectually stimulating and emotionally fulfilling. Engaging with powerful stakeholders in the decision-making process, such as donor organizations, non-profits, and government officials, and sharing my learnings from the field and from the data is very rewarding professionally and helps me feel like I'm doing my bit to impact policy design and implementation toward an efficient and equitable solution."

For students interested in a research career, Sruthi recommends developing the ability to focus on details and the flexibility to examine the big picture of a project. She explains, "Research can at times be painstakingly slow and frustrating, so patience and single-minded focus on the end goal can help one through the tough times. Being aware of competing methodologies and research studies in relevant fields can also prove to be quite useful in understanding the advantages and pitfalls in your own research. If you are inspired to take up research, make sure you choose a field close to your heart since this will be personally and professionally rewarding. If you are unsure, take up an internship or a short-term project to see how much you may enjoy it."

MAIN POINTS

- The mode, the median, and the mean are measures of central tendency—numbers that describe what is average or typical about the distribution.
- The mode is the category or score with the largest frequency (or percentage) in the distribution. It is often used to describe the most commonly occurring category of a nominal-level variable.
- The median is a measure of central tendency that represents the exact middle of the distribution. It is calculated for variables measured on at least an ordinal level of measurement.
- The mean is typically used to describe central tendency in interval-ratio variables, such as income, age, or education. We obtain the mean by summing all the scores and dividing by the total (N) number of scores.
- In a symmetrical distribution, the frequencies at the right and left tails of the distribution are identical. In skewed distributions, there are either a few extremely high (positive skew) or a few extremely low (negative skew) values.
- Measures of variability are numbers that describe how much variation or diversity there is in a distribution.

- The range measures variation in interval ratio and ordinal variables and is the difference between the highest (maximum) and the lowest (minimum) scores in the distribution. To find the range, subtract the lowest from the highest score in a distribution. For an ordinal variable, report the lowest and the highest values without subtracting.
- The interquartile range measures the width of the middle 50% of the distribution for interval-ratio and ordinal variables. It is defined as the difference between the lower and upper quartiles (Q_1 and Q_3). In some instances, reporting the full range (the values of Q_1 and Q_3) may provide more information than the single interquartile range value.
- The box plot is a graphical device that visually presents the range, the interquartile range, the median, the lowest (minimum) score, and the highest (maximum) score. The box plot provides us with a way to visually examine the center, the variation, and the shape of a distribution.
- The variance and the standard deviation are two closely related measures of variation for interval-ratio and ordinal variables that increase or decrease based on how closely the scores cluster around the mean. The variance is the average of the squared deviations from the center (mean) of the distribution; the standard deviation is the square root of the variance.

KEY TERMS

interquartile range (IQR) 95
mean 78
measures of central tendency 70
measures of variability 70
median 72
mode 70
negatively skewed distribution 89
percentile 77
positively skewed distribution 88
range 94
skewed distribution 87
standard deviation 102
symmetrical distribution 86
variance 102

DIGITAL RESOURCES

Access key study tools at https://edge.sagepub.com/ssdsess4e

- eFlashcards of the glossary terms
- Datasets and codebooks
- SPSS and Excel walk-through videos
- SPSS and Excel demonstrations and problems to accompany each chapter
- Appendix F: Basic Math Review

CHAPTER EXERCISES

1. The following GSS 2018 frequency distribution presents information about people's self-evaluations of their lives, based on three categories: (1) exciting, (2) routine, and (3) dull.

Respondent's Assessment of Life	Frequency	Percentage	Cumulative Percentage
Exciting	348	46.6	46.6
Routine	356	47.7	94.3
Dull	42	5.6	99.9
Total	746	99.9	

   a. Find the mode.
   b. Find the median.
   c. Interpret the mode and the median.
   d. Why would you not want to report the mean for this variable?

2. GSS 2014 and GSS 2018 respondents were asked their opinion on whether homosexuals should have the right to marry. (The 2014 data were collected before the 2015 U.S. Supreme Court decision granting same-sex couples the right to marry.) Responses were measured on a 5-point scale from *strongly agree* (1) to *strongly disagree* (5).

Homosexuals Should Have Right to Marry (2014)	Frequency	Percentage	Cumulative Percentage
Strongly agree	251	32.5	32.5
Agree	195	25.2	57.5
Neither agree nor disagree	74	9.6	67.3
Disagree	92	11.9	79.2
Strongly disagree	161	20.8	100.0
Total	773	100.0	

Homosexuals Should Have Right to Marry (2018)	Frequency	Percentage	Cumulative Percentage
Strongly agree	293	39.5	39.5
Agree	200	27.0	66.5

Homosexuals Should Have Right to Marry (2018)	Frequency	Percentage	Cumulative Percentage
Neither agree nor disagree	70	9.4	75.9
Disagree	67	9.0	84.9
Strongly disagree	111	15.0	99.9
Total	741	99.9	

a. What is the level of measurement for this variable?

b. What is the mode for 2014? For 2018?

c. Calculate the median for this variable. In general, how would you characterize the public's attitude about same-sex marriage? Is there a difference in attitudes between the 2 years?

3. This frequency distribution contains information on the number of hours worked last week for a sample of 32 Latino adults from the GSS 2010.

Hours Worked Last Week	Frequency	Percentage	Cumulative Percentage
20	3	9.4	9.4
25	2	6.3	15.6
28	1	3.1	18.8
29	1	3.1	21.9
30	3	9.4	31.3
32	1	3.1	34.4
40	14	43.8	78.1
50	2	6.3	84.4
52	1	3.1	87.5
55	1	3.1	90.6
60	1	3.1	93.8
64	1	3.1	96.9
70	1	3.1	100.0
Total	32	100.0	

a. What is the level of measurement, mode, and median for "hours worked last week"?

b. Construct quartiles for weeks worked last year. What is the 25th percentile? The 50th percentile? The 75th percentile? Why don't you need to calculate the 50th percentile to answer this question?

4. GSS 2018 respondents were asked to rate their agreement to the following statement: "Even if it brings no immediate benefits, scientific research that advances the frontiers of knowledge is necessary and should be supported by the federal government." Responses were measured on a 4-point scale: *strongly agree*, *agree*, *disagree*, and *strongly disagree*.

	Frequency
Strongly agree	243
Agree	468
Disagree	101
Strongly disagree	23
Total	835

a. What is the modal category?

b. Calculate the median category for this variable.

c. Identify which categories contain the 20th and 80th percentiles.

5. Monitoring the Future is a longitudinal study of the behaviors and attitudes of American secondary school students. Data from the 2014 survey are presented, measuring the frequency of eating breakfast (an indicator of a healthy lifestyle) by student race. The table features a different format, presenting frequencies and percentages in a single column.

	Black	White	Hispanic
	f	*f*	*f*
	%	%	%
Never	4	14	5
	9.3%	6.4%	7.0%
Seldom	11	22	9
	25.6%	10.1%	12.7%
Sometimes	10	43	21
	23.3%	19.7%	29.6%

	Black	White	Hispanic
	f	*f*	*f*
	%	%	%
Most days	2	25	9
	4.7%	11.5%	12.7%
Nearly every day	7	30	2
	16.3%	13.8%	2.8%
Everyday	9	84	25
	20.9%	38.5%	35.2%
Total	43	218	71
	100%	100%	100%

a. Calculate the median and mode for each racial group.

b. Use this information to describe how teens' breakfast habits vary by race. In your opinion, which statistic provides a better description of the data—the median or mode?

6. U.S. households have become smaller over the years. The following table from the GSS contains information on the number of people currently aged 18 years or older living in a respondent's household. Calculate the mean number of people living in a U.S. household.

Household Size	Frequency
1	381
2	526
3	227
4	200
5	96
6	42
7	19
8	5
9	2

(*Continued*)

(Continued)

Household Size	Frequency
10	2
Total	1,500

7. The U.S. Census Bureau collects information about divorce rates. The following table summarizes the divorce rate for 10 U.S. states in 2017. Use the table to answer the questions that follow.

State	Divorce Rate per 1,000 Population
Alaska	3.6
Florida	3.6
Idaho	3.9
Maine	3.2
Maryland	2.5
Nevada	4.5
New Jersey	2.6
Texas	2.2
Vermont	2.9
Wisconsin	2.4

Source: National Center for Health Statistics, *Divorce Rates by State, 1999–2017*.

a. Calculate and interpret the range and the IQR. Which is a better measure of variability? Why?
b. Calculate and interpret the mean and standard deviation.
c. Identify two possible explanations for the variation in divorce rates across the 10 states.

8. Individuals with higher levels of education tend to delay parenting, having children at an older age in comparison with individuals with lower levels of education. We examine this relationship based on GSS 2018 data. In the following table, the mean, standard deviation, and variance for respondent's age when first child was born (AGEKDBRN) are reported for five categories (DEGREE).

a. Identify the level of measurement for DEGREE and AGEKDBRN.
b. Describe the relationship between respondent degree and age when first child was born.

	Less Than High School	High School	Some College	Bachelor's Degree	Graduate Degree
Mean	21.33	23.08	24.83	27.44	28.59
Standard deviation	5.055	5.498	4.581	5.314	5.446
Variance	25.556	30.232	20.984	28.234	29.654

9. You are interested in studying the variability of violent crimes committed and police expenditures in the eastern and midwestern United States. The U.S. Census Bureau collected the following statistics on these two variables for 21 states in the East and Midwest in 2014–2015.

State	Number of Violent Crimes per 100,000 Population	Police Protection Expenditures per Capita
Maine	130.1	194
New Hampshire	199.3	304
Vermont	118.0	312
Massachusetts	390.9	349
Rhode Island	242.5	371
Connecticut	218.5	323
New York	379.7	477
New Jersey	255.4	325
Pennsylvania	315.1	276
Ohio	291.9	282
Indiana	387.5	183
Illinois	383.8	388
Michigan	415.5	242
Wisconsin	305.8	305
Minnesota	242.6	321
Iowa	286.1	233
Missouri	497.4	281
North Dakota	239.4	249

(*Continued*)

(Continued)

State	Number of Violent Crimes per 100,000 Population	Police Protection Expenditures per Capita
South Dakota	383.1	212
Nebraska	274.9	220
Kansas	389.9	260

Source: ProQuest, *Statistical Abstract of the United States 2018: The National Data Book*, Tables 334 and 361.

The SPSS output showing the mean and the standard deviation for both variables is presented below.

Descriptive Statistics

	N	Mean	Std. Deviation
Number of Crimes	21	302.2571	97.12543
Police Protection Expenditures	21	290.8095	70.32256
Valid N (listwise)	21		

a. What are the means? The standard deviations?

b. Compare the mean with the standard deviation for each variable. Does there appear to be more variability in the number of crimes or in police expenditures per capita in these states? Which states contribute more to this greater variability?

c. Suggest why one variable has more variability than the other. In other words, what social forces would cause one variable to have a relatively large standard deviation?

10. The following table summarizes the racial differences in education and the number of children for GSS 2018 Hispanic and American Indian respondents. Based on the means and standard deviations (in parentheses), what conclusions can be drawn about differences in the years of education and the number of children?

	Hispanic	American Indian
Education (years)	10.97 (4.17)	13.00 (3.82)
Number of children	2.37 (2.13)	2.29 (1.65)

11. The following output depicts data for population change in midwestern and western states between 2010 and 2016 from Table 3.5.

West	Mean		6.600	.7478
	95% Confidence Interval for Mean	Lower Bound	4.971	
		Upper Bound	8.229	
	5% Trimmed Mean		6.694	
	Median		6.800	
	Variance		7.270	
	Std. Deviation		2.6963	
	Minimum		1.1	
	Maximum		10.4	
	Range		9.3	
	Interquartile Range		3.9	
	Skewness		-.387	.616
	Kurtosis		-.212	1.191

Midwest	Mean		3.233	1.0118
	95% Confidence Interval for Mean	Lower Bound	1.006	
		Upper Bound	5.460	
	5% Trimmed Mean		2.898	
	Median		2.100	
	Variance		12.286	
	Std. Deviation		3.5051	
	Minimum		-.2	
	Maximum		12.7	
	Range		12.9	
	Interquartile Range		3.4	
	Skewness		1.998	.637
	Kurtosis		4.718	1.232

a. Compare the range for the western states with that of the Midwest. Which region had a greater range?

b. Examine the IQR for each region. Which is greater?

c. Use the statistics to characterize the variability in population increase in the two regions. Does one region have more variability than another? If yes, why do you think that is?

12. The 2018 average life expectancy for the total population is reported for 10 countries. Calculate the appropriate measures of central tendency and variability for both European countries and non-European countries. Is there more variability in life expectancy for European countries or non-European countries? If so, what might explain these differences?

Country	Life Expectancy at Birth
European countries	
France	82
Germany	80.9
Netherlands	81.5
Spain	81.8
Turkey	75.3
Non-European countries	
Japan	85.5
Australia	82.4
Mexico	76.3
Iceland	83.1
Israel	82.7

Source: *CIA Workbook*, 2018.

13. We examine education (measured in years), age (measured in years), and frequency of religious services (measured on an ordinal scale: 0 = never, 1 = once a month to 8 = more than once a week) for males and females who voted for Hillary Clinton and Donald Trump in the 2016 presidential election. Using the GSS 2018 statistics, describe the characteristics of Clinton and Trump supporters. *Note:* The mean for each variable is presented alongside the standard deviation (in parentheses).

	Clinton		Trump	
	Males (N = 144)	Females (N = 217)	Males (N = 132)	Females (N = 125)
Education	14.69 (2.73)	14.40 (2.92)	13.80 (2.75)	14.00 (2.57)
Age	50.49 (18.84)	51.93 (18.21)	55.39 (17.03)	56.19 (15.94)
Frequency of religious services	2.50 (2.74)	3.53 (2.80)	3.72 (2.93)	3.82 (2.75)

4

THE NORMAL DISTRIBUTION

We have learned some important things about distributions: how to organize them into frequency distributions, how to display them using graphs, and how to describe their central tendencies and variation using measures such as the mean and the standard deviation. The distributions that we have described so far are all empirical distributions—that is, they are all based on real data.

On the other hand, the focus of this chapter is the distribution known as the normal curve or the **normal distribution**. The normal distribution is a theoretical distribution, similar to an empirical distribution in that it can be organized into frequency distributions, displayed using graphs, and described by its central tendency and variation using measures such as the mean and the standard deviation. However, unlike an empirical distribution, a theoretical distribution is based on theory rather than on real data. The value of the theoretical normal distribution lies in the fact that many empirical distributions that we study seem to approximate it. We can often learn a lot about the characteristics of these empirical distributions based on our knowledge of the theoretical normal distribution.

Chapter Learning Objectives

1. Explain the importance and use of the normal distribution in statistics.
2. Describe the properties of the normal distribution.
3. Transform a raw score into standard (Z) score and vice versa.
4. Transform a Z score into proportion (or percentage) and vice versa.
5. Calculate and interpret the percentile rank of a score.

PROPERTIES OF THE NORMAL DISTRIBUTION

The normal curve (Figure 4.1) is bell shaped. One of the most striking characteristics of the normal distribution is its perfect symmetry. If you fold Figure 4.1 exactly in the middle, you have two equal halves, each the mirror image of the other. This means that precisely half the observations fall on each side of the middle of the distribution. In addition, the midpoint of the normal curve is the point having the maximum frequency. This is also the point at which three measures coincide: (1) the mode (the point of the

Normal distribution: A bell-shaped and symmetrical theoretical distribution with the mean, the median, and the mode all coinciding at its peak and with the frequencies gradually decreasing at both ends of the curve.

Figure 4.1 The Normal Curve

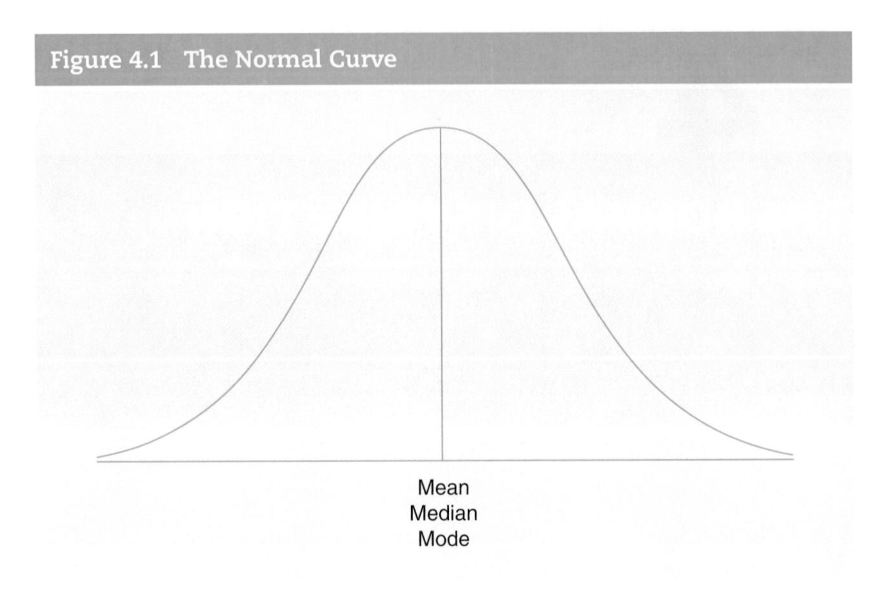

highest frequency), (2) the median (the point that divides the distribution into two equal halves), and (3) the mean (the average of all the scores). Notice also that most of the observations are clustered around the middle, with the frequencies gradually decreasing at both ends of the distribution.

Empirical Distributions Approximating the Normal Distribution

The normal curve is a theoretical ideal, and real-life distributions never match this model perfectly. However, researchers study many variables (e.g., standardized tests such as the SAT, ACT, or GRE) that closely resemble this theoretical model. When we say that a variable is "normally distributed," we mean that the graphic display will reveal an approximately bell-shaped and symmetrical distribution closely resembling the idealized model shown in Figure 4.1. This property makes it possible for us to describe many empirical distributions based on our knowledge of the normal curve.

Areas Under the Normal Curve

In all normal or nearly normal curves, we find a constant proportion of the area under the curve lying between the mean and any given distance from the mean when measured in standard deviation units.

The area under the normal curve may be conceptualized as a proportion or percentage of the number of observations in the sample. Thus, the entire area under the curve is equal to 1.00 or 100% (1.00 × 100) of the observations. Because the normal curve is perfectly symmetrical, exactly 0.50 or 50% of the

Figure 4.2 Percentages Under the Normal Curve

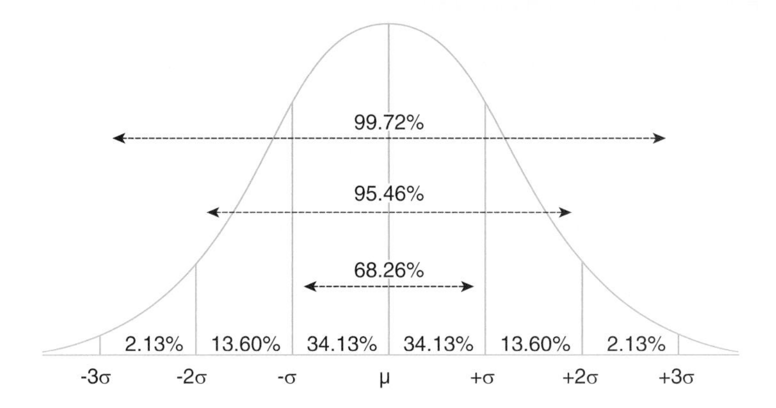

observations lie above or to the right of the center, which is the mean of the distribution, and 50% lie below or to the left of the mean.

In Figure 4.2, note the percentage of cases that will be included between the mean and 1, 2, and 3 standard deviations above and below the mean. The mean of the distribution divides it exactly into half; 34.13% is included between the mean and 1 standard deviation to the right of the mean, and the same percentage is included between the mean and 1 standard deviation to the left of the mean. The plus signs indicate standard deviations above the mean; the minus signs denote standard deviations below the mean. Thus, between the mean and ±1 standard deviation, 68.26% of all the observations in the distribution occur; between the mean and ±2 standard deviations, 95.46% of all observations in the distribution occur; and between the mean and ±3 standard deviations, 99.72% of the observations occur.

LEARNING CHECK 4.1

Review the properties of the normal curve. What is the area underneath the curve equal to? What percentage of the distribution is within 1 standard deviation? Within 2 and 3 standard deviations? Verify the percentage of cases by summing the percentages in Figure 4.2.

Interpreting the Standard Deviation

The fixed relationship between the distance from the mean and the areas under the curve represents a property of the normal curve that has highly practical applications. As long as a distribution is normal and we know the mean and the standard deviation, we can determine the proportion or percentage of cases that fall between any score and the mean.

This property provides an important interpretation for the standard deviation of empirical distributions that are approximately normal. For such distributions, when we know the mean and the standard deviation, we can determine the percentage or proportion of scores that are within any distance, measured in standard deviation units, from that distribution's mean.

Not every empirical distribution is normal. We've learned that the distributions of some common variables, such as income, are skewed and therefore not normal. The fixed relationship between the distance from the mean and the areas under the curve applies only to distributions that are normal or approximately normal.

AN APPLICATION OF THE NORMAL CURVE

For the rest of this chapter discussion, we rely on the results of the 2018 SAT examination. You may have taken the SAT exam as part of your college admission process. Although there is much debate on the predictive value of SAT scores on college success and some schools have revised their SAT requirements, the exam is still widely regarded as the standardized assessment test to measure college readiness and student quality.

The current SAT includes two components: (1) Evidence-Based Reading and Writing (ERW) and (2) Mathematics. The perfect score for each component is 800, for a total possible of 1,600. Table 4.1 presents mean and standard deviation statistics for all 2018 high school graduates who took the SAT. The results of the SAT exam, combined or for each component, are assumed to be normally distributed. Throughout this chapter, we will use the normal (theoretical) curve to describe and better understand the characteristics of the SAT ERW empirical (real data) distribution.

Table 4.1 2018 SAT Component Means and Standard Deviations for High School Graduates Who Took the SAT During High School

Number of Total Test Takers	Evidence-Based Reading and Writing (ERW)		Mathematics	
	Mean	Standard Deviation	Mean	Standard Deviation
2,136,539	536	102	531	114

Source: *SAT Suite of Assessments Annual Report: Total Group*. Copyright © 2018. The College Board. www.collegeboard.org

Transforming a Raw Score Into a *Z* Score

We can express the difference between any score in a distribution and the mean in terms of standard scores, also known as *Z* scores. A **standard (*Z*) score** is the number of standard deviations that a given raw score (or the observed score) is above or below the mean. A raw score can be transformed into a *Z* score to find how many standard deviations it is above or below the mean.

Standard (*Z*) score: The number of standard deviations that a given raw score is above or below the mean.

To transform a raw score into a *Z* score, we divide the difference between the score and the mean by the standard deviation. For example, if we want to transform a 638 SAT ERW score into a *Z* score, we subtract the mean ERW score of 536 from 638 and divide the difference by the standard deviation of 102 (mean and standard deviation reported in Table 4.1).

This calculation gives us a method of standardization known as transforming a raw score into a *Z* score (also known as a standard score). The *Z* score formula is

$$Z = \frac{Y - \bar{Y}}{s} \qquad (4.1)$$

Thus, the *Z* score of 638 is

$$\frac{638 - 536}{102} = \frac{48}{102} = 1.00$$

or 1 standard deviation above the mean. Similarly, for a 434 SAT ERW score, the *Z* score is

$$\frac{434 - 536}{102} = \frac{-102}{102} = -1.00$$

or 1 standard deviation below the mean. The negative sign indicates that this score is below (on the left side of) the mean.

A *Z* score allows us to represent a raw score in terms of its relationship to the mean and to the standard deviation of the distribution. It represents how far a given raw score is from the mean in standard deviation units. A positive *Z* indicates that a score is larger than the mean, and a negative *Z* indicates that it is smaller than the mean. The larger the *Z* score, the larger the difference between the score and the mean.

THE STANDARD NORMAL DISTRIBUTION

When a normal distribution is represented in standard scores (*Z* scores), we call it the **standard normal distribution**. Standard scores, or *Z* scores, are the numbers that tell us the distance between an actual score and the mean in terms of standard deviation units. The standard normal distribution has a mean of 0.0 and a standard deviation of 1.0.

Standard normal distribution: A normal distribution represented in standard (*Z*) scores.

Figure 4.3 The Standard Normal Distribution

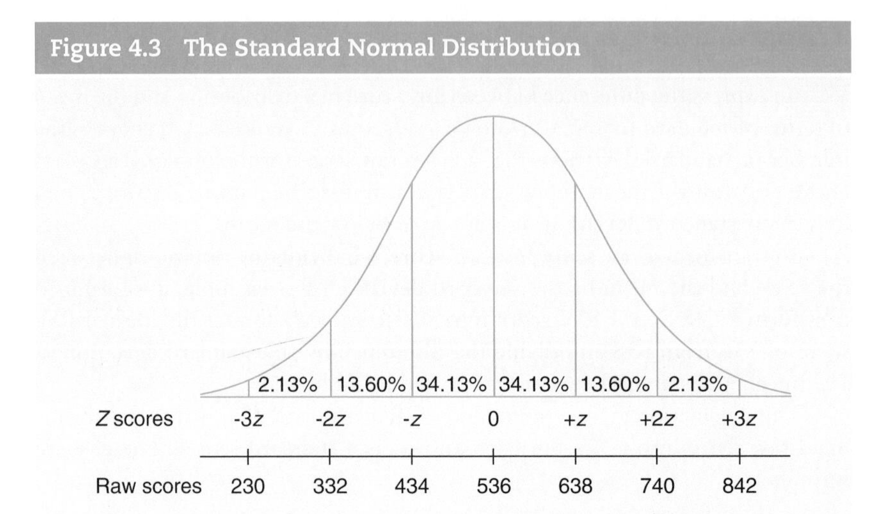

Source: Adapted from College Board, SAT *Suite of Assessments Annual Report: Total Group,* 2018. The College Board. www.collegeboard.org.

Figure 4.3 shows a standard normal distribution with areas under the curve associated with one, two, and three standard scores above and below the mean. To help you understand the relationship between raw scores of a distribution and their respective standard *Z* scores, we also show the SAT ERW scores that correspond to these standard scores. For example, notice that the mean for the SAT ERW distribution is 536 and the corresponding *Z* score—the mean of the standard normal distribution—is 0. As we've already calculated, the score of 638 is 1 standard deviation above the mean (536 + 102 = 638); therefore, its corresponding *Z* score is +1. Similarly, the score of 434 is 1 standard deviation below the mean (536 – 102 = 434), and its *Z* score equivalent is –1.

THE STANDARD NORMAL TABLE

We can use *Z* scores to determine the proportion of cases that are included between the mean and any *Z* score in a normal distribution. The areas or proportions under the standard normal curve, corresponding to any *Z* score or its fraction, are organized into a special table called the **standard normal table**. The table is presented in Appendix B. In this section, we discuss how to use this table.

Standard normal table: A table showing the area (as a proportion, which can be translated into a percentage) under the standard normal curve corresponding to any *Z* score or its fraction.

Table 4.2 reproduces a small part of the standard normal table. Note that the table consists of three columns (rather than having one long table, we have moved the second half of the table next to the first half, so the three columns are presented in a six-column format).

Column A lists positive *Z* scores. (Note that *Z* scores are presented with two decimal places. In our chapter calculations, we will do the same.) Because the normal curve is symmetrical, the proportions that correspond to positive *Z* scores are identical to the proportions corresponding to negative *Z* scores.

Table 4.2 The Standard Normal Table

(A)	(B)	(C)	(A)	(B)	(C)
Z	Area Between Mean and Z	Area Beyond Z	Z	Area Between Mean and Z	Area Beyond Z
0.00	0.0000	0.5000	0.21	0.0832	0.4168
0.01	0.0040	0.4960	0.22	0.0871	0.4129
0.02	0.0080	0.4920	0.23	0.0910	0.4090
0.03	0.0120	0.4880	0.24	0.0948	0.4052
0.04	0.0160	0.4840	0.25	0.0987	0.4013
0.05	0.0199	0.4801	0.26	0.1026	0.3974
0.06	0.0239	0.4761	0.27	0.1064	0.3936
0.07	0.0279	0.4721	0.28	0.1103	0.3897
0.08	0.0319	0.4681	0.29	0.1141	0.3859
0.09	0.0359	0.4641	0.30	0.1179	0.3821
0.10	0.0398	0.4602	0.31	0.1217	0.3783

Figure 4.4 Areas Between Mean and Z (B) and Beyond Z (C)

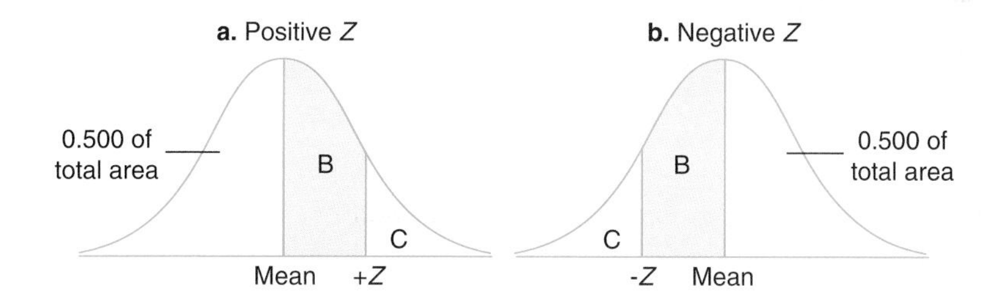

Column B shows the area included between the mean and the Z score listed in column A. Note that when Z is positive, the area is located on the right side of the mean (see Figure 4.4a), whereas for a negative Z score, the same area is located left of the mean (Figure 4.4b).

Column C shows the proportion of the area that is beyond the Z score listed in column A. Areas corresponding to positive Z scores are on the right side of the curve (see Figure 4.4a). Areas corresponding to negative Z scores are identical except that they are on the left side of the curve (Figure 4.4b).

In Sections 1–4, we present examples of how to transform Z scores into proportions or percentages to describe different areas of the empirical distribution of SAT ERW scores.

1. Finding the Area Between the Mean and a Positive or Negative *Z* Score

We can use the standard normal table to find the area between the mean and specific Z scores. To find the area between 536 and 736, follow these steps.

1. Convert 736 to a Z score:

$$\frac{736-536}{102}=\frac{200}{102}=1.96$$

2. Look up 1.96 in column A (in Appendix B) and find the corresponding area in column B, 0.4750. We can translate this proportion into a percentage ($0.4750 \times 100 = 47.50\%$) of the area under the curve included between the mean and a Z score of 1.96 (Figure 4.5).

3. Thus, 47.50% of the total area lies between 536 and 736.

To find the actual number of students who scored between 536 and 736, multiply the proportion 0.4750 by the total number of students. Thus, 1,014,856 students ($0.4750 \times 2{,}136{,}539 = 1{,}014{,}856$) obtained a score between 536 and 736.

For a score lower than the mean, such as 305, we can use the standard normal table and the following steps.

Figure 4.5 Finding the Area Between the Mean and a Specified Positive Z Score

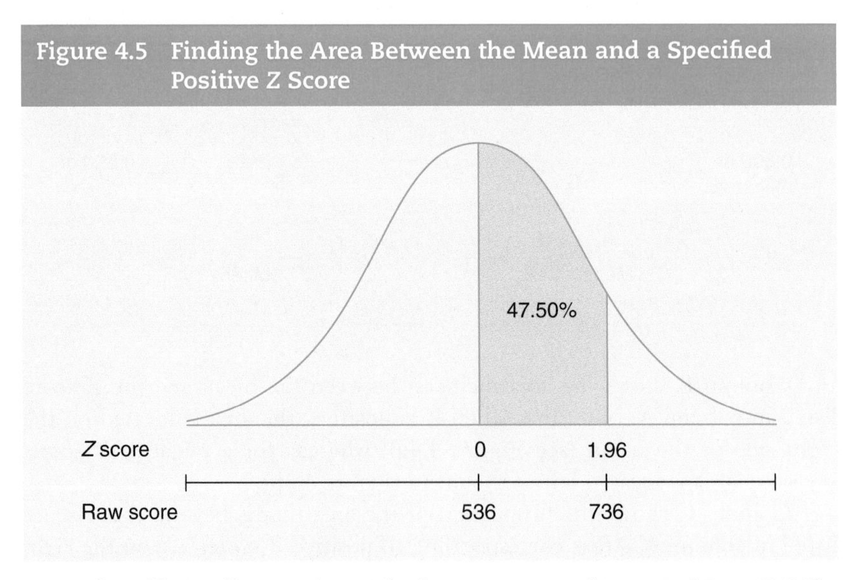

Source: Adapted from College Board, *SAT Suite of Assessments Annual Report: Total Group*, 2018. The College Board. www.collegeboard.org.

Figure 4.6 Finding the Area Between the Mean and a Specified Negative Z Score

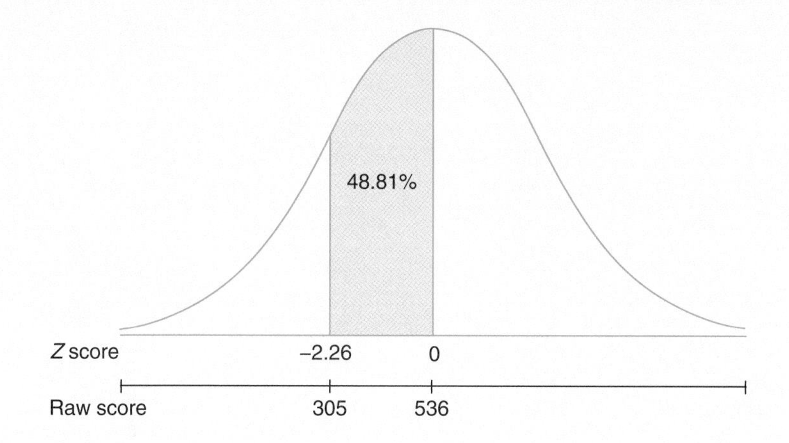

Source: Adapted from College Board, *SAT Suite of Assessments Annual Report: Total Group*, 2018. The College Board. www.collegeboard.org.

1. Convert 305 to a Z score:

$$\frac{305-536}{102}=\frac{-231}{102}=-2.26$$

2. Because the proportions that correspond to positive Z scores are identical to the proportions corresponding to negative Z scores, we ignore the negative sign of Z and look up 2.26 in column A. The area corresponding to a Z score of 2.26 is .4881. This indicates that 0.4881 of the area under the curve is included between the mean and a Z of −2.26 (Figure 4.6). We convert this proportion to a percentage, 48.81%.
3. Thus, 48.81% of the distribution lies between the scores 305 and 536.

LEARNING CHECK 4.2

How many students obtained a score between 305 and 536?

2. Finding the Area Above a Positive *Z* Score or Below a Negative *Z* Score

The normal distribution table can also be used to find the area beyond a Z score, SAT scores that lie at the tip of the positive or negative sides of the distribution (Figure 4.7).

Figure 4.7 Finding the Area Above a Positive Z Score or Below a Negative Z Score

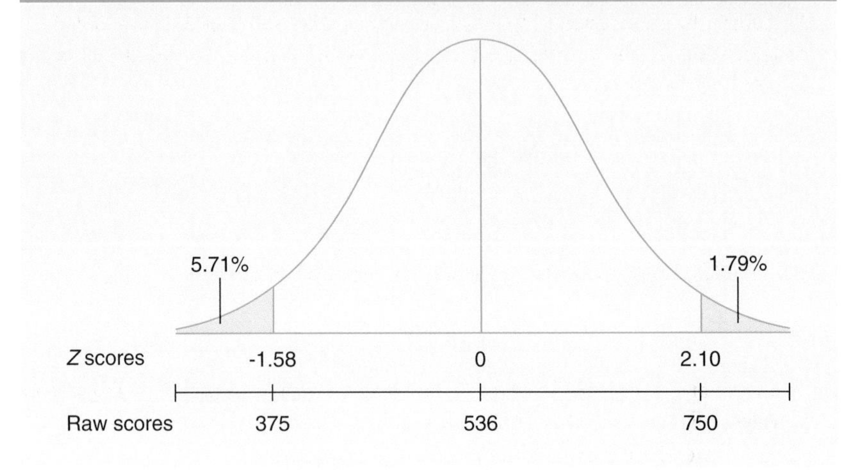

Source: Adapted from College Board, *SAT Suite of Assessments Annual Report: Total Group*, 2018. The College Board. www.collegeboard.org.

For example, what is the area below a score of 750? The Z score corresponding to a final SAT ERW score of 750 is equal to 2.10:

$$\frac{750-536}{102}=\frac{214}{102}=2.10$$

The area beyond a Z of 2.10 includes all students who scored above 750. This area is shown in Figure 4.7. To find the proportion of students whose scores fall in this area, refer to the entry in column C that corresponds to a Z of 2.10, 0.0179. This means that 1.79% ($0.0179 \times 100 = 1.79\%$) of the students scored above 750, a very small percentage. To find the actual number of students in this group, multiply the proportion 0.0179 by the total number of students. Thus, there were $2{,}136{,}539 \times 0.0179$, or about 38,244 students, who scored above 750.

A similar procedure can be applied to identify the number of students on the opposite end of the distribution. Let's first convert a score of 375 to a Z score:

$$\frac{375-536}{102}=\frac{-161}{102}=-1.58$$

The Z score corresponding to a final score of 375 is equal to -1.58. The area beyond a Z of -1.58 includes all students who scored below 375. This area is also shown in Figure 4.7. Locate the proportion of students in this area in column C in the entry corresponding to a Z of -1.58. Remember that the proportions corresponding to positive or negative Z scores are identical. This proportion is equal to .0571. Thus, 5.71% (0.0571×100) of the group, or about 121,996 ($.0571 \times 2{,}136{,}539$) students, performed poorly on the ERW portion of the SAT exam.

LEARNING CHECK 4.3

Calculate the proportion of test takers who earned a SAT ERW score of 400 or less. What is the proportion of students who earned a score of 800 or higher?

3. Transforming Proportions and Percentages Into *Z* Scores

We can also convert proportions or percentages into *Z* scores.

Finding a *Z* Score That Bounds an Area Above It

Let's say we are interested in identifying the score that corresponds to the top 10% of SAT ERW test takers. We will need to identify the cutoff point for the top 10% of the class. This problem involves two steps:

1. Find the *Z* score that bounds the top 10% or 0.1000 ($0.1000 \times 100 = 10\%$) of all the students who took the SAT ERW (Figure 4.8).

Refer to the areas under the normal curve shown in Appendix B. First, look for an entry of 0.1000 (or the value closest to it) in column C. The entry closest to 0.1000 is 0.1003. Then, locate the *Z* in column A that corresponds to this proportion. The *Z* score associated with the proportion 0.1003 SAT ERW is 1.28.

Figure 4.8 Finding a Z Score That Bounds an Area Above It

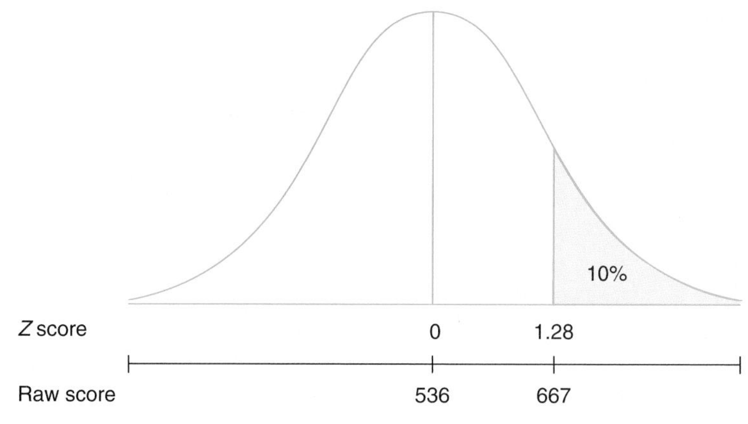

Source: Adapted from College Board, *SAT Suite of Assessments Annual Report: Total Group*, 2018. The College Board. www.collegeboard.org.

2. Find the score associated with a Z of 1.28.

This step involves transforming the Z score into a raw score. To transform a Z score into a raw score, we multiply the score by the standard deviation and add that product to the mean (Formula 4.2):

$$Y = \bar{Y} + Z(s) \tag{4.2}$$

Thus,

$$Y = 536 + 1.28(102) = 536 + 130.56 = 667$$

The cutoff point for the top 10% of SAT ERW exam test takers is 667.

Finding a *Z* Score That Bounds an Area Below It

Now, let's identify the score that corresponds to the bottom 5% of test takers. This problem involves two steps:

1. Find the Z score that bounds the lowest 5% or 0.0500 of all the students who took the class (Figure 4.9).

Refer to the areas under the normal curve, and look for an entry of 0.0500 (or the value closest to it) in column C. The entry closest to 0.0500 is 0.0495. Then, locate the Z in column A that corresponds to this proportion, 1.65. Because the area we are looking for is on the left side of the curve—that is,

Figure 4.9 Finding a Z Score That Bounds an Area Below It

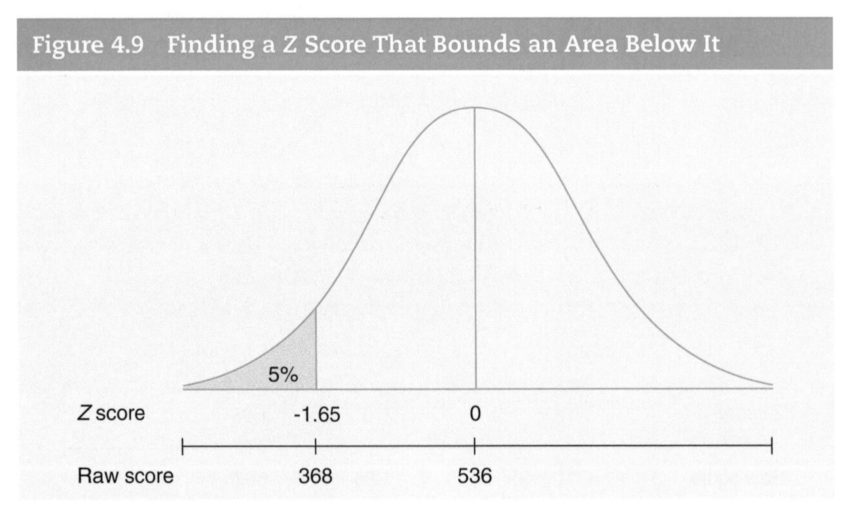

Source: Adapted from College Board, *SAT Suite of Assessments Annual Report: Total Group*, 2018. The College Board. www.collegeboard.org.

below the mean—the Z score is negative. Thus, the Z associated with the lowest 0.0500 (or 0.0495) is −1.65.

2. To find the final SAT ERW score associated with a Z of −1.65, convert the Z score to a raw score:

$$Y = 536 + (-1.65)(102) = 536 - 168.3 = 368$$

The cutoff for the lowest 5% of SAT ERW scores is 368.

LEARNING CHECK 4.4

Which score corresponds to the top 5% of SAT ERW writing test takers?

4. Working With Percentiles in a Normal Distribution

In Chapter 2, we defined percentiles as scores below which a specific percentage of the distribution falls. For example, the 95th percentile is a score that divides the distribution so that 95% of the cases are below it and 5% of the cases are above it. How are percentile ranks determined? How do you convert a percentile rank to a raw score? To determine the percentile rank of a raw score requires transforming Z scores into proportions or percentages. Converting percentile ranks to raw scores is based on transforming proportions or percentages into Z scores. In the following examples, we illustrate both procedures based on the SAT ERW scores data.

Finding the Percentile Rank of a Score Higher Than the Mean

Suppose you took the SAT ERW exam during the same year. You recall that your final score was 680, but how well did you do relative to the other students who took the exam? To evaluate your performance, you need to translate your raw score into a percentile rank. Figure 4.10 illustrates this problem.

To find the percentile rank of a score higher than the mean, follow these steps:

1. Convert the raw score to a Z score:

$$\frac{680-536}{102} = \frac{144}{102} = 1.41$$

The Z score corresponding to a raw score of 680 is 1.41.

Figure 4.10 Finding the Percentile Rank of a Score Higher Than the Mean

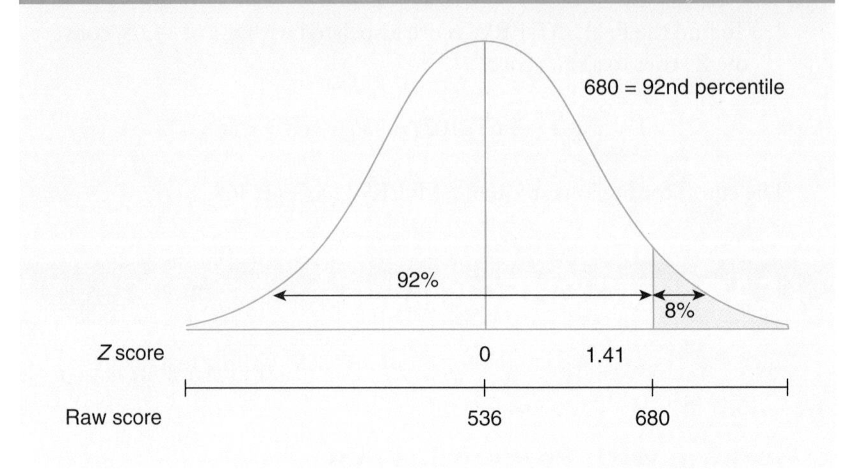

Source: Adapted from College Board, *SAT Suite of Assessments Annual Report: Total Group*, 2018. The College Board. www.collegeboard.org.

2. Find the area beyond Z in Appendix B, column C. The area beyond a Z score of 1.41 is 0.0793.

3. Subtract the area from 1.00 and multiply by 100 to obtain the percentile rank:

$$\text{Percentile rank} = (1.0000 - .0793 = 0.9207) \times 100 = 92.07\% = 92\%$$

Being in the 92nd percentile means that 92% of all test takers scored lower than 680 and 8% scored higher than 680.

Finding the Percentile Rank of a Score Lower Than the Mean

If your SAT ERW score was 450, what was your percentile rank? Figure 4.11 illustrates this problem.

To find the percentile rank of a score lower than the mean, follow these steps:

1. Convert the raw score to a Z score:

$$\frac{450-536}{102} = \frac{-86}{102} = -0.84$$

The Z score corresponding to a raw score of 450 is –0.84.

2. Find the area beyond Z in Appendix B, column C. The area beyond a Z score of –0.84 is 0.2005.

Figure 4.11 Finding the Percentile Rank of a Score Lower Than the Mean

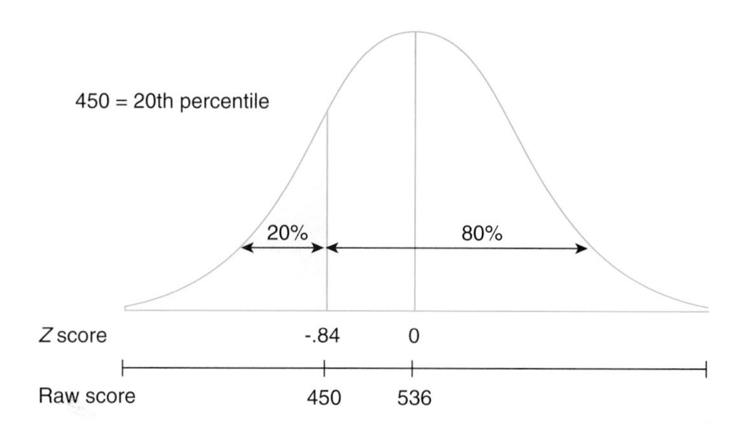

Source: Adapted from College Board, *SAT Suite of Assessments Annual Report: Total Group*, 2018. The College Board. www.collegeboard.org.

3. Multiply the area by 100 to obtain the percentile rank:

$$\text{Percentile rank} = 0.2005(100) = 20.05\% = 20\%$$

The 20th percentile rank means that 20% of all test takers scored lower than you (i.e., 20% scored lower, but 80% scored the same or higher).

Finding the Raw Score Associated With a Percentile Higher Than 50

Now, let's assume that for an honors English program, your university will only admit students who scored at or above the 95th percentile in the SAT ERW exam. What is the cutoff point required for acceptance? Figure 4.12 illustrates this problem.

To find the score associated with a percentile higher than 50, follow these steps:

1. Divide the percentile by 100 to find the area below the percentile rank:

$$\frac{95}{100} = 0.95$$

2. Subtract the area below the percentile rank from 1.00 to find the area above the percentile rank:

$$1.00 - 0.95 = 0.05$$

3. Find the Z score associated with the area above the percentile rank.

Figure 4.12 Finding the Raw Score Associated With a Percentile Higher Than 50

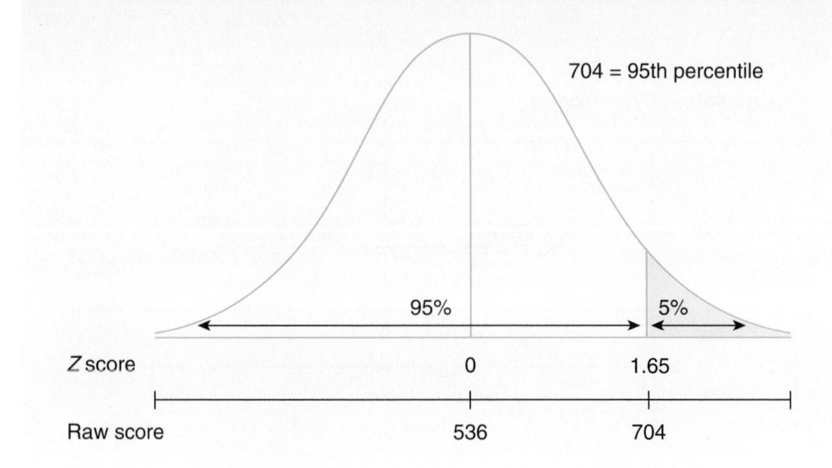

Source: Adapted from College Board, *SAT Suite of Assessments Annual Report: Total Group*, 2018. The College Board. www.collegeboard.org.

Refer to the area under the normal curve shown in Appendix B. First, look for an entry of 0.0500 (or the value closest to it) in column C. The entry closest to 0.0500 is 0.0495. Now, locate the Z in column A that corresponds to this proportion, 1.65.

4. Convert the Z score to a raw score:

$$Y = 536 + 1.65(102) = 536 + 168.3 = 704.3 = 704$$

The final SAT ERW score associated with the 95th percentile is 704. This means that you will need a score of 704 or higher to be admitted into the honors English program.

LEARNING CHECK 4.5

In a normal distribution, how many standard deviations from the mean is the 95th percentile?

Finding the Raw Score Associated With a Percentile Lower Than 50

Finally, what is the SAT ERW score associated with the 40th percentile? To find the percentile rank of a score lower than 50, follow these steps (Figure 4.13).

Figure 4.13 Finding the Raw Score Associated With a Percentile Lower Than 50

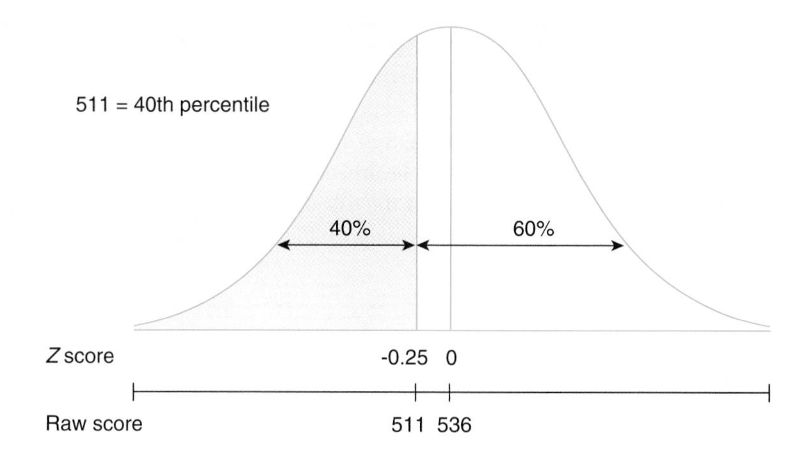

Source: Adapted from College Board, *SAT Suite of Assessments Annual Report: Total Group*, 2018. The College Board. www.collegeboard.org.

1. Divide the percentile by 100 to find the area below the percentile rank:

$$\frac{40}{100} = 0.40$$

2. Find the Z score associated with this area.

Refer to the area under the normal curve shown in Appendix B. First, look for an entry of 0.4000 (or the value closest to it) in column C. The entry closest to 0.4000 is 0.4013. Now, locate the Z in column A that corresponds to this proportion. The Z score associated with the proportion 0.4013 is –0.25.

3. Convert the Z score to a raw score:

$$Y = 536 + (-0.25)(102) = 536 - 25.5 = 510.5 = 511$$

The SAT ERW score associated with the 40th percentile is 511. This means that 40% of the students scored below 511 and 60% scored above it.

LEARNING CHECK 4.6

What is the raw SAT ERW score associated with the 50th percentile?

READING THE RESEARCH LITERATURE: CHILD HEALTH AND ACADEMIC ACHIEVEMENT

Margot Jackson (2015) relied on data from the British National Child Development Study 1958–1974 to examine the intersection of economic disadvantage, poor health, and academic achievement for children aged 7 to 16 years. The longitudinal data set is comprehensive, tracking child health, educational progress, income, and family relationships. Jackson suggests that the role of health in producing academic inequality depends on when, and for how long, children are in poor health.[1]

A CLOSER LOOK 4.1

Percentages, Proportions, and Probabilities

We take a moment to note the relationship between the theoretical normal curve and the estimation of probabilities, a topic that we'll explore in more detail in Chapter 5.

We consider probabilities in many instances. What is the probability of winning the lottery? Of selecting the Queen of Hearts out of a deck of 52 cards? Of getting your favorite parking spot on campus? But we rarely make the connection to some statistical computation or a normal curve.

A probability is a quantitative measure that a particular event will occur. Probability can be calculated as follows:

p = Number of times an event will occur / Total number of events

Probabilities range in value from 0 (the event will not occur) to 1 (the event will certainly occur). In fact, the shape of the normal curve and the percentages beneath it (refer to Figure 4.2) can be used to determine probabilities. By definition, we know that the majority of cases are near the mean, within 1 to 3 standard deviation units. Thus, it is common—or a high-probability occurrence—to find a case near the mean. Rare events—with smaller corresponding probabilities—are further a way from the mean, toward the tail ends of the normal curve.

In our chapter example of SAT ERW scores, we've determined the proportion or percentage of cases that fall within a certain area of the empirical distribution. But these same calculations tell us the probability of the occurrence of a specific score or set of scores based on their distance from the mean. We calculated the proportion of cases between the mean of 536 and a test score of 736 as .4750. We can say that the probability of earning a score between 536 and 736 is .4750 (at least 47 times out of 100 events). What percentage of scores were 375 or less on the SAT ERW exam? According to our calculations, it is 5.71%. The probability of earning a score of 375 or less is 5.71% (6 times out of 100 events). Notice how the calculations are unchanged, but we are shifting our interpretation from a percentage or a proportion to a probability.

The relationship between probabilities and sampling methods will be the focus of Chapter 5.

Jackson examined how standardized reading and math scores vary by the child's health condition at ages 7, 11, and 16 (the presence of a slight, moderate, severe, or no health condition impeding the child's normal functioning; asthma was the most common health condition), low child birth weight (weight below 5.5 pounds), and maternal smoking (amount of smoking by the mother after the fourth month of pregnancy).[2]

A portion of her study analyses is presented in Table 4.3. Jackson converts reading and math scores into *Z* scores, measuring the distance from the overall mean test score in standard deviation units. A positive *Z* score indicates that the group's test score is higher than the mean; a negative score indicates that the test score is lower.

Jackson confirms her hypothesis about the cumulative effects of health on a child's academic performance:

> Table 4.3, which disaggregates average achievement by health status and age, reveals clear variation in reading and math achievement across health categories. Respondents with no childhood health conditions score highest on reading and math assessments at all ages. In contrast, low-birth-weight respondents, those exposed to heavy prenatal smoking late in utero, and those with early

Table 4.3 Academic Achievement by Health, Birth to Age 16 Sample: National Child Development Study, 1958–1974 (*N* = 9,252)

	No Health Condition	Low Birth Weight	Medium/ Variable Smoking	Heavy Prenatal Smoking	Condition at Age 7	Condition at Ages 7, 11, and 16
Age 7						
Mean reading *Z* score	0.139	–0.304	–0.046	–0.084	–0.661	–1.561
Mean math *Z* score	0.087	–0.321	–0.053	–0.085	–0.439	–1.026
Age 11						
Mean reading *Z* score	0.102	–0.337	–0.101	–0.144	–0.454	–1.196
Mean math *Z* score	0.101	–0.367	–0.081	–0.160	–0.411	–0.952
Age 16						
Mean reading *Z* score	0.115	–.317	–0.115	–0.160	–0.432	–1.206
Mean math *Z* score	0.084	–.321	–0.149	–0.196	–0.388	–0.706
N	7,806	445	419	1,080	586	126

Source: Margot Jackson, "Cumulative Inequality in Child Health and Academic Achievement," *Journal of Health and Social Behavior* 56, no. 2 (2015): 269.

school-age health limitations perform more poorly, ranging from .10 to .5 of a standard deviation below average. Children with health conditions at all school ages perform nearly a full standard deviation lower in math and reading. (p. 269)[3]

LEARNING CHECK 4.7

Review the mean math Z scores for the variable "conditions at age 7, 11, and 16" (the last column of Table 4.3). From ages 7, 11, and 16, was there an improvement in their math scores? Explain.

DATA AT WORK

Claire Wulf Winiarek: Director of Collaborative Policy Engagement

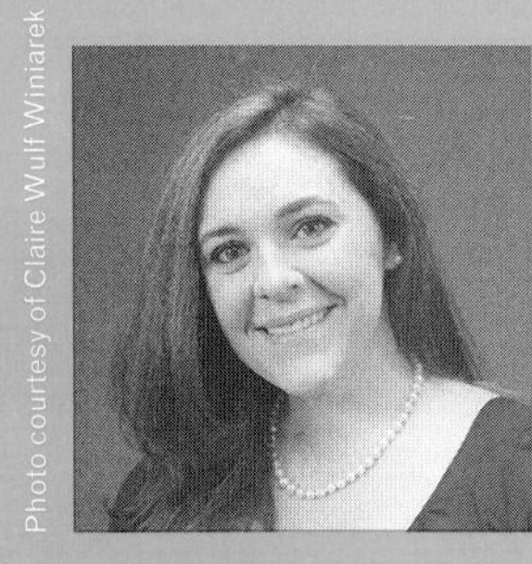

Photo courtesy of Claire Wulf Winiarek

Claire has had an impressive career in public policy. She's worked for a member of Congress, coordinated international human rights advocacy initiatives, and led a public policy team. Her experiences have led her to her current position with a *Fortune* 50 health insurance company. Research is a constant in her work. "The critical and analytic thinking a strong foundation in research methods allows informs my work—from analyzing draft legislation and proposed regulation to determining next year's department budget, from estimating potential growth to making the case for a new program. Research is part of my every day."

Claire was drawn to a career in research because she wanted to make a difference in the public sector. "Early in my career, the frequency with which I returned to and leveraged research methods surprised me—these were not necessarily research positions. However, as government and its private sector partners increasingly rely on data and evidence-based decision making, research methods and the ability to analyze their output have become more critical in public affairs."

"Whether you pursue a research career or not, the importance of research methods—especially with regard to data—will be necessary for success. The information revolution is impacting all industries and sectors, as well as government and our communities. With this ever growing and ever richer set of information, today's professionals must have the know-how to understand and apply this data in a meaningful way. Research methods will create the critical and analytical foundation to meet the challenge, but internships or special research projects in your career field will inform that foundation with practical experience. Always look for that connection between research and reality."

MAIN POINTS

- The normal distribution is central to the theory of inferential statistics. It also provides a model for many empirical distributions that approximate normality.
- In all normal or nearly normal curves, we find a constant proportion of the area under the curve lying between the mean and any given distance from the mean when measured in standard deviation units.
- The standard normal distribution is a normal distribution represented in standard scores, or *Z* scores, with mean = 0 and standard deviation = 1. *Z* scores express the number of standard deviations that a given score is located above or below the mean. The proportions corresponding to any *Z* score or its fraction are organized into a special table called the standard normal table.

KEY TERMS

normal distribution 123
standard normal distribution 127
standard normal table 128
standard (*Z*) score 127

DIGITAL RESOURCES

Access key study tools at https://edge.sagepub.com/ssdsess4e

- eFlashcards of the glossary terms
- Datasets and codebooks
- SPSS and Excel walk-through videos
- SPSS and Excel demonstrations and problems to accompany each chapter
- Appendix F: Basic Math Review

CHAPTER EXERCISES

1. We discovered that 1,014 GSS18SSDS-A respondents watched television for an average of 2.97 hours per day, with a standard deviation of 3.00 hours. Answer the following questions assuming the distribution of the number of television hours is normal:

a. What is the Z score for a person who watches more than 8 hours per day?

b. What proportion of people watch television less than 5 hours per day? How many does this correspond to in the sample?

c. What number of television hours per day corresponds to a Z score of +1?

d. What is the percentage of people who watch between 1 and 6 hours of television per day?

2. You are asked to do a study of shelters for abused and battered women to determine the necessary capacity in your city to provide housing for most of these women. After recording data for a whole year, you find that the mean number of women in shelters each night is 250, with a standard deviation of 75. Fortunately, the distribution of the number of women in the shelters each night is normal, so you can answer the following questions posed by the city council:

   a. If the city's shelters have a capacity of 350, will that be enough places for abused women on 95% of all nights? If not, what number of shelter openings will be needed?

   b. The current capacity is only 220 openings, because some shelters have been closed. What is the percentage of nights that the number of abused women seeking shelter will exceed current capacity?

3. Based on GSS data, we find the mean number of years of education is 13.71 with a standard deviation of 3.04. A total of 1,498 GSS18SSDS-B respondents were included in the survey. Assuming that years of education is normally distributed, answer the following questions:

   a. If you have 13.71 years of education—that is, the mean number of years of education—what is your Z score?

   b. If your friend is in the 60th percentile, how many years of education does she have?

   c. How many people have between your years of education (13.71) and your friend's years of education?

4. A criminologist developed a test to measure recidivism, where low scores indicated a lower probability of repeating the undesirable behavior. The test is normed so that it has a mean of 140 and a standard deviation of 40.

   a. What is the percentile rank of a score of 172?

   b. What is the Z score for a test score of 200?

   c. What percentage of scores falls between 100 and 160?

   d. What proportion of respondents should score above 190?

   e. Suppose an individual is in the 67th percentile in this test. What is his or her corresponding recidivism score?

5. We report the average years of education for GSS18SSDS-B respondents by their social class—lower, working, middle, and upper. Standard deviations are also reported for each class.

	Mean	Standard Deviation	*N*
Lower class	12.03	3.24	136
Working class	13.05	2.77	636
Middle class	14.56	2.95	661
Upper class	15.48	2.76	56

a. Assuming that years of education is normally distributed in the population, what proportion of working-class respondents have 12 to 16 years of education? What proportion of upper-class respondents have 12 to 16 years of education?

b. What is the probability that a working-class respondent, drawn at random from the population, will have more than 16 years of education? What is the equivalent probability for a middle-class respondent drawn at random?

c. What is the probability that a lower-class respondent will have less than 10 years of education?

6. As reported in Table 4.1, the mean SAT Evidence-Based Reading and Writing (ERW) score was 536 with a standard deviation of 102 in 2018.

a. What percentage of students scored above 625?

b. What percentage of students scored between 400 and 625?

c. A college decides to adjust its admission policy. As a first step, the admissions committee decides to exclude student applicants scoring below the 20th percentile on the ERW SAT exam. Translate this percentile into a Z score. Then, calculate the equivalent SAT ERW test score.

7. The standardized IQ test is described as a normal distribution with 100 as the mean score and a 15-point standard deviation.

a. What is the Z score for a score of 150?

b. What percentage of scores are above 150?

c. What percentage of scores fall between 85 and 150?

d. Explain what is meant by scoring in the 95th percentile. What is the corresponding score?

8. We'll examine the results of the 2018 SAT math exam with a mean of 531 and standard deviation of 114, as reported in Table 4.1.

   a. What percentage of test takers scored lower than 400 on the math SAT?

   b. What percentage scored between 600 and 700 points?

   c. Your score is 725. What is your percentile rank?

9. The Hate Crime Statistics Act of 1990 requires the attorney general to collect national data about crimes that manifest evidence of prejudice based on race, religion, sexual orientation, or ethnicity, including the crimes of murder and nonnegligent manslaughter, forcible rape, aggravated assault, simple assault, intimidation, arson, and destruction, damage, or vandalism of property. The Hate Crime Data collected in 2007 reveal, based on a randomly selected sample of 300 incidents, that the mean number of victims in a particular type of hate crime was 1.28, with a standard deviation of 0.82. Assuming that the number of victims was normally distributed, answer the following questions:

   a. What proportion of crime incidents had more than two victims?

   b. What is the probability that there was more than one victim in an incident?

   c. What proportion of crime incidents had fewer than four victims?

10. The number of hours people work each week varies widely for many reasons. Using the GSS18SSDS-A, you find that the mean number of hours worked last week was 41.32, with a standard deviation of 15.28 hours, based on a sample size of 875.

    a. Assume that hours worked are approximately normally distributed in the sample. What is the probability that someone in the sample will work 60 hours or more in a week? How many people in the sample should have worked 60 hours or more?

    b. What is the probability that someone will work 30 hours or fewer in a week (i.e., work part-time)? How many people does this represent in the sample?

    c. What number of hours worked per week corresponds to the 60th percent?

11. The National Collegiate Athletic Association has a public access database on each Division I sports team in the United States, which contains data on team-level Academic Progress Rates (APRs), eligibility rates, and retention rates. The APR score combines team rates for academic eligibility and retention. The mean APR of all reporting men's and women's teams for the 2013–2014 academic year was 981 (based on a 1,000-point scale), with a standard deviation of 27.3. Assuming that the distribution of APRs for the teams is approximately normal:

a. Would a team be at the upper quartile (the top 25%) of the APR distribution with an APR score of 990?

b. What APR score should a team have to be more successful than 75% of all the teams?

c. What is the Z value for this score?

12. According to the same National Collegiate Athletic Association data, the means and standard deviations of eligibility and retention rates (based on a 1,000-point scale) for the 2013–2014 academic year are presented, along with the fictional scores for two basketball teams, A and B. Assume that rates are normally distributed.

	Mean	Standard Deviation	Team A	Team B
Eligibility	983	33	971	987
Retention	976	34.9	958	970

a. On which criterion (eligibility or retention) did Team A do better than Team B? Calculate appropriate statistics to answer this question.

b. What proportion of the teams have retention rates below Team B?

c. What is the percentile rank of Team A's eligibility rate?

13. We present data from the 2014 International Social Survey Programme for five European countries. The average number of completed years of education, standard deviations, and sample size are reported in the table. Assuming that the data are normally distributed, for each country, calculate the number of years of education that corresponds to the 95th percentile.

Country	Mean	Standard Deviation	N
Hungary	11.76	2.91	501
Czech Republic	12.82	2.29	914
Denmark	13.93	5.83	651
France	14.12	5.73	975
Ireland	15.15	3.90	581

5

SAMPLING AND SAMPLING DISTRIBUTIONS

Chapter Learning Objectives

1. Describe the aims of sampling and basic principles of probability.
2. Explain the relationship between a sample and a population.
3. Identify and apply different sampling designs.
4. Apply the concept of the sampling distribution.
5. Describe the central limit theorem.

Until now, we have ignored the question of who or what should be observed when we collect data or whether the conclusions based on our observations can be generalized to a larger group of observations. In truth, we are rarely able to study or observe everyone or everything we are interested in. Although we have learned about various methods to analyze observations, remember that these observations represent a fraction of all the possible observations we might have chosen. Consider the following research examples.

Example 1: The Academic Advising Office is trying to determine how to better address the parking needs of more than 15,000 commuter students but determines that it has only enough time and money to survey 500 students.

Example 2: The Black Lives Matter student organization on your campus is interested in conducting a study of experiences with police brutality. They have enough funds to survey 300 students from the more than 20,000 enrolled students at your school.

Example 3: Environmental activists would like to assess recycling practices in 2-year and 4-year colleges and universities. There are more than 4,700 colleges and universities nationwide.[1]

The primary problem in each situation is that there is too much information and not enough resources to collect and analyze it.

AIMS OF SAMPLING

Researchers in the social sciences rarely have enough time or money to collect information about the entire group that interests them. Known as the population, this group includes all the cases (individuals, groups, or objects) in which the researcher is interested. For example, in our first illustration, the population is 15,000 commuter students; the population in the second illustration is more than 20,000 students; and in the third illustration, our population consists of 4,700 colleges and universities.

Fortunately, we can learn a lot about a population if we carefully select a subset of it. This subset is called a sample. Through the process of sampling—selecting a subset of observations from the population—we attempt to generalize the characteristics of the larger group (population) based on what we learn from the smaller group (the sample). This is the basis of inferential statistics—making predictions or inferences about a population from observations based on a sample. Thus, it is important how we select our sample.

Parameter: A measure (e.g., mean or standard deviation) used to describe the population distribution.

The term **parameter**, associated with the population, refers to measures used to describe the population we are interested in. For instance, the average commuting time for the 15,000 commuter students on your campus is a population parameter because it refers to a population characteristic. In previous chapters, we have learned the many ways of describing a distribution, such as a proportion, a mean, or a standard deviation. When used to describe the population distribution, these measures are referred to as parameters. Thus, a population mean, a population proportion, and a population standard deviation are all parameters.

Statistic: A specific measure used to describe the sample distribution.

We use the term **statistic** when referring to a corresponding characteristic calculated for the sample. For example, the average commuting time for a sample of commuter students is a sample statistic. Similarly, a sample mean, a sample proportion, and a sample standard deviation are all statistics.

In this chapter and in the remaining chapters of this text, we discuss some of the principles involved in generalizing results from samples to the population. We will use different notations when referring to sample statistics and population parameters in our discussion. Table 5.1 presents the sample notation and the corresponding population notation.

The distinctions between a sample and a population and between a parameter and a statistic are illustrated in Figure 5.1. We've included for illustration the population parameter of 0.60—the proportion of white respondents in the population. However, since we almost never have enough resources to collect information about the population, it is rare that we know the value of a parameter. The goal of most research is to find the population parameter. Researchers usually select a sample from the population to obtain an estimate of the population parameter. Thus, the major objective of sampling theory and statistical inference is to provide estimates of unknown parameters from sample statistics that can be easily obtained and calculated.

LEARNING CHECK 5.1

Take a moment to review the definitions of population, sample, parameter, and statistic mean. Use your own words so that the concepts make sense to you. Also review the sample and population notations. These concepts and notations will be used throughout the rest of the text in our discussion of inferential statistics.

Table 5.1 Sample and Population Notations

Measure	Sample Notation	Population Notation
Mean	$\bar{Y}$	μ
Proportion	p	π
Standard deviation	s	σ
Variance	s^2	σ^2

Figure 5.1 The Proportion of White Respondents in a Population and in a Sample

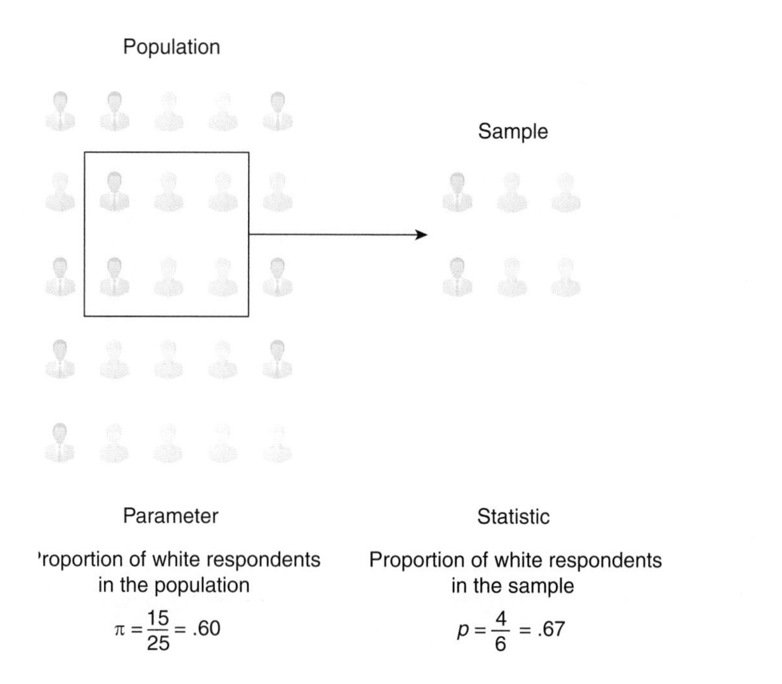

BASIC PROBABILITY PRINCIPLES

We all use the concept of probability in everyday conversation. You might ask, "What is the probability that it will rain tomorrow?" or "What is the likelihood that I will do well on a test?" In everyday conversations, our answers to these questions are rarely systematic, but in the study of statistics, probability has a far more precise meaning.

Probability: A quantitative measure that a particular event will occur.

A **probability** is a quantitative measure that a particular event will occur. It is expressed as a ratio of the number of times an event will occur relative to the set of all possible and equally likely outcomes. Probability is represented by a lowercase p.

p = Number of times an event will occur/total number of events.

Probabilities range in value from 0 (the event will not occur) to 1 (the event will certainly occur). As first discussed in A Closer Look 4.1, probabilities can be expressed as proportions or percentages. Consider, for example, the outcome of rolling a 3 with a six-sided, equally weighted die. The probability of rolling a 3 is 1/6 or 0.17, because this outcome can occur only once out of a total of six possible outcomes: 1, 2, 3, 4, 5, or 6.

p (*rolling* a 3) = 1/6 = .1667 = .17

Over many rolls of the die, the chances of rolling a 3 is .17. So for every 100 rolls, a 3 would come up 17 times. The other 83 times we would see the other values of the die. We can also convert the proportion to a percentage (.17 × 100) and conclude that the probability of rolling a 3 is 17%.

Sometimes we use information from past events to help us predict the likelihood of future events. Such a method is called the relative frequency method. Let's consider, for example, a sample of 1,017 respondents from the 2018 General Social Survey. Respondents were asked how often they read the newspaper; their responses are summarized in Table 5.2.

The ratio of respondents who read the newspaper every day is 210:1,017 or, when reduced, approximately 1:5. To convert a ratio to a proportion, we divide the numerator (210) by the denominator (1,017), as shown in Table 5.2. Thus, the probability 210/1,067 is equivalent to .21. Now, imagine that we wrote down each of the 1,017 respondents' names and placed them in a hat. Because the proportion of .21 is closer to 0 than it is to 1, there is a low likelihood that we would select a respondent who reads the newspaper every day.

Table 5.2 How Often Respondent Reads the Newspaper, GSS18SSDS-B

	Frequency	Proportion (rounded)
Every day	210	.21
Few times a week	117	.12
Once a week	115	.11
Less than once a week	140	.14
Never	435	.43
Total	1,017	1.01

The observed relative frequencies are just an approximation of the true probability of identifying how often a respondent uses the media for political news. The true probabilities can only be determined if we were to repeat the study many times under the same conditions. Then, our long-run relative frequency (or probability) will approximate the true probability.

In Chapter 4, we converted the areas under the normal distribution into proportions or percentages of the number of observations in a sample based on standard deviation units from the mean. These proportions make it possible to estimate the probability of occurrence of these observations. For example, a study of 200 teen girls on the prevalence of texting found the average number of messages a teen girl texts per day to be 70 with a standard deviation of 10. We can estimate that the probability of randomly selecting a teen girl who texts between 70 and 80 messages per day is approximately .3413 (based on the definition of the normal distribution discussed in Chapter 4). We can also say that there is a 34.13% chance that any teen girl drawn randomly from the sample of 200 girls would text between 70 and 80 messages per day.

LEARNING CHECK 5.2

What is the probability of drawing an ace out of a normal deck of 52 playing cards? It's not 1/52. There are four aces, so the probability is 4/52 or 1/13. The proportion is .08. The probability of drawing the specific ace, like the ace of spades, is 1/52 or .02.

PROBABILITY SAMPLING

Social researchers are systematic in their efforts to obtain samples that are representative of the population. Such researchers have adopted a number of approaches for selecting samples from populations. Only one general approach, probability sampling, allows the researcher to use the principles of statistical inference to generalize from the sample to the population.[3]

Probability sampling: A method of sampling that enables the researcher to specify for each case in the population the probability of its inclusion in the sample.

Probability sampling is a method that enables the researcher to specify for each case in the population the probability of its inclusion in the sample. The purpose of probability sampling is to select a sample that is as representative as possible of the population. The sample is selected in such a way as to allow the use of the principles of probability to evaluate the generalizations made from the sample to the population. A probability sample design enables the researcher to estimate the extent to which the findings based on one sample are likely to differ from what would be found by studying the entire population.

Although accurate estimates of sampling error can be made only from probability samples, social scientists often use nonprobability samples because they are more convenient and cheaper to collect. Nonprobability samples are useful under many circumstances for a variety of research purposes. Their main limitation is that they do not allow the use of the method of inferential statistics to generalize from the sample to the population.

The Simple Random Sample

The simple random sample is the most basic probability sampling design, and it is incorporated into even more elaborate probability sampling designs. A **simple random sample** is a sample design chosen in such a way as to ensure that (a) every member of the population has an equal chance of being chosen and (b) every combination of *N* members has an equal chance of being chosen.

Simple random sample: A sample designed in such a way as to ensure that (a) every member of the population has an equal chance of being chosen and (b) every combination of *N* members has an equal chance of being chosen.

Let's take a very simple example to illustrate. Suppose we are conducting a cost-containment study of the 10 hospitals in our region, and we want to draw a sample of two hospitals to study intensively. We can put into a hat 10 slips of paper, each representing one of the 10 hospitals, and mix the slips carefully. We select one slip out of the hat and identify the hospital it represents. We then make the second draw and select another slip out of the hat and identify it. The two hospitals we identified on the two draws become the two members of our sample:

(1) Assuming that we made sure the slips were really well mixed, pure chance determined which hospital was selected on each draw. The sample is a simple random sample because every hospital had the same chance of being selected as a member of our sample of two and (2) every combination of ($N = 2$) hospitals was equally likely to be chosen.

Researchers usually use computer programs or tables of random numbers in selecting random samples. An abridged table of random numbers is reproduced in Appendix A. To use a random number table, list each member of the population and assign the member a number. Begin anywhere on the table and read each digit that appears in the table in order—up, down, or sideways; the direction does not matter, as long as it follows a consistent path. Whenever we come across a digit in the table of random digits that corresponds to the number of a member in the population of interest, that member is selected for the sample. Continue this process until the desired sample size is reached.

THE CONCEPT OF THE SAMPLING DISTRIBUTION

We began this chapter with a few examples illustrating why researchers in the social sciences almost never collect information on the entire population that interests them. Instead, they usually select a sample from that population and use the principles of statistical inference to estimate the characteristics, or parameters, of that population based on the characteristics, or statistics, of the sample. In this section, we describe one of the most important concepts in statistical inference—sampling

Table 5.3 The Population: Personal Income (in Dollars) for 20 Individuals (Hypothetical Data)

Individual	Income (Y)
Case 1	11,350 (Y_1)
Case 2	7,859 (Y_2)
Case 3	41,654 (Y_3)
Case 4	13,445 (Y_4)
Case 5	17,458 (Y_5)
Case 6	8,451 (Y_6)
Case 7	15,436 (Y_7)
Case 8	18,342 (Y_8)
Case 9	19,354 (Y_9)
Case 10	22,545 (Y_{10})
Case 11	25,345 (Y_{11})
Case 12	68,100 (Y_{12})
Case 13	9,368 (Y_{13})
Case 14	47,567 (Y_{14})
Case 15	18,923 (Y_{15})
Case 16	16,456 (Y_{16})
Case 17	27,654 (Y_{17})
Case 18	16,452 (Y_{18})
Case 19	23,890 (Y_{19})
Case 20	25,671 (Y_{20})
Mean (μ) = 22,766	Standard deviation (σ) = 14,687

distribution. The sampling distribution helps estimate the likelihood of our sample statistics and, therefore, enables us to generalize from the sample to the population.

The Population

To illustrate the concept of the sampling distribution, let's consider as our population the 20 individuals listed in Table 5.3.[4] Our variable, Y, is the income (in dollars) of these 20 individuals, and the parameter we are trying to estimate is the mean income.

We use the symbol μ to represent the population mean. Using Formula 3.1, we can calculate the population mean:

$$\mu = \frac{\Sigma Y}{N} = \frac{Y_1 + Y_2 + Y_3 + Y_4 + Y_5 + \ldots + Y_{20}}{20}$$

$$= \frac{11,350 + 7,859 + 41,654 + 13,445 + 17,458 + \ldots + 25,671}{20}$$

$$= 22,766$$

Using Formula 3.3, we can also calculate the standard deviation for this population distribution. We use the Greek symbol sigma (σ) to represent the population's standard deviation:

$$\sigma = 14,687$$

Of course, most of the time, we do not have access to the population. So instead, we draw one sample, compute the mean—the statistic—for that sample, and use it to estimate the population mean—the parameter.

The Sample

Let's assume that μ is unknown and that we estimate its value by drawing a random sample of 3 individuals ($N = 3$) from the population of 20 individuals and calculate the mean income for that sample. The incomes included in that sample are as follows:

Case 8	18,342
Case 16	16,456
Case 17	27,654

Now let's calculate the mean for that sample:

$$\bar{Y} = \frac{18,342 + 16,456 + 27,654}{3} = 20,817$$

Sampling error: The discrepancy between a sample estimate of a population parameter and the real population parameter.

Note that our sample mean, $\bar{Y} = \$20,817$, differs from the actual population parameter, \$22,766. This discrepancy is due to sampling error. **Sampling error** is the discrepancy between a sample estimate of a population parameter and the real population parameter. By comparing the sample statistic with the population parameter, we can determine the sampling error. The sampling error for our example is 1,949 (22,766 – 20,817 = 1,949).

Now let's select another random sample of three individuals. This time, the incomes included are as follows:

Case 15	18,923
Case 5	17,458
Case 17	27,654

The mean for this sample is

$$\bar{Y} = \frac{18,923 + 17,458 + 27,654}{3} = 21,345$$

The sampling error for this sample is 1,421(22,766 – 21,345 = 1,421), somewhat less than the error for the first sample we selected.

The Dilemma

Although comparing the sample estimates of the average income with the actual population average is a perfect way to evaluate the accuracy of our estimate, in practice, we rarely have information about the actual population parameter. If we did, we would not need to conduct a study! Moreover, few, if any, sample estimates correspond exactly to the actual population parameter. This, then, is our dilemma: If sample estimates vary and if most estimates result in some sort of sampling error, how much confidence can we place in the estimate? On what basis can we infer from the sample to the population?

The Sampling Distribution

The answer to this dilemma is to use a device known as the sampling distribution. The **sampling distribution** is a theoretical probability distribution of all possible sample values for the statistic in which we are interested. If we were to draw all possible random samples of the same size from our population of interest, compute the statistic for each sample, and plot the frequency distribution for that statistic, we would obtain an approximation of the sampling distribution. Every statistic—for example, a proportion, a mean, or a variance—has a sampling distribution. Because it includes all possible sample values, the sampling distribution enables us to compare our sample result with other sample values and determine the likelihood associated with that result.[5]

Sampling distribution: The sampling distribution is a theoretical probability distribution of all possible sample values for the statistics in which we are interested.

THE SAMPLING DISTRIBUTION OF THE MEAN

Sampling distributions are theoretical distributions, which means that they are never really observed. Constructing an actual sampling distribution would involve taking all possible random samples of a fixed size from the population. This process would be very tedious because it would involve a very large number of samples. However, to help grasp the concept of the sampling distribution, let's illustrate how one could be generated from a limited number of samples.

An Illustration

Sampling distribution of the mean: A theoretical probability distribution of sample means that would be obtained by drawing from the population all possible samples of the same size.

For our illustration, we use one of the most common sampling distributions—the sampling distribution of the mean. The **sampling distribution of the mean** is a theoretical distribution of sample means that would be obtained by drawing from the population all possible samples of the same size.

Let's go back to our example in which our population is made up of 20 individuals and their incomes. From that population (Table 5.3), we now randomly draw 50 possible samples of size 3 ($N = 3$), computing the mean income for each sample and replacing it before drawing another.

In our first sample of $N = 3$, we draw three incomes: \$8,451, \$41,654, and \$18,923. The mean income for this sample is

$$\bar{Y} = \frac{8,451 + 41,654 + 18,923}{3} = 23,009$$

Now we restore these individuals to the original list and select a second sample of three individuals. The mean income for this sample is

$$\bar{Y} = \frac{15,436 + 25,345 + 16,456}{3} = 19,079$$

We repeat this process 48 more times, each time computing the sample mean and restoring the sample to the original list. Table 5.4 lists the means of the first five and the 50th samples of $N = 3$ that were drawn from the population of 20 individuals. (Note that $\Sigma\bar{Y}$ refers to the sum of all the means computed for each of the samples and M refers to the total number of samples that were drawn.)

The grouped frequency distribution for all 50 sample means ($M = 50$) is displayed in Table 5.5; Figure 5.2 is a histogram of this distribution. This distribution is an example of a sampling distribution of the mean. Note that in its structure, the sampling distribution resembles a frequency distribution of raw scores, except that here each score is a sample mean, and the corresponding frequencies are the number of samples with that particular mean value. For example, the third interval in Table 5.5 ranges from \$19,500 to \$23,500, with a corresponding frequency of 14, or 28%. This means that we drew 14 samples (28%) with means ranging between \$19,500 and \$23,500.

Table 5.4 Mean Income of 50 Samples of Size 3

Sample	Mean ($\bar{Y}$)
First	23,009
Second	19,079
Third	18,873
Fourth	26,885
Fifth	21,847
⋮	⋮
Fiftieth	26,645
Total (M) = 50	$\Sigma\bar{Y} = 1,237,48$[illegible]

Table 5.5 Sampling Distribution of Sample Means for Sample Size N = 3 Drawn From the Population of 20 Individuals' Incomes

Sample Mean Intervals	Frequency	Percentage
11,500–15,500	6	12
15,500–19,500	7	14
19,500–23,500	14	28
23,500–27,500	4	8
27,500–31,500	9	18
31,500–35,500	7	14
35,500–39,500	1	2
39,500–43,500	2	4
Total (M)	50	100

Remember that the distribution depicted in Table 5.5 and Figure 5.2 is an empirical distribution, whereas the sampling distribution is a theoretical distribution. In reality, we never really construct a sampling distribution. However, even this simple empirical example serves to illustrate some of the most important characteristics of the sampling distribution.

Figure 5.2 Sampling Distribution of Sample Means for Sample Size N = 3 Drawn From the Population of 20 Individuals' Incomes

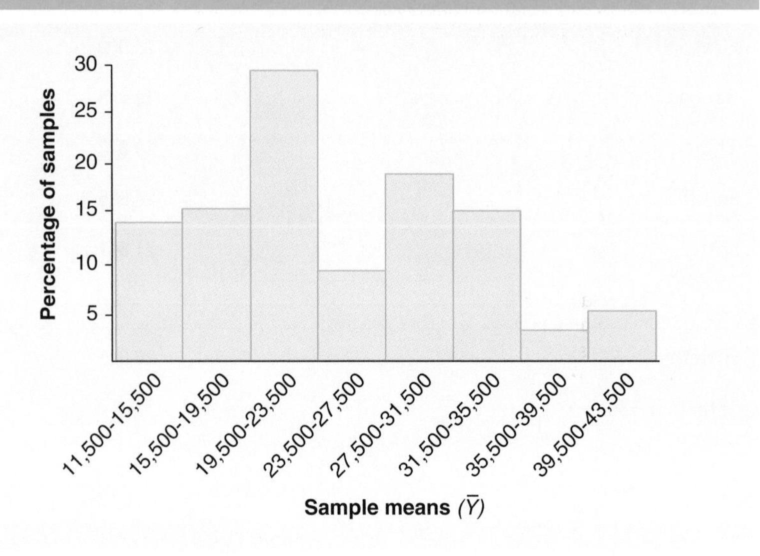

Review

Before we continue, let's take a moment to review the three distinct types of distribution.

The Population: We began with the population distribution of 20 individuals. This distribution actually exists. It is an empirical distribution that is usually unknown to us. We are interested in estimating the mean income for this population.

The Sample: We drew a sample from that population. The sample distribution is an empirical distribution that is known to us and is used to help us estimate the mean of the population. We selected 50 samples of $N = 3$ and calculated the mean income. We generally use the sample mean $(\bar{Y})$ as an estimate of the population mean (μ).

The Sampling Distribution of the Mean: For illustration, we generated an approximation of the sampling distribution of the mean, consisting of 50 samples of $N = 3$. The sampling distribution of the mean does not really exist. It is a theoretical distribution.

To help you understand the relationship among the population, the sample, and the sampling distribution, we have illustrated in Figure 5.3 the process of generating an empirical sampling distribution of the mean. From a population of raw scores (*Y*s), we draw *M* samples of size *N* and calculate the mean of each sample. The resulting sampling distribution of the mean, based on *M* samples of size *N*, shows the values that the mean could take and the frequency (number of samples) associated with each value. Make sure you understand these relationships. The concept of the sampling distribution is crucial to understanding statistical inference. In

Figure 5.3 Generating the Sampling Distribution of the Mean

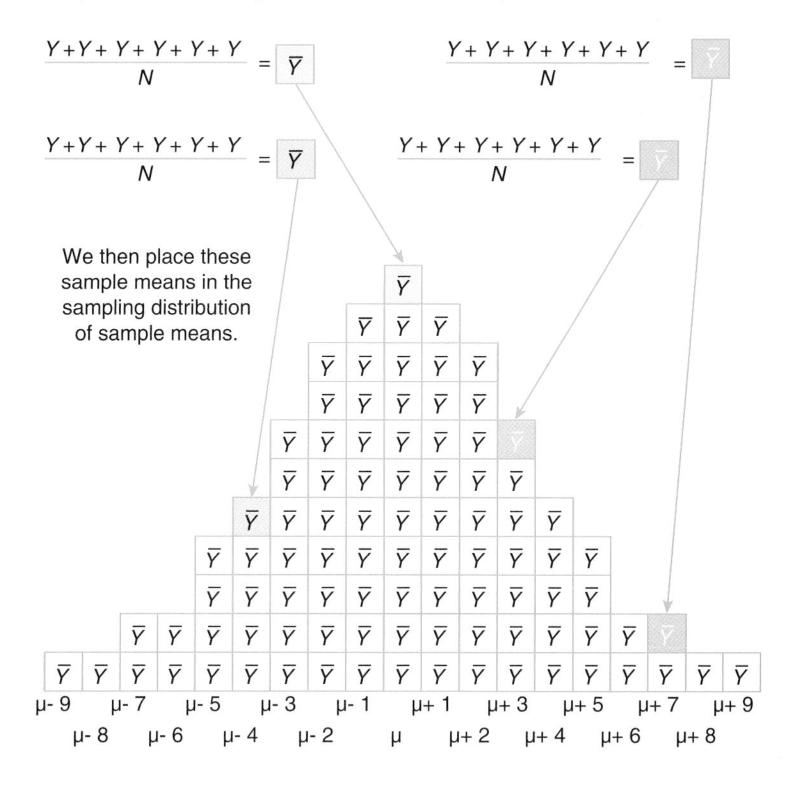

this and the following chapter, we learn how to employ the sampling distribution to draw inferences about the population on the basis of sample statistics.

The Mean of the Sampling Distribution

Like the population and sample distributions, the sampling distribution can be described in terms of its mean and standard deviation. We use the symbol $\mu_{\bar{Y}}$ to represent the mean of the sampling distribution. The subscript indicates the specific variable of this sampling distribution. To obtain the mean of the sampling distribution, add all the individual sample means ($\Sigma\bar{Y} = 1{,}237{,}482$) and divide by the number of samples ($M = 50$). Thus, the mean of the sampling distribution of the mean is actually the mean of means:

$$\mu_{\bar{Y}} = \frac{\Sigma\bar{Y}}{M} = \frac{1{,}237{,}482}{50} = 24{,}750$$

The Standard Error of the Mean

Standard error of the mean: The standard deviation of the sampling distribution of the mean. It describes how much dispersion there is in the sampling distribution of the mean.

The standard deviation of the sampling distribution is also called the **standard error of the mean**. The standard error of the mean describes how much dispersion there is in the sampling distribution, or how much variability there is in the value of the mean from sample to sample:

$$\sigma_{\bar{Y}} = \frac{\sigma}{\sqrt{N}}$$

This formula tells us that the standard error of the mean is equal to the standard deviation of the population σ divided by the square root of the sample size (N). For our example, because the population standard deviation is 14,687 and our sample size is 3, the standard error of the mean is

$$\sigma_{\bar{Y}} = \frac{14,687}{\sqrt{3}} = 8,480$$

THE CENTRAL LIMIT THEOREM

In Figure 5.4a and b, we compare the histograms for the population and sampling distributions of Tables 5.3 and 5.4. Figure 5.4a shows the population distribution of 20 incomes, with a mean $\mu = 22,766$ and a standard deviation $\sigma = 14,687$. Figure 5.4b shows the sampling distribution of the means from 50 samples of $N = 3$ with a mean $\mu_{\bar{Y}} = 24,750$ and a standard deviation (the standard error of the mean) $\sigma_{\bar{Y}} = 8,480$. These two figures illustrate some of the basic properties of sampling distributions in general and the sampling distribution of the mean in particular.

First, as can be seen from Figure 5.4a and b, the shapes of the two distributions differ considerably. Whereas the population distribution is skewed to the right, the sampling distribution of the mean is less skewed—that is, it is closer to a symmetrical, normal distribution.

Second, whereas only a few of the sample means coincide exactly with the population mean, $22,766, the sampling distribution centers on this value. The mean of the sampling distribution is a pretty good approximation of the population mean.

In the discussions that follow, we make frequent references to the mean and standard deviation of the three distributions. To distinguish among the different distributions, we use certain conventional symbols to refer to the means and standard deviations of the sample, the population, and the sampling distribution. Note that we use Greek letters to refer to both the sampling and the population distributions.

Third, the variability of the sampling distribution is considerably smaller than the variability of the population distribution. Note that the standard deviation for the sampling distribution $\sigma_{\bar{Y}} = 8,480$ is almost half that for the population ($\sigma = 14,687$).

Figure 5.4 Three Income Distributions

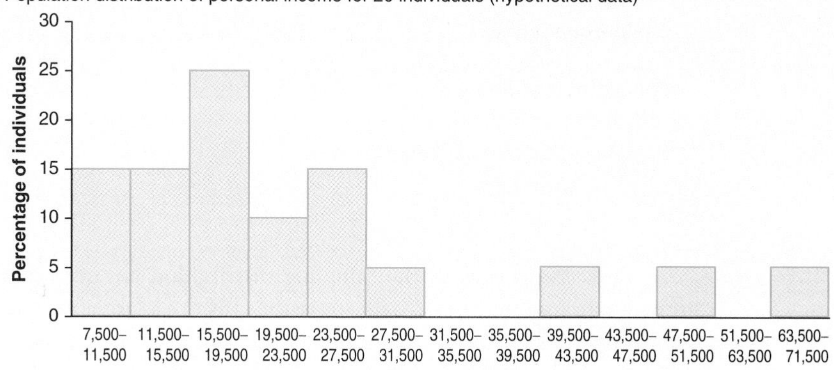

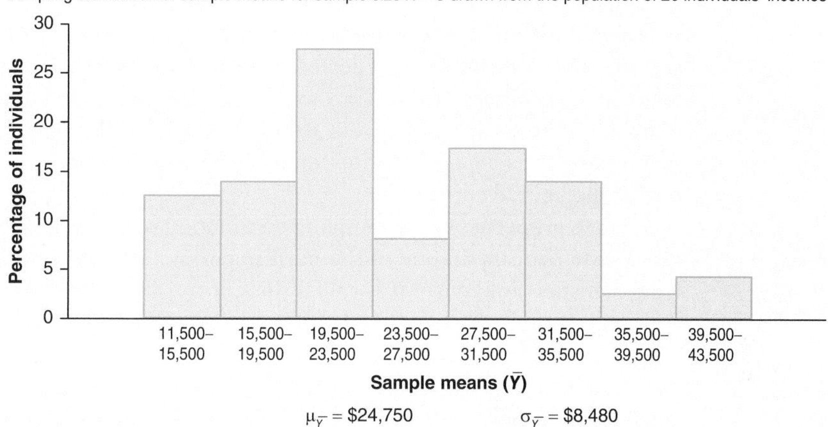

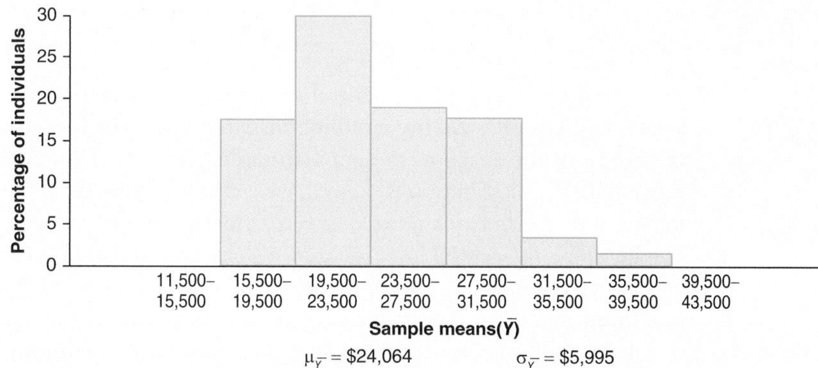

Table 5.6 Mean and Standard Deviation Notations

	Mean	Standard Deviation
Sample distribution	$\bar{Y}$	s
Population distribution	μ	σ
Sampling distribution of the mean	$\mu_{\bar{Y}}$	$\mu_{\bar{Y}}$

These properties of the sampling distribution are even more striking as the sample size increases. To illustrate the effect of a larger sample on the shape and properties of the sampling distribution, we went back to our population of 20 individual incomes and drew 50 additional samples of $N = 6$. We calculated the mean for each sample and constructed another sampling distribution. This sampling distribution is shown in Figure 5.4c. It has a mean $\mu_{\bar{Y}} = 24{,}064$ and a standard deviation $\sigma_{\bar{Y}} = 5{,}995$. Note that as the sample size increased, the sampling distribution became more compact. This decrease in the variability of the sampling distribution is reflected in a smaller standard deviation: With an increase in sample size from $N = 3$ to $N = 6$, the standard deviation of the sampling distribution decreased from 8,480 to 5,995. Furthermore, with a larger sample size, the sampling distribution of the mean is an even better approximation of the normal curve.

These properties of the sampling distribution of the mean are summarized more systematically in one of the most important statistical principles underlying statistical inference. It is called the **central limit theorem**, and it states that if all possible random samples of size N are drawn from a population with a mean μ and a standard deviation σ, then as N becomes larger, the sampling distribution of sample means becomes approximately normal, with mean $\mu_{\bar{Y}}$ equal to the population mean and a standard deviation equal to

$$\sigma_{\bar{Y}} = \frac{\sigma}{\sqrt{N}}$$

Central limit theorem: If all possible random samples of size N are drawn from a population with a mean μ_Y and a standard deviation, σ_Y then as N becomes larger, the sampling distribution of sample means becomes approximately normal, with mean $\mu_{\bar{Y}}$ and standard deviation, $\sigma_{\bar{Y}} = \sigma_Y / \sqrt{N}$

The significance of the central limit theorem is that it tells us that with a sufficient sample size, the sampling distribution of the mean will be normal regardless of the shape of the population distribution. Therefore, even when the population distribution is skewed, we can still assume that the sampling distribution of the mean is normal, given random samples of large enough size. Furthermore, the central limit theorem also assures us that (a) as the sample size gets larger, the mean of the sampling distribution becomes equal to the population mean, and (b) as the sample size gets larger, the standard error of the mean (the standard deviation of the sampling distribution of the mean) decreases in size. The standard error of the mean tells how much variability in the sample estimates there is from sample to sample. The smaller the standard error of the mean, the closer (on average) the sample means will be to the population mean. Thus, the larger the sample, the more closely the sample statistic clusters around the population parameter.

The Size of the Sample

Although there is no hard-and-fast rule, a general rule of thumb is that when N is 50 or more, the sampling distribution of the mean will be approximately normal regardless of the shape of the distribution. However, we can assume that the sampling distribution will be normal even with samples as small as 30 if we know that the population distribution approximates normality.

LEARNING CHECK 5.3

What is a normal population distribution? If you can't answer this question, go back to Chapter 4. You must understand the concept of a normal distribution before you can understand the techniques involved in inferential statistics.

The Significance of the Sampling Distribution and the Central Limit Theorem

In the preceding sections, we have covered a lot of abstract material. You may have a number of questions at this time. Why is the concept of the sampling distribution so important? What is the significance of the central limit theorem? To answer these questions, let's go back and review our 20 incomes example.

To estimate the mean income of a population of 20 individuals, we drew a sample of three cases and calculated the mean income for that sample. Our first sample mean, $\bar{Y} = 20{,}817$, differs from the actual population parameter, $\mu = 22{,}766$. When we selected different samples, we found each time that the sample mean differed from the population mean. These discrepancies are due to sampling errors. Had we taken a number of additional samples, we probably would have found that the mean was different each time because every sample differs slightly. Few, if any, sample means would correspond exactly to the actual population mean. Usually, we have only one sample statistic as our best estimate of the population parameter.

So now let's restate our dilemma: If sample estimates vary and if most result in some sort of sampling error, how much confidence can we place in the estimate? On what basis can we infer from the sample to the population?

The solution lies in the sampling distribution and its properties. Because the sampling distribution is a theoretical distribution that includes all possible sample outcomes, we can compare our sample outcome with it and estimate the likelihood of its occurrence.

Since the sampling distribution is theoretical, how can we know its shape and properties so that we can make these comparisons? Our knowledge is based on what the central limit theorem tells us about the properties of the sampling distribution of the mean. We know that if our sample size is large enough (at least 50 cases), most sample means will be quite close to the true population mean. It is highly unlikely that our sample mean would deviate much from the actual population mean.

In Chapter 4, we saw that in all normal curves, a constant proportion of the area under the curve lies between the mean and any given distance from the mean when measured in standard deviation units, or *Z* scores. We can find this proportion in the standard normal table (Appendix B).

Knowing that the sampling distribution of the means is approximately normal, with a mean $\mu_{\bar{Y}}$ and a standard deviation $\sigma / \sqrt{N}$ (the standard error of the mean), we can use Appendix B to determine the probability that a sample mean will fall within a certain distance—measured in standard deviation units, or *Z* scores—of $\mu_{\bar{Y}}$ or μ. For example, we can expect approximately 68% (or we can say the probability is approximately 0.68) of all sample means to fall within ±1 standard error ($\sigma / \sqrt{N}$), or the standard deviation of the sampling distribution of the mean of $\mu_{\bar{Y}}$ or μ. Similarly, the probability is about 0.95 that the sample mean will fall within ±2 standard errors of $\mu_{\bar{Y}}$ or μ. In the next chapter, we will see how this information helps us evaluate the accuracy of our sample estimates.

LEARNING CHECK 5.4

Suppose a population distribution has a mean $\mu = 150$ and a standard deviation $s = 30$, and you draw a simple random sample of $N = 100$ cases. What is the probability that the mean is between 147 and 153? What is the probability that the sample mean exceeds 153? Would you be surprised to find a mean score of 159? Why? (*Hint:* To answer these questions, you need to apply what you learned in Chapter 4 about Z scores and areas under the normal curve [Appendix B].) To translate a raw score into a Z score, we used this formula:

$$Z = \frac{Y - \bar{Y}}{s}$$

However, because here we are dealing with a sampling distribution, replace Y with the sample mean $\bar{Y}$, $\bar{Y}$ with the sampling distribution's mean μ Y, and σ with the standard error of the mean.

$$Z = \frac{\bar{Y} - \mu_{\bar{Y}}}{\sigma / \sqrt{N}}$$

STATISTICS IN PRACTICE: THE 2016 U.S. PRESIDENTIAL ELECTION

There are numerous applications of the central limit theorem in research, business, medicine, and popular media. As varied as these applications may be, what they have in common is that the data are derived from relatively small random samples taken from considerably larger and often varied populations. And the data have consequence—informing our understanding of the social world, influencing decisions, shaping social policy, and predicting social behavior.

On Tuesday, November 8, 2016, Republican nominee Donald J. Trump was elected the 45th president of the United States. Although he didn't win the popular vote,[6] he won the election in what many people around the world, including political pundits, viewed as an unimaginable outcome. According to a wide range of election forecasters, the odds of the Democratic nominee, Hillary R. Clinton, winning the election over Trump ranged from 70% to 99%! So, what happened on November 8? Were research methodologists who worked for the various polling outlets as surprised as the general public seemed to be? Probably not, given, through their experience and training, they understand the complexities of sampling that drive public opinion polls.

The day after the election, Andrew Mercer, Claudia Deane, and Kyley McGeeney from the Pew Research Center ("a nonpartisan fact tank that . . . conduct[s] public opinion polling, demographic research, content analysis and other data-driven social science research"[7]) published an article on their website entitled, "Why 2016 Election Polls Missed Their Mark." They acknowledge that "across the board, polls underestimated Trump's level of support," and in the process, they offer three possible explanations for the faulty predictions.

The first explanation has to do with nonresponse bias. They explain that nonresponse bias "occurs when certain kinds of people systematically do not respond to surveys despite equal opportunity outreach to all parts of the electorate." This hypothesis rests on the possibility that pollsters weren't able to successfully reach a key group of Trump supporters: working-class and less-educated Americans. Or maybe, the Pew Research Center scientists conclude, "the frustration and anti-institutional feelings that drove the Trump campaign may also have aligned with an unwillingness [for Trump supporters] to respond to polls."

The second explanation has to do with a lack of honesty by survey respondents. The Pew Research Center scientists wonder how many respondents were "shy Trumpers" who didn't share their intent to vote for Trump with the pollsters given they might have worried that their support for Trump was socially undesirable—viewed unfavorably. This certainly is a plausible hypothesis given the public criticism of Trump—especially after a 2005 audio recording surfaced before the election where Trump is heard saying to TV host Billy Bush that he can "grab [women he is attracted to] by the pussy."[8] Although Trump minimized the leaked audio recording as "locker room banter"[9] between men, many people around the world were horrified that someone who would aggressively speak about women was the Republican nominee for the U.S. presidency.

The third explanation has to do with the possibility that pollsters inaccurately predicted which likely voters would actually vote. The Pew Research Center scientists explain, "This is a notoriously difficult task, and small differences in assumptions can produce sizable differences in election predictions. We may find that the voters that pollsters were expecting, particularly in the Midwestern and Rust Belt states that so defied expectations, were not the ones that showed up."

While we may never know for sure which hypothesis, or combination of hypotheses, explains "Why 2016 Election Polls Missed Their Mark," it's safe to assume research methodologists will continue to study what happened in the 2016 U.S. presidential election in an attempt to improve their predictions in future elections.

DATA AT WORK

Emily Treichler: Postdoctoral Fellow

Photo courtesy of Emily Treichler

As an undergraduate, Dr. Treichler wanted to figure out a way to make mental health treatment more effective and more accessible. Having completed her PhD, currently she is a postdoctoral fellow conducting research on schizophrenia and other related disorders in a Veterans Affairs (VA) hospital. "I divide my time between research, clinical work, and other kinds of training, including learning new methods. I conduct research in clinical settings working with people who are experiencing mental health problems, and use the results of my research and other research literature to try to improve mental health treatment."

"I use statistics and methods constantly. I read research literature in order to learn more about my area, to apply it in clinical situations, and to apply it to my own research. I conduct clinical research, using both qualitative and quantitative methodology, and conducting statistics on quantitative data. I collect data in settings I work in as a clinician and conduct statistics on that data in order to understand how our services are working, and how to improve clinical services."

According to Treichler, "Quantitative research can be an incredibly fun area." For students interested in the field, she advises, "Get a wide range of training in statistics and methods so you can understand the literature in your area, and have access to multiple methods for your own studies. Choosing appropriate methods and statistics given your research question and the literature in your area is key to creating a successful project."

MAIN POINTS

- Through the process of sampling, researchers attempt to generalize the characteristics of a large group (the population) from a subset (sample) selected from that group. The term parameter, associated with the population, refers to the information we are interested in finding out. Statistic refers to a corresponding calculated sample statistic.
- A probability sample design allows us to estimate the extent to which the findings based on one sample are likely to differ from what we would find by studying the entire population.
- A simple random sample is chosen in such a way as to ensure that every member of the population and every combination of N members have an equal chance of being chosen.
- The sampling distribution is a theoretical probability distribution of all possible sample values for the statistic in which we are interested. The sampling distribution of the mean is a frequency distribution of all possible sample means of the same size that can be drawn from the population of interest.
- According to the central limit theorem, if all possible random samples of size N are drawn from a population with a mean μ and a standard deviation σ, then as N becomes larger, the sampling distribution of sample means becomes approximately normal, with mean μ and standard deviation $\sigma / \sqrt{N}$.
- The central limit theorem tells us that with sufficient sample size, the sampling distribution of the mean will be normal regardless of the shape of the population distribution. Therefore, even when the population distribution is skewed, we can still assume that the sampling distribution of the mean is normal, given a large enough randomly selected sample size.

KEY TERMS

central limit theorem 164
parameter 150
probability 152
probability sampling 153
sampling distribution 157
sampling distribution of the mean 158
sampling error 156
simple random sample 154
standard error of the mean 162
statistic 150

DIGITAL RESOURCES

Access key study tools at https://edge.sagepub.com/ssdsess4e

- eFlashcards of the glossary terms
- Datasets and codebooks
- SPSS and Excel walk-through videos
- SPSS and Excel demonstrations and problems to accompany each chapter
- Appendix F: Basic Math Review

CHAPTER EXERCISES

1. Explain which of the following is a statistic and which is a parameter.
   a. The mean age of Americans from the 2020 decennial census
   b. The unemployment rate for the population of U.S. adults, estimated by the government from a large sample
   c. The percentage of students at your school who receive an athletic scholarship
   d. The percentage of Arizonans opposed to increasing border control from a poll of 1,000 residents
   e. The mean salaries of employees at your school (e.g., administrators, faculty, maintenance)
2. The mayor of your city has been talking about the need for a tax hike. The city's newspaper uses letters sent to the editor to judge public opinion about this possible hike, reporting on their results in an article.
   a. Do you think that these letters represent a random sample? Why or why not?
   b. What alternative sampling method would you recommend to the mayor?
3. A social scientist gathers a carefully chosen group of 20 people whom she has selected to represent a broad cross section of the population in New York City. She interviews them in depth for a study she is doing on race relations in the city. Is this a probability sample? Explain your answer.
4. An upper-level sociology class has 120 registered students: 34 seniors, 57 juniors, 22 sophomores, and 7 freshmen.
   a. Imagine that you choose one random student from the classroom (perhaps by using a random number table). What is the probability that the student will be a junior?
   b. What is the probability that the student will be a freshman?

5. Can the standard error of a variable ever be larger than, or even equal in size to, the standard deviation for the same variable? Justify your answer by means of both a formula and a discussion of the relationship between these two concepts.

6. When taking a random sample from a very large population, how does the standard error of the mean change when

   a. the sample size is increased from 100 to 1,600?

   b. the sample size is decreased from 300 to 150?

   c. the sample size is multiplied by 4?

7. Facebook now allows users to create a poll of their own to assess their FB friends' views on any given issue. In light of the Black Lives Matter movement, you could create a Facebook poll asking your FB friends if they feel police brutality is a serious issue offering the response options "Yes," "No," and "Not sure."

   a. Is your FB poll a probability sample? Why or why not?

   b. Specify the population from which the sample is drawn.

8. The following table shows the number of active military personnel in 2009, by region (including the District of Columbia).

Pacific	229,634	Mountain	89,816	West South Central	177,336
West North Central	64,564	East North Central	26,384	East South Central	68,440
South Atlantic	376,034	Middle Atlantic	41,441	New England	8,579

Source: U.S. Census Bureau, *Statistical Abstract of the United States: 2012*, Table 508 (data) and U.S. Census Bureau, *Census Regions and Divisions of the United States* (regions).

   a. Calculate the mean and standard deviation for the population.

   b. Now take 10 samples of size 3 from the population. Use simple random sampling to identify your samples. Calculate the mean for each sample.

   c. Once you have calculated the mean for each sample, calculate the mean of means (i.e., add up your 10 sample means and divide by 10). How does this mean compare with the mean for all states?

   d. How does the value of the standard deviation that you calculated in Exercise 8a compare with the value of the standard error (i.e., the standard deviation of the sampling distribution)?

   e. Construct two histograms, one for the distribution of values in the population and the other for the various sample means taken from Exercise 8b. Describe and explain any differences you observe between the two distributions.

f. It is important that you have a clear sense of the population that we are working with in this exercise. What is the population?

9. You've been asked to determine the percentage of students at your university who support open border policies that would allow folks to freely move between international jurisdictions with minimal or no restrictions. You want to take a sample of 10% of the students enrolled on your campus to make the estimate using a survey you created. Explain whether each of the following scenarios describes a reliable sample reflective of all students at your university.

    a. You distribute your survey to all students eating lunch in your campus student union on a Tuesday at 12:30 p.m.

    b. You attend a rally on your campus in support of DACA (Deferred Action for Childhood Arrivals) and ask everyone in attendance to complete your survey.

    c. What sampling procedure would you recommend to complete your study?

10. Imagine that you are working with a total population of 21,473 respondents. You assign each respondent a value from 0 to 21,473 and proceed to select your sample using the random number table in Appendix A. Starting at column 7, line 1 in Appendix A and going down, which are the first five respondents that will be included in your sample?

11. A small population of $N = 10$ has values of 4, 7, 2, 11, 5, 3, 4, 6, 10, and 1.

    a. Calculate the mean and standard deviation for the population.

    b. Take 10 simple random samples of size 3, and calculate the mean for each.

    c. Calculate the mean and standard deviation of all these sample means. How closely does the mean of all sample means match the population mean? How is the standard deviation of the means related to the standard deviation for the population?

12. The following table presents the number of parolees (per 100,000 people) for 12 of the most populous states as of July 2015.

State	Parolees (per 100,000 people)
California	292
Texas	556
New York	288
Florida	28
Illinois	299

State	Parolees (per 100,000 people)
Pennsylvania	1,035
Ohio	193
Georgia	334
Michigan	239
North Carolina	130
New Jersey	214
Virginia	27

Source: National Institute of Corrections, *Correction Statistics by State*, 2016.

a. Assume that $\sigma = 226.83$ for the entire population of 50 states. Calculate and interpret the standard error. (Consider the formula for the standard error. Since we provided the population standard deviation, calculating the standard error requires only minor calculations.)

b. Write a brief statement on the following: the standard error compared with the standard deviation of the population, the shape of the sampling distribution, and suggestions for reducing the standard error.

6 ESTIMATION

In this chapter, we discuss the procedures involved in estimating population means and proportions based on the principles of sampling and statistical inference discussed in Chapter 5. Knowledge about the sampling distribution allows us to estimate population means and proportions from sample outcomes and to assess the accuracy of these estimates. Consider three examples of information derived from samples.

Example 1: Based on a random sample of 1,505 U.S. adults, a January 2019 Pew Research poll found that the percentage of Americans who believe immigrants strengthen rather than burden the country was 62%, compared with 28% of Americans who believe immigrants are a burden because they take jobs and health care. In its report, Pew Research noted a partisan division in attitudes: 83% of Democrats believe immigrants strengthen the country compared to 38% of Republicans who believe the same.[1]

Example 2: Every other year, the National Opinion Research Center conducts the General Social Survey (GSS) on a representative sample of about 1,500 respondents. The GSS, from which many of the examples in this book are selected, is designed to provide social science researchers with a readily accessible database of socially relevant attitudes, behaviors, and attributes of a cross section of the U.S. adult population. For example, in analyzing the responses to the 2018 GSS, researchers found that the average respondent's education was about 13.71 years. This average probably differs from the average of the population from which the GSS sample was drawn. However, we can establish that in most cases, the sample mean (in this case, 13.71 years) is fairly close to the actual true average in the population.

Example 3: In 2018, a CBS News poll of 1,012 Americans revealed that 65% of those surveyed support stricter gun sale laws.

Chapter Learning Objectives

1. Explain the concept of estimation, point estimates, confidence level, and confidence interval.
2. Calculate and interpret confidence intervals for means.
3. Describe the concept of risk and how to reduce it.
4. Calculate and interpret confidence intervals for proportions.

Fifty-nine percent were pessimistic that President Donald Trump and the U.S. Congress would enact any such laws in the near future.[2]

The percentage of Americans who believe immigrants are a strength to our country as approximated by the Pew Research Center, the average level of education in the United States as calculated from the GSS, and the percentage of Americans who support stricter restrictions on the sale of firearms based on the CBS News poll are all sample estimates of population parameters. Population parameters are the actual percentage of Americans who believe immigrants are a strength to our country or the actual average level of education in the United States. The 65% who support stricter gun sale laws can be used to estimate the actual percentage of all Americans who support such restrictions.

Estimation: A process whereby we select a random sample from a population and use a sample statistic to estimate a population parameter.

These are illustrations of estimation. **Estimation** is a process whereby we select a random sample from a population and use a sample statistic to estimate a population parameter. We can use sample proportions as estimates of population proportions, sample means as estimates of population means, or sample variances as estimates of population variances.

Why estimate? The goal of most research is to find the population parameter. Yet we hardly ever have enough resources to collect information about the entire population. We rarely know the actual value of the population parameter. On the other hand, we can learn a lot about a population by randomly selecting a sample from that population and obtaining an estimate of the population parameter. The major objective of sampling theory and statistical inference is to provide estimates of unknown population parameters from sample statistics.

POINT AND INTERVAL ESTIMATION

Point estimate: A sample statistic used to estimate the exact value of a population parameter.

Confidence interval (CI): A range of values defined by the confidence level within which the population parameter is estimated to fall.

Estimates of population characteristics can be divided into two types: (1) point estimates and (2) interval estimates. **Point estimates** are sample statistics used to estimate the exact value of a population parameter. When the Pew Research organization reports that 62% of Americans believe immigrants strengthen the country, they are using a point estimate. Similarly, if we reported the average level of education of the population of adult Americans to be exactly 13.71 years, we would be using a point estimate.

The problem with point estimates is that sample statistics vary, usually resulting in some sort of sampling error. Thus, we never really know how accurate they are. As a result, we rarely rely on them as estimators of population parameters such as average income or percentage of the population who are in favor of gay and lesbian relations.

One method of increasing accuracy is to use an interval estimate rather than a point estimate. In interval estimation, we identify a range of values within which the population parameter may fall. This range of values is called a **confidence interval (CI)**. Instead of using a single value, 13.71 years, as an estimate of the mean education of adult Americans, we could say that the population mean is somewhere between 12 and 14 years.

When we use confidence intervals to estimate population parameters, such as mean educational levels, we can also evaluate the accuracy of this estimate by assessing the likelihood that any given interval will contain the mean. This likelihood, expressed as a percentage or a probability, is called a confidence level. Confidence intervals are defined in terms of confidence levels. Thus, by selecting a 95% confidence level, we are saying that there is a .95 probability—or 95 chances out of 100—that a specified interval will contain the population mean. Confidence intervals can be constructed for any level of confidence, but the most common ones are the 90%, 95%, and 99% levels. You should also know that confidence intervals are sometimes referred to in terms of margin of error. In short, margin of error is simply the radius of a confidence interval. If we select a 95% confidence level, we would have a 5% chance of our interval being incorrect.

Confidence level: The likelihood, expressed as a percentage or a probability, that a specified interval will contain the population parameter.

Margin of error: The radius of a confidence interval.

Confidence intervals can be constructed for many different parameters based on their corresponding sample statistics. In this chapter, we describe the rationale and the procedure for the construction of confidence intervals for means and proportions.

LEARNING CHECK 6.1

What is the difference between a point estimate and a confidence interval?

CONFIDENCE INTERVALS FOR MEANS

To illustrate the procedure for establishing confidence intervals for means, we'll reintroduce one of the research examples mentioned in Chapter 5—assessing the needs of commuter students.

Recall that we have been given enough money to survey a random sample of 500 students. One of our tasks is to estimate the average commuting time of all 15,000 commuters on our campus—the population parameter. To obtain this estimate, we calculate the average commuting time for the sample. Suppose the sample average is $\bar{Y} = 7.5$ hours/week, and we want to use it as an estimate of the true average commuting time for the entire population of commuting students.

Because it is based on a sample, this estimate is subject to sampling error. We do not know how close it is to the true population mean. However, based on what the central limit theorem tells us about the properties of the sampling distribution of the mean, we know that with a large enough sample size, most sample means will tend to be close to the true population mean. Therefore, it

A CLOSER LOOK 6.1

Estimation as a Type Inference

Using inferential statistics, a researcher is able to describe a population based entirely on information from a sample of that population. A confidence interval is an example of this—by knowing a sample mean, sample size, and sample standard deviation, we are able to say something about the population from which that sample was drawn.

Combining this information gives us a range within which we can confidently say that the population mean falls.

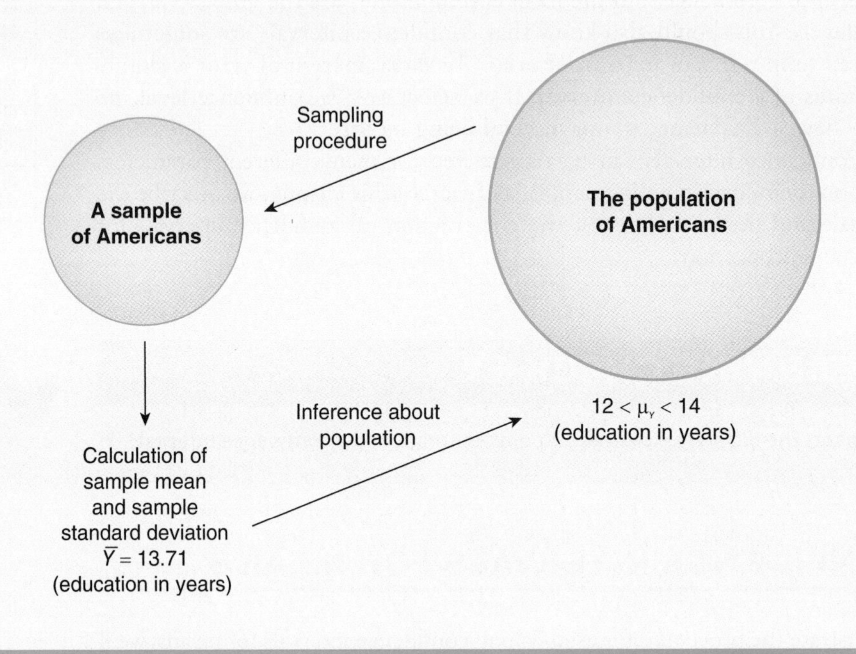

is unlikely that our sample mean, $\bar{Y}$ = 7.5 hours/week, deviates much from the true population mean.

We know that the sampling distribution of the mean is approximately normal with a mean equal to the population mean μ and a standard error $\bar{Y}$ (standard deviation of the sampling distribution) as follows:

$$\sigma_{\bar{Y}} = \frac{\sigma}{\sqrt{N}} \tag{6.1}$$

This information allows us to use the normal distribution to determine the probability that a sample mean will fall within a certain distance—measured in standard deviation (standard error) units or Z scores—of μ or $\mu_{\bar{Y}}$. We can make the following assumptions:

- A total of 68% of all random sample means will fall within ±1 standard error of the true population mean.
- A total of 95% of all random sample means will fall within ±1.96 standard errors of the true population mean.
- A total of 99% of all random sample means will fall within ±2.58 standard errors of the population mean.

On the basis of these assumptions and the value of the standard error, we can establish a range of values—a confidence interval—that is likely to contain the actual population mean. We can also evaluate the accuracy of this estimate by assessing the likelihood that this range of values will actually contain the population mean.

The general formula for constructing a confidence interval (CI) for any level is

$$CI = \bar{Y} \pm Z(\sigma_{\bar{Y}}) \tag{6.2}$$

Note that to calculate a confidence interval, we take the sample mean and add to or subtract from it the product of a Z value and the standard error.

The Z score we choose depends on the desired confidence level. For example, to obtain a 95% confidence interval, we would choose a Z of 1.96 because we know (from Appendix B) that 95% of the area under the curve lies between ±1.96. Similarly, for a 99% confidence level, we would choose a Z of 2.58. The relationship between the confidence level and Z is illustrated in Figure 6.1 for the 95% and 99% confidence levels.

Figure 6.1 Relationship Between Confidence Level and Z for 95% and 99% Confidence Intervals

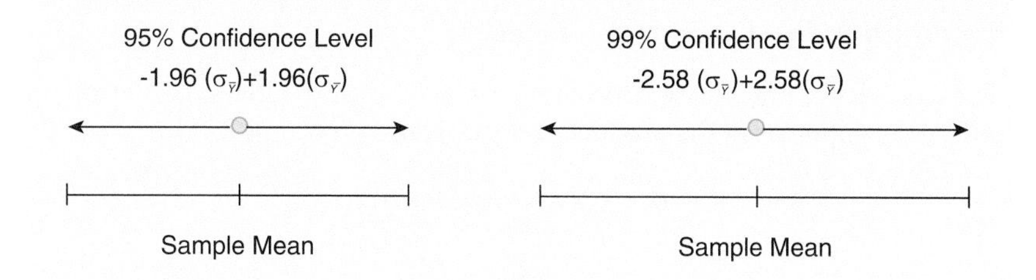

Source: Based on David Freedman, Robert Pisani, Roger Purves, and Ani Akhikari, *Statistics*, 2nd ed. (New York: Norton, 1991).

LEARNING CHECK 6.2

To understand the relationship between the confidence level and Z, review the material in Chapter 4. What would be the corresponding Z value for a 98% confidence interval?

Determining the Confidence Interval

To determine the confidence interval for means, follow these steps:

1. Calculate the standard error of the mean.
2. Decide on the level of confidence, and find the corresponding Z value.
3. Calculate the confidence interval.
4. Interpret the results.

Let's return to the problem of estimating the mean commuting time of the population of students on our campus. How would you find the 95% confidence interval?

Calculating the Standard Error of the Mean

Let's suppose that the standard deviation for our population of commuters is $\sigma = 1.5$. We calculate the standard error for the sampling distribution of the mean:

$$\sigma_{\bar{Y}} = \frac{\sigma}{\sqrt{N}} = \frac{1.5}{\sqrt{500}} = 0.07$$

Deciding on the Level of Confidence and Finding the Corresponding Z Value

We decide on a 95% confidence level. The Z value corresponding to a 95% confidence level is 1.96.

Calculating the Confidence Interval

The confidence interval is calculated by adding and subtracting from the observed sample mean the product of the standard error and Z:

$$95\%\ \text{CI} = 7.5 \pm 1.96(0.07)$$

$$= 7.5 \pm 0.14$$

$$= 7.36 \text{ to } 7.64$$

The 95% CI for the mean commuting time is illustrated in Figure 6.2.

Figure 6.2 Ninety-Five Percent Confidence Interval for the Mean Commuting Time (N = 500)

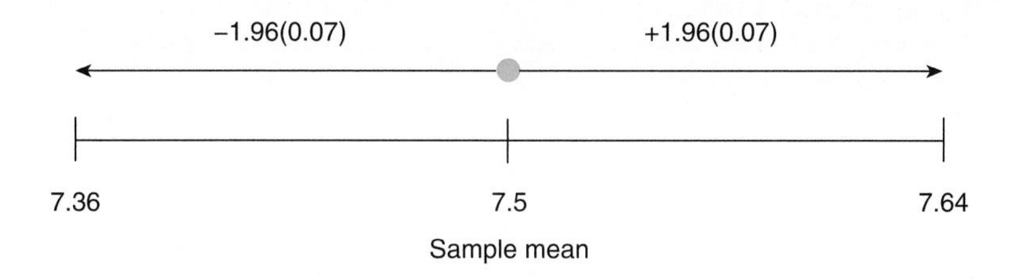

Interpreting the Results

We can be 95% confident that the actual mean commuting time—the true population mean—is not less than 7.36 hours and not greater than 7.64 hours. In other words, if we collected a large number of samples ($N = 500$) from the population of commuting students, 95 times out of 100, the true population mean would be included within our computed interval. With a 95% confidence level, there is a 5% risk that we are wrong. Five times out of 100, the true population mean will not be included in the specified interval.

Remember that we can never be sure whether the population mean is actually contained within the confidence interval. Once the sample is selected and the confidence interval defined, the confidence interval either does or does not contain the population mean—but we will never be sure.

LEARNING CHECK 6.3

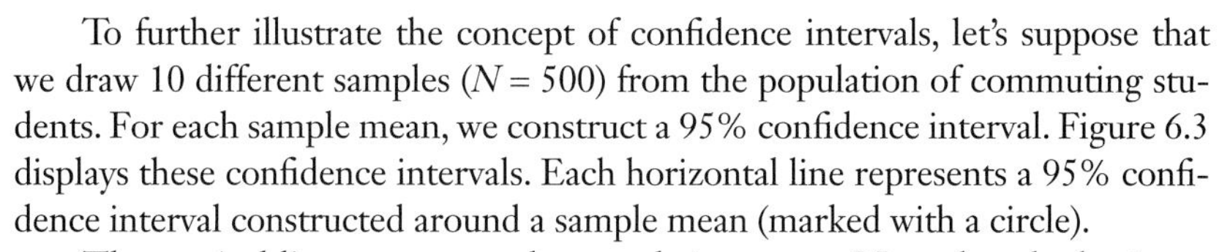

What is the 90% confidence interval for the mean commuting time? First, find the Z value associated with a 90% confidence level.

To further illustrate the concept of confidence intervals, let's suppose that we draw 10 different samples ($N = 500$) from the population of commuting students. For each sample mean, we construct a 95% confidence interval. Figure 6.3 displays these confidence intervals. Each horizontal line represents a 95% confidence interval constructed around a sample mean (marked with a circle).

The vertical line represents the population mean. Note that the horizontal lines that intersect the vertical line are the intervals that contain the true population mean. Only 1 out of the 10 confidence intervals does not intersect the vertical line, meaning it does not contain the population mean. What would happen if we continued to draw samples of the same size from this population and constructed a 95% confidence interval for each sample? For about 95% of all samples, the specified interval would contain the true population mean, but for 5% of all samples, it would not.

Figure 6.3 Ninety-Five Percent Confidence Intervals for 10 Samples

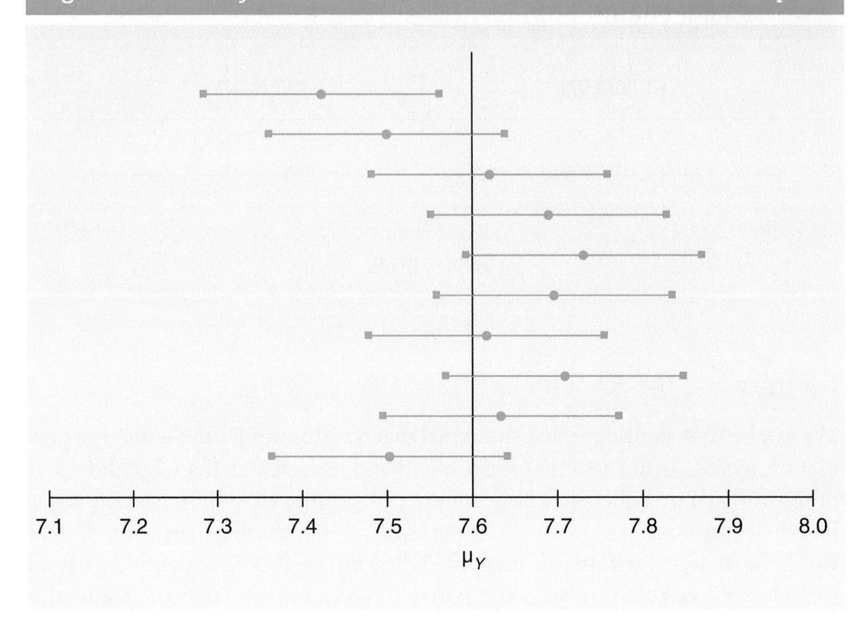

Reducing Risk

One way to reduce the risk of being incorrect is by increasing the level of confidence. For instance, we can increase our confidence level from 95% to 99%. The 99% confidence interval for our commuting example is

$$99\%\ \text{CI} = 7.5 \pm 2.58(0.07)$$

$$= 7.5 \pm 0.18$$

$$= 7.32 \text{ to } 7.68$$

When using the 99% confidence interval, there is only a 1% risk that we are wrong and the specified interval does not contain the true population mean. We can be almost certain that the true population mean is included in the interval ranging from 7.32 to 7.68 hours/week. Note that by increasing the confidence level, we have also increased the width of the confidence interval from 0.28 (7.36 – 7.64) to 0.36 hours (7.32 – 7.68), thereby making our estimate less precise.

This is the trade-off between achieving greater confidence in an estimate and the precision of that estimate. Although using a higher level of confidence increases our confidence that the true population mean is included in our confidence interval, the estimate becomes less precise as the width of the interval increases. Although we are only 95% confident that the interval ranging between 7.36 and 7.64 hours includes the true population mean, it is a more precise estimate than the 99% interval ranging from 7.32 to 7.68 hours. The relationship between the confidence level and the precision of the confidence interval is illustrated in Figure 6.4. Table 6.1 lists three commonly used confidence levels along with their corresponding *Z* values.

Figure 6.4 Confidence Intervals, 95% Versus 99% (Mean Commuting Time)

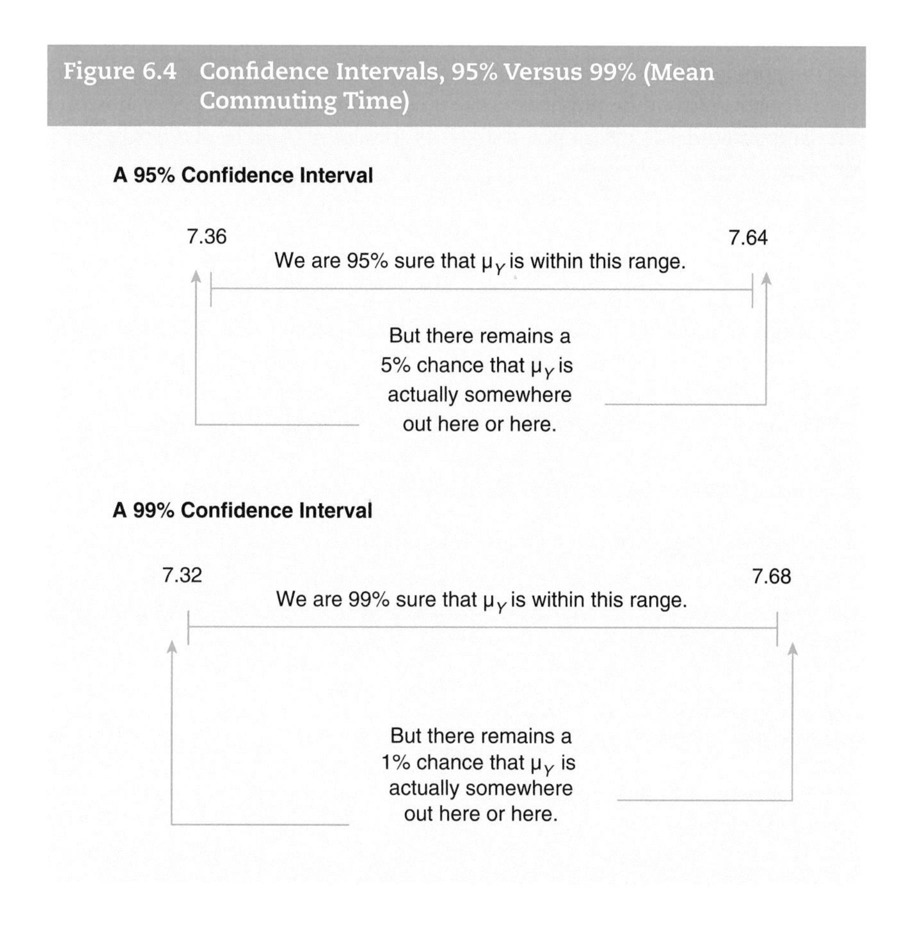

Table 6.1 Confidence Levels and Corresponding Z Values

Confidence Level	*Z* Value
90%	1.65
95%	1.96
99%	2.58

Estimating Sigma

To calculate confidence intervals, we need to know the standard error of the sampling distribution. The standard error is a function of the population standard deviation and the sample size:

$$\sigma_{\bar{Y}} = \frac{\sigma}{\sqrt{N}}$$

In our commuting example, we have been using a hypothetical value, $\sigma = 1.5$, for the population standard deviation. Typically, both the mean (μ) and the standard deviation (σ) of the population are unknown to us. When $N \geq 50$, however, the sample standard deviation, s, is a good estimate of $\sigma_{\bar{Y}}$. The standard error is then calculated as follows:

$$s_{\bar{Y}} = \frac{s}{\sqrt{N}} \tag{6.3}$$

As an example, we'll estimate the mean hours per day that Americans spend watching television based on the 2018 GSS. The mean hours per day spent watching television for a sample of $N = 778$ is 2.95, and the standard deviation is 2.94 hours. Let's determine the 95% confidence interval for these data.

Calculating the Estimated Standard Error of the Mean

The estimated standard error for the sampling distribution of the mean is

$$s_{\bar{Y}} = \frac{2.94}{\sqrt{778}} = 0.11$$

Deciding on the Level of Confidence and Finding the Corresponding *Z* Value

We decide on a 95% confidence level. The Z value corresponding to a 95% confidence level is 1.96.

Calculating the Confidence Interval

The confidence interval is calculated by adding to and subtracting from the observed sample mean the product of the standard error and Z:

$$95\%\ \text{CI} = 2.95 \pm 1.96(0.11)$$

$$= 2.95 \pm 0.22$$

$$= 2.73 \text{ to } 3.17$$

Interpreting the Results

We can be 95% confident that the actual mean hours spent watching television by Americans from which the GSS sample was taken is not less than 2.73 hours and not greater than 3.17 hours. In other words, if we drew a large number of samples ($N = 778$) from this population, then 95 times out of 100, the true population mean would be included within our computed interval.

Sample Size and Confidence Intervals

Researchers can increase the precision of their estimate by increasing the sample size. In Chapter 5, we learned that larger samples result in smaller standard errors and, therefore, sampling distributions are more clustered around the population mean (Figure 5.6). A more tightly clustered sampling distribution means that our confidence intervals will be narrower and more precise. To illustrate the relationship between sample size and the standard error, and thus the confidence interval, let's calculate the 95% confidence interval for our GSS data with (a) a sample of $N = 150$ and (b) a sample of $N = 1{,}556$ (double our original sample size).

With a sample size $N = 150$, the estimated standard error for the sampling distribution is

$$s_{\bar{Y}} = \frac{s}{\sqrt{N}} = \frac{2.94}{\sqrt{150}} = 0.24$$

and the 95% confidence interval is

$$\begin{aligned} 95\%\ \text{CI} &= 2.95 \pm 1.96(0.24) \\ &= 2.95 \pm 0.47 \\ &= 2.48 \text{ to } 3.42 \end{aligned}$$

With a sample size $N = 1{,}556$, the estimated standard error for the sampling distribution is

$$s_{\bar{Y}} = \frac{s}{\sqrt{N}} = \frac{2.94}{\sqrt{1{,}556}} = 0.07$$

and the 95% confidence interval is

$$\begin{aligned} 95\%\ \text{CI} &= 2.95 \pm 1.96(0.07) \\ &= 2.95 \pm 0.14 \\ &= 2.81 \text{ to } 3.09 \end{aligned}$$

Table 6.2 Ninety-Five Percent Confidence Interval and Width for Mean Number of Hours per Day Watching Television for Three Different Sample Sizes

Sample Size (N)	Confidence Interval	Interval Width	s	$s_{\bar{Y}}$
150	2.48–3.42	0.94	2.94	0.24
778	2.73–3.17	0.44	2.94	0.11
1,556	2.81–3.09	0.28	2.94	0.07

In Table 6.2, we summarize the 95% confidence intervals for the mean number of hours watching television for these three sample sizes: $N = 150$, $N = 778$, and $N = 1{,}556$.

Note that there is an inverse relationship between sample size and the width of the confidence interval. The increase in sample size is linked with increased precision of the confidence interval. The 95% confidence interval for the GSS sample of 150 cases is 0.94 hours. But the interval widths decrease to 0.44 and 0.28 hours, respectively, as the sample sizes increase to $N = 778$ and then to $N = 1{,}556$. We had to double the size of the sample to reduce the confidence interval by almost one half (from 0.44 to 0.28 hours). In general, although the precision of estimates increases steadily with sample size, the gains would appear to be rather modest after N reaches 1,556. An important factor to keep in mind is the increased cost associated with a larger sample. Researchers have to consider at what point the increase in precision is too small to justify the additional cost associated with a larger sample.

LEARNING CHECK 6.4

Why do smaller sample sizes produce wider confidence intervals? (See Figure 6.5.) Compare the standard errors of the mean for the three sample sizes.

STATISTICS IN PRACTICE: HISPANIC MIGRATION AND EARNINGS

There were nearly 58.9 million people of Latino ethnicity in the United States in 2017. Mexican-origin Hispanics comprise the single largest group of Hispanics in the country. But the origins of the Hispanic population have

Figure 6.5 The Relationship Between Sample Size and Confidence Interval Width

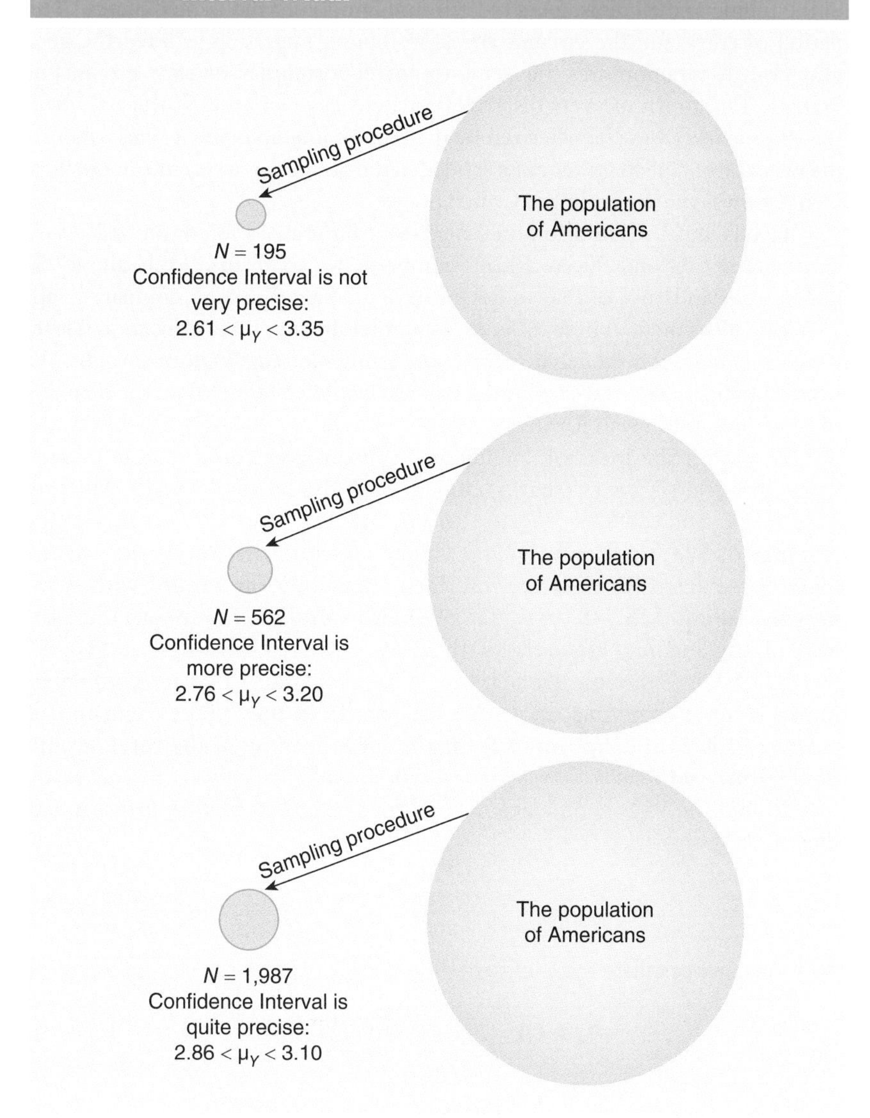

diversified with the addition of immigrants from other Latin American nations and Puerto Rico.[3]

During the past few decades, numerous studies have noted the discrepancy in earnings between these groups.[4] The gap in earnings has been attributed mainly to differences in migration status and in level of education. In this classic study,

Marta Tienda and Franklin Wilson argued that Mexicans, Puerto Ricans, and Cubans varied markedly in socioeconomic characteristics because of differences in the timing and circumstances of their immigration to the United States.[5] The period of entry and the circumstances prompting migration affected the geographical distribution and the employment opportunities of each group. For example, Puerto Ricans were disproportionately located in the Northeast, where the labor market was characterized by the highest unemployment rates, whereas the majority of Cuban immigrants resided in the Southeast, where the unemployment rate was the lowest in the United States.

Tienda and Wilson also noted persistent differences in educational levels among Mexicans and Puerto Ricans compared with Cubans.[6] Only about 9% of Mexicans and 16% of Puerto Ricans have graduated college, compared with 25% of Cuban men.[7] These differences in migrant status and educational level were likely to be reflected in disparities in earnings among the three groups. We would anticipate that the earnings of Cubans would be higher than the earnings of Mexicans and Puerto Ricans.

We tested the ideas of Tienda and Wilson based on a sample of men from the 2000 Census that included 29,233 Cubans, 34,620 Mexican Americans, and 66,933 Puerto Ricans. As hypothesized, with average earnings of \$24,018 (s = \$36,298), Cubans were at the top of the income hierarchy. Puerto Ricans were intermediate among the groups with earnings averaging \$18,748 (s = \$25,694). Mexican men were at the bottom of the income hierarchy with average annual earnings of \$16,537 (s = \$23,502). Although Tienda and Wilson did not calculate confidence intervals for their estimates, we use the data from the 2000 Census to calculate a 95% confidence interval for the mean income of the three groups of Hispanic men.[8]

To find the 95% confidence interval for Cuban income, we first estimate the standard error:

$$s_{\bar{Y}} = \frac{36{,}298}{\sqrt{29{,}233}} = 212.30$$

Then, we calculate the confidence interval:

$$\begin{aligned} 95\%\ \text{CI} &= 24{,}018 \pm 1.96(212.30) \\ &= 24{,}018 \pm 416 \\ &= 23{,}602 \text{ to } 24{,}434 \end{aligned}$$

For Puerto Rican income, the estimated standard error is

$$s_{\bar{Y}} = \frac{25{,}694}{\sqrt{66{,}933}} = 99.31$$

and the 95% confidence interval is

$$95\%\ \text{CI} = 18{,}748 \pm 1.96(99.31)$$

A CLOSER LOOK 6.2

What Affects Confidence Interval Width?

"Holding other factors constant . . . "

If the sample size goes up	↑	the confidence interval becomes more precise.	→←
If the level of confidence goes down (from 99% to 95%)	↓	the confidence interval becomes less precise.	←→
If the sample size goes down	↓	the confidence interval becomes less precise.	←→
If the value of the sample standard deviation goes up	↑	the confidence interval becomes more precise.	→←
If the value of the sample standard deviation goes down	↓	the confidence interval becomes less precise.	←→
If the level of confidence goes up (from 95% to 99%)	↑	the confidence interval becomes more precise	→←

$$= 18{,}748 \pm 195$$
$$= 18{,}533 \text{ to } 18{,}943$$

Finally, for Mexican income, the estimated standard error is

$$s_{\bar{Y}} = \frac{23{,}502}{\sqrt{34{,}620}} = 126.31$$

and the 95% confidence interval is

$$\begin{aligned} 95\%\ \text{CI} &= 16{,}537 \pm 1.96(126.31) \\ &= 16{,}537 \pm 248 \\ &= 16{,}287 \text{ to } 16{,}785 \end{aligned}$$

The confidence intervals for mean annual income of Cuban, Puerto Rican, and Mexican immigrants are illustrated in Figure 6.6. We can say with 95% confidence that the true income mean for each Hispanic group lies somewhere within the corresponding confidence interval. Note that the confidence intervals do not overlap, thus revealing great disparities in earnings among the three groups. Highest interval estimates are for Cubans, followed by Mexicans and then Puerto Ricans.

Figure 6.6 Ninety-Five Percent Confidence Intervals for the Mean Income of Puerto Ricans, Mexicans, and Cubans

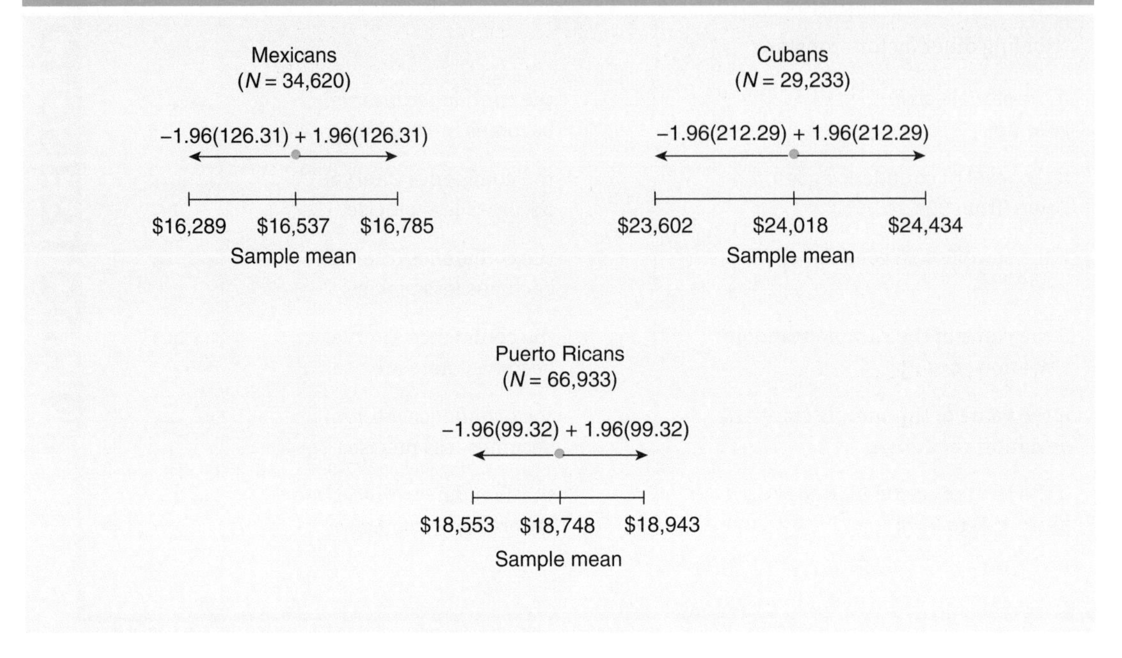

Source: Original material, based on Marta Tienda, "The Ghetto Underclass: Social Science Perspectives," *Annals of the American Academy of Political and Social Science* 501 (January 1989): 105–119.

CONFIDENCE INTERVALS FOR PROPORTIONS

You may already be familiar with confidence intervals, having seen them applied in opinion and election polls like those conducted by the Pew Research Center or CBS News. Pollsters interview a random sample representative of a defined population to assess their opinion on a certain issue or their voting preference. Sample proportions or percentages are usually reported along with a margin error, plus or minus a particular value. The margin of error is the confidence interval and is used to estimate the population proportions or percentages. The same conceptual foundations of sampling and statistical inference that are central to the estimation of population means—the selection of random samples and the special properties of the sampling distribution—are also central to the estimation of population proportions.

Earlier, we saw that the sampling distribution of the means underlies the process of estimating population means from sample means. Similarly, the sampling distribution of proportions underlies the estimation of population proportions from sample proportions. Based on the central limit theorem, we know that with sufficient sample size, the sampling distribution of proportions is approximately normal, with mean μ_p equal to the population proportion π and with a standard error of proportions (the standard deviation of the sampling distribution of proportions) equal to

$$\sigma_p = \sqrt{\frac{(\pi)(1-\pi)}{N}} \qquad (6.4)$$

where

σ_p = the standard error of proportions

π = the population proportion

N = the population size

However, since the population proportion, π, is unknown to us (that is what we are trying to estimate), we can use the sample proportion, p, as an estimate of π. The estimated standard error then becomes

$$s_p = \sqrt{\frac{(p)(1-p)}{N}} \qquad (6.5)$$

where

s_p = the estimated standard error of proportions

p = the sample proportion

N = the sample size

Let's calculate the estimated standard error for the 2019 Pew Research survey about attitudes about immigrants. Based on a random sample of 1,505 adults, the percentage who believe immigrants are a strength was reported at 62%. Based on Formula 6.5, with $p = 0.62$, $1 - p = (1 - 0.62) = 0.38$, and $N = 1,505$, the standard error is $s_p = \sqrt{(.62)(1-.62)/1,505} = 0.01$. We will have to consider two factors to meet the assumption of normality with the sampling distribution of proportions: (1) the sample size N and (2) the sample proportions p and $1 - p$. When p and $1 - p$ are about 0.50, a sample size of at least 50 is sufficient. But when >0.50 (or $1 - p < 0.50$), a larger sample is required to meet the assumption of normality. Usually, a sample of 100 or more is adequate for any single estimate of a population proportion.

Determining the Confidence Interval

Because the sampling distribution of proportions is approximately normal, we can use the normal distribution to establish confidence intervals for proportions in the same manner that we used the normal distribution to establish confidence intervals or means.

The general formula for constructing confidence intervals for proportions for any level of confidence is

$$\text{CI} = p \pm Z(s_p) \qquad (6.6)$$

where

CI = the confidence interval

p = the observed sample proportion

Z = the Z corresponding to the confidence level

s_p = the estimated standard error of proportions

Let's examine this formula in more detail. Note that to obtain a confidence interval at a certain level, we take the sample proportion and add to or subtract from it the product of a Z value and the standard error. The Z value we choose depends on the desired confidence level. We want the area between the mean and the selected $\pm Z$ to be equal to the confidence level.

For example, to obtain a 95% confidence interval, we would choose a Z of 1.96 because we know (from Appendix B) that 95% of the area under the curve is included between ±1.96. Similarly, for a 99% confidence level, we would choose a Z of 2.58. (The relationship between confidence level and Z values is illustrated in Figure 6.1.)

To determine the confidence interval for a proportion, we follow the same steps that were used to find confidence intervals for means:

1. Calculate the estimated standard error of the proportion.
2. Decide on the desired level of confidence, and find the corresponding Z value.
3. Calculate the confidence interval.
4. Interpret the results.

To illustrate these steps, we use the Gallup survey results about the percentage of Americans who identify as environmentalists.

Calculating the Estimated Standard Error of the Proportion

The standard error of the proportion 0.62 (62%) with a sample N = 1,505 is 0.01.

Deciding on the Desired Level of Confidence and Finding the Corresponding *Z* Value

We choose the 95% confidence level. The Z corresponding to a 95% confidence level is 1.96.

Calculating the Confidence Interval

We calculate the confidence interval by adding to and subtracting from the observed sample proportion the product of the standard error and Z:

$$\begin{aligned} 95\%\ \text{CI} &= 0.62 \pm 1.96(0.01) \\ &= 0.62 \pm 0.02 \\ &= 0.60 \text{ to } 0.64 \end{aligned}$$

Interpreting the Results

We are 95% confident that the true population proportion is somewhere between 0.60 and 0.64. In other words, if we drew a large number of samples from the population of adults, then 95 times out of 100, the confidence interval we obtained would contain the true population proportion. We can also express this result in percentages and say that we are 95% confident that the true population percentage of Americans who believe immigrants strengthen the country is between 60% and 64%.

Note that with a 95% confidence level, there is a 5% risk that we are wrong. If we continued to draw large samples from this population, in 5 out of 100 samples, the true population proportion would not be included in the specified interval.

We can decrease our risk by increasing the confidence level from 95% to 99%.

$$\begin{aligned} 99\%\ \text{CI} &= 0.62 \pm 2.58(0.01) \\ &= 0.62 \pm 0.03 \\ &= .59 \text{ to } .65 \end{aligned}$$

When using the 99% confidence interval, we can be almost certain (99 times out of 100) that the true population proportion is included in the interval ranging from .59 to .65 (or 59% to 65%). However, as we saw earlier, there is a trade-off between achieving greater confidence in making an estimate and the precision of that estimate. Although using a 99% level increased our confidence level from 95% to 99% (thereby reducing our risk of being wrong from 5% to 1%), the estimate became less precise as the width of the interval increased.[9]

LEARNING CHECK 6.5

Calculate the 95% confidence interval for the CBS News poll results for those who support stricter laws for gun sales.

READING THE RESEARCH LITERATURE: WOMEN VICTIMS OF INTIMATE VIOLENCE

Janet Fanslow and Elizabeth Robinson (2010)[10] studied help-seeking behavior and motivation among women victims of intimate partner violence in New Zealand. The researchers argue that

> historically, responses to victims have been developed on the basis of identified need and active advocacy work rather than driven by data. . . . Questions remain, however, in terms of whether the responses that have been developed are the most appropriate ways to deliver help to victims of intimate partner violence according to victims' personal perceptions of who they would like to receive help from and the types and nature of help they would like to receive. (p. 930)

Fanslow and Robinson relied on data from the New Zealand Violence Against Women Study to document the reasons why victims sought help or left their partner due to domestic violence.

We present one of the tables featured in Fanslow and Robinson's study (see Table 6.3). Note that 95% confidence intervals are reported for each reason category. The confidence intervals are based on the confidence level, the standard error of the proportion (which can be estimated from *p*), and the sample size. For example, we know that based on their sample of 486 women, 48.5% reported that "could not endure more" was a reason for seeking help. Based on the 95% confidence interval, we can say that the actual proportion of women victims who identify that they "could not endure more" as a reason for seeking help lies somewhere between 43.6% and 53.3%.

Fanslow and Robinson identify the categories with the highest percentages for each sample: for seeking help (could not endure more, encouraged by friends or family, children suffering and badly injured) and for leaving (could not endure more, he threatened to kill her, children suffering and badly injured). They write as follows:

> The majority of women who sought help, and the majority of women who left violent relationships reported doing so because they "could not endure more." Where and when this line is drawn will undoubtedly differ from person to person . . . but factors that may affect it include realistic assessment of the man's behavior and his likelihood to change, and recognition of the seriousness of the violence they experienced. Other commonly reported reasons for leaving and/or seeking help included because women experienced serious injury or feared for their lives. This emphasizes that the

Table 6.3 Percentage of Women With Lifetime Experience of Physical and/or Sexual Intimate Partner Violence Who Reported Reasons for Asking for Help With and for Leaving Violent Relationships

	Reasons That Made You Go for Help (n = 486), (95% CI)	Reasons for Leaving the Last Time (n = 508), (95% CI)
Could not endure more	48.5 (43.6–53.3)	64.2 (59.8–68.6)
Encouraged by friends/family	17.7 (14.0–21.5)	6.7 (4.4–9.0)
Badly injured	15.4 (12.0–18.7)	7.1 (4.2–9.9)
He threatened to kill her	11.3 (8.1–14.5)	10.2 (7.0–13.3)
Afraid he would kill her	11.4 (8.3–14.6)	5.9 (3.8–7.9)
Children suffering	17.2 (13.5–20.9)	8.6 (5.9–11.3)
He threatened or hit children	7.9 (5.2–10.5)	5.0 (3.0–7.0)
She was afraid she would kill him	2.1 (0.8–3.5)	1.2 (0.2–2.2)
She was thrown out of the home	2.0 (0.6–3.3)	2.2 (0.7–3.7)
For her mental health/save sanity	8.2 (5.5–10.9)	—
To get information/legal help	8.1 (5.5–10.6)	—
Encouraged by organization	—	1.7 (0.1–3.3)
No particular incident	—	5.3 (3.3–7.4)
He was unfaithful	—	3.8(2.1–5.5)
She was pregnant	—	1.4 (0.4–2.3)
To have time out/break from relationships	—	1.4 (0.3–2.4)
Other	22.4 (18.5–26.2)	8.3 (5.6–10.9)

Source: Adapted from Janet Fanslow and Elizabeth Robinson, "Help Seeking Behaviors and Reasons for Help Seeking Reported by a Representative Sample of Women Victims of Intimate Partner Violence in New Zealand," *Journal of Interpersonal Violence* 25, no. 5 (2010): 929–951.

Note: Percentages sum to greater than 100% because individuals could provide multiple responses.

> violence experienced by many women in the context of intimate relationships is not trivial. (p. 946)[11]

Concern for children suffering was also identified as an important reason for female victims to seek help and/or to leave their abuser.

When estimates are reported for subgroups, the confidence intervals are likely to vary. Even when a confidence interval is reported only for the overall sample, we can easily compute separate confidence intervals for each of the subgroups if the confidence level and the size of each of the subgroups are included.

DATA AT WORK

Laurel Person Mecca: Research Specialist

Photo courtesy of Laurel Person Mecca

Laurel works as a research specialist in a university center for social and urban research. The center provides support and consultation to behavioral and social science investigators. She first connected with the center while she was recruiting subjects for her master's thesis research.

"Interacting with human subjects offers a unique view into people's lives," says Laurel, "thereby providing insights into one's own life and a richer understanding of the human condition." She's consulted on an array of projects: testing prototypes of technologies designed to enable older adults and persons with disabilities to live independently, facilitating focus groups with low-income individuals to explore their barriers to Supplemental Nutrition Assistance Program participation, and evaluating an intervention designed to improve parent–adolescent communication about sexual behaviors to reduce sexually transmitted diseases and unintended pregnancies among teens.

For students interested in a career in research and data analysis, she offers this advice: "Gain on-the-job experience while in college, even if it is an unpaid internship. Find researchers who are conducting studies that interest you, and inquire about working for them. Even if they are not posting an available position, they may bring you on board (as I have done with students). Persistence pays off! You are much more likely to be selected for a position if you demonstrate a genuine interest in the work and if you continue to show your enthusiasm by following up. Definitely check out the National Science Foundation's Research Experience for Undergraduates program. Though most of these internships are in the 'hard' sciences, there are plenty of openings in social sciences disciplines. These internships include a stipend (YES!), and oftentimes, assistance with travel and housing. They are wonderful opportunities to work directly on a research project, and may provide the additional benefit of a conference presentation and/or publication."

MAIN POINTS

- The goal of most research is to find population parameters. The major objective of sampling theory and statistical inference is to provide estimates of unknown parameters from sample statistics.
- Researchers make point estimates and interval estimates. Point estimates are sample statistics used to estimate the exact value of a population parameter. Interval estimates are ranges of values

within which the population parameter may fall.

- Confidence intervals can be used to estimate population parameters such as means or proportions. Their accuracy is defined with the confidence level. The most common confidence levels are 90%, 95%, and 99%.
- To establish a confidence interval for a mean or a proportion, add or subtract from the mean or the proportion the product of the standard error and the Z value corresponding to the confidence level.

KEY TERMS

confidence interval (CI) 176
confidence level 177
estimation 176
margin of error 177
point estimate 176

DIGITAL RESOURCES

Access key study tools at https://edge.sagepub.com/ssdsess4e

- eFlashcards of the glossary terms
- Datasets and codebooks
- SPSS and Excel walk-through videos
- SPSS and Excel demonstrations and problems to accompany each chapter
- Appendix F: Basic Math Review

CHAPTER EXERCISES

1. In the 2017 National Crime Victimization Study, the Federal Bureau of Investigation found that 21% of Americans age 12 or older had been victims of crime during a 1-year period. This result was based on a sample of 239,541 persons.[12]
   a. Estimate the percentage of U.S. adults who were victims at the 90% confidence level. Provide an interpretation of the confidence interval.
   b. Estimate the percentage of victims at the 99% confidence level. Provide an interpretation of the confidence interval.

2. Average years of education are reported by social class based on data from the GSS 2018.

	Mean	Standard Deviation	*N*
Lower class	12.19	3.08	102
Working class	13.16	2.93	523
Middle class	14.60	2.88	498
Upper class	15.21	3.01	34

   a. Construct the 95% confidence interval for the mean number of years of education for lower- and working-class respondents.
   b. Construct the 99% confidence interval for the mean number of years of education for lower-class and middle-class respondents.
   c. As our confidence in the result increases, how does the size of the confidence interval Change? Explain why this is so.

3. In 2016, the Pew Research Center conducted a survey of 1,004 Canadians and 1,003 Americans to assess their opinion of climate change. The data show that 51% of Canadians and 45% of Americans believe climate change is a very serious problem.[13]
   a. Estimate the proportion of all Canadians who believe climate change is a very serious problem at the 95% confidence interval.
   b. Estimate the proportion of all Americans who believe climate change is a very serious problem at the 95% confidence interval.
   c. Would you be confident in reporting that the majority (51% or higher) of Canadians and Americans believe climate change is a very serious problem? Explain.

4. Although 70% of women with children younger than 18 years participate in the labor force,[14] society still upholds the stay-at-home mother as the traditional model. Some believe that employment distracts mothers from their parenting role, affecting the well-being of children.
   a. In the GSS 2018, respondents were asked to indicate their level of agreement to the statement, "A working mother hurts children." Of the 337 male respondents who answered the question, 21% strongly agreed that a working mother does not hurt children. Construct a 90% confidence interval for this statistic.
   b. Of the 441 female respondents who answered the question, 37% strongly agreed that a working mother does not hurt children. Construct a 90% confidence interval for this statistic.
   c. Why do you think there is a difference between men and women on this issue?

5. Gallup conducted a survey from April 1 to 25, 2010, to determine the congressional vote preference of the American voters.[15] They found that 51% of the male voters preferred a Republican candidate to a Democratic candidate in a sample of 5,490 registered voters. Gallup asks you, their statistical consultant, to tell them whether you could declare the Republican candidate as the likely winner of the votes coming from men if there was an election today. What is your advice? Why?

6. You have been doing research for your statistics class on the prevalence of binge drinking among teens.

   a. According to 2011 Monitoring the Future data, the average binge drinking score, for this sample of 914 teens, is 1.27, with a standard deviation of 0.80. Binge drinking is defined as the number of times the teen drank five or more alcoholic drinks during the past week. Construct the 95% confidence interval for the true average severe binge drinking score.

   b. Your roommate is concerned about your confidence interval calculation, arguing that severe binge drinking scores are not normally distributed, which in turn makes the confidence interval calculation meaningless. Assume your roommate is correct about the distribution of severe binge drinking scores. Does that imply that the calculation of a confidence interval is not appropriate? Why or why not?

7. From the GSS 2018 subsample, we find that 83% of respondents believe in life after death ($N = 1{,}054$).

   a. What is the 95% confidence interval for the percentage of the U.S. population who believe in life after death?

   b. Without doing any calculations, estimate the lower and upper bounds of 90% and 99% confidence intervals.

8. A social service agency plans to conduct a survey to determine the mean income of its clients. The director of the agency prefers that you measure the mean income very accurately, to within ±$500. From a sample taken 2 years ago, you estimate that the standard deviation of income for this population is about $5,000. Your job is to figure out the necessary sample size to reduce sampling error to ±$500.

   a. Do you need to have an estimate of the current mean income to answer this question? Why or why not?

   b. What sample size should be drawn to meet the director's requirement at the 95% level of confidence? (*Hint:* Use the formula for a confidence interval and solve for N, the sample size.)

   c. What sample size should be drawn to meet the director's requirement at the 99% level of confidence?

9. Education data (measured in years) from the ISSP 2014 are presented below for four countries. Calculate the 90% confidence interval for each country.

Country	Mean	Standard Deviation	*N*
France	14.12	5.73	975
Japan	12.48	2.53	528
Croatia	12.18	2.71	480
Turkey	9.15	11.98	783

10. Throughout the 2016 presidential election primaries, Millennials (those aged 20 to 36 years) consistently supported Senator Bernie Sanders over Secretary Hillary Clinton. According to the 2016 Gallup poll of 1,754 Millennials, 55% had a favorable opinion of Sanders compared with Hillary Clinton (38%).[16] Calculate the 90% confidence interval for both reported percentages.

11. According to a 2014 survey by the Pew Research Center, 18% of registered Republicans and 15% of registered Democrats follow political candidates on social media.[17] These data are based on a national survey of 446 registered Republicans and 522 registered Democrats. What is the 95% confidence interval for the percentage of Republicans who follow political candidates on social media? The 95% confidence interval for Democrats?

12. According to a report published by the Pew Research Center in February 2010, 61% of Millennials (Americans in their teens and 20s) think that their generation has a unique and distinctive identity ($N = 527$).[18]

    a. Calculate the 95% confidence interval to estimate the percentage of Millennials who believe that their generation has a distinctive identity as compared with the other generations (Generation X, baby boomers, or the Silent Generation).

    b. Calculate the 99% confidence interval.

    c. Are both these results compatible with the conclusion that the majority of Millennials believe that they have a unique identity that separates them from the previous generations?

13. In 2018, GSS respondents ($N = 1,117$) were asked about their political views. The data show that 29% identified as liberal, while 32% were identified as conservative.

    a. For each reported percentage, calculate the 95% confidence interval.

    b. Approximately 39% of GSS respondents identified as moderate politically. Calculate the 95% confidence interval.

    c. Based on your calculations, what conclusions can you draw about the public's political views?

7

TESTING HYPOTHESES

According to economist Ethan Harris, "People may not remember too many numbers about the economy, but there are certain signposts they do pay attention to. As a short hard way to assess how the economy is doing, everybody notices the price of gas."[1]

The impact of high and volatile fuel prices is felt across the nation, affecting consumer spending and the economy, but the burden remains greater among distinct social economic groups and geographic areas. Lower-income Americans spend eight times more of their disposable income on gasoline than wealthier Americans do.[2] For example, in Wilcox, Alabama, individuals spend 12.72% of their income to fuel one vehicle, while in Hunterdon Co., New Jersey, people spend 1.52%. Nationally, Americans spend 3.8% of their income fueling one vehicle. The first state to reach the $5 per gallon milestone was California in 2012. California's drivers were especially hit hard by the rising price of gas, due in part to their reliance on automobiles, especially for work commuters.

Let's say we drew a random sample of California gas stations ($N = 100$) and calculated the mean price for a gallon of regular gas, $3.05. Based on consumer information,[3] we also know that nationally, the mean price of a gallon was $2.62 with a standard deviation of 0.21 for the same week. We can thus compare the mean price of gas in California with the mean price of all gas stations in March 2019. By comparing these means, we are asking whether it is reasonable to consider our random sample of California gas as representative of the population of gas stations in the United States. Actually, we expect to find that the average price of gas from a sample of California gas stations will be unrepresentative of the population of gas stations because we assume higher gas prices in California.

The sample mean of $3.05 is higher than the population mean of $2.62, but it is an estimate based on a single sample. Thus, it could mean one of two things: (1) The average price of gas in

Chapter Learning Objectives

1. Describe the assumptions of statistical hypothesis testing.
2. Define and apply the components in hypothesis testing.
3. Explain what it means to reject or fail to reject a null hypothesis.
4. Calculate and interpret a test for two sample cases with means or proportions.
5. Determine the significance of t-test and Z-test statistics.

California is indeed higher than the national average, or (2) the average price of gas in California is about the same as the national average, and this sample happens to show a particularly high mean.

How can we decide which explanation makes more sense? Because most estimates are based on single samples and different samples may result in different estimates, sampling results cannot be used directly to make statements about a population. We need a procedure that allows us to evaluate hypotheses about population parameters based on sample statistics. In Chapter 6, we saw that population parameters can be estimated from sample statistics. In this chapter, we will learn how to use sample statistics to make decisions about population parameters. This procedure is called **statistical hypothesis testing**.

Statistical hypothesis testing: A procedure that allows us to evaluate hypotheses about population parameters based on sample statistics.

ASSUMPTIONS OF STATISTICAL HYPOTHESIS TESTING

Statistical hypothesis testing requires several assumptions. These assumptions include considerations of the level of measurement of the variable, the method of sampling, the shape of the population distribution, and the sample size. The specific assumptions may vary, depending on the test or the conditions of testing. However, without exception, all statistical tests assume random sampling. Tests of hypotheses about means also assume interval-ratio level of measurement and require that the population under consideration be normally distributed or that the sample size be larger than 50.

Based on our data, we can test the hypothesis that the average price of gas in California is higher than the average national price of gas. The test we are considering meets these conditions:

1. The sample of California gas stations was randomly selected.
2. The variable *price per gallon* is measured at the interval-ratio level.
3. We cannot assume that the population is normally distributed. However, because our sample size is sufficiently large ($N > 50$), we know, based on the central limit theorem, that the sampling distribution of the mean will be approximately normal.

STATING THE RESEARCH AND NULL HYPOTHESES

Hypotheses are usually defined in terms of interrelations between variables and are often based on a substantive theory. Earlier, we defined hypotheses as tentative answers to research questions. They are tentative because we can find evidence for them only after being empirically tested. The testing of hypotheses is an important step in this evidence-gathering process.

The Research Hypothesis (H_1)

Our first step is to formally express the hypothesis in a way that makes it amenable to a statistical test. The substantive hypothesis is called the **research hypothesis** and is symbolized as H_1. Research hypotheses are always expressed in terms of population parameters because we are interested in making statements about population parameters based on our sample statistics.

Research hypothesis (H_1): A statement reflecting the substantive hypothesis. It is always expressed in terms of population parameters, but its specific form varies from test to test.

In our research hypothesis (H_1), we state that the average price of gas in California is higher than the average price of gas nationally. Symbolically, we use μ to represent the population mean; our hypothesis can be expressed as

$$H_1: \mu > \$2.62$$

In general, the research hypothesis (H_1) specifies that the population parameter is one of the following:

1. Not equal to some specified value: $\mu \neq$ some specified value
2. Greater than some specified value: $\mu >$ some specified value
3. Less than some specified value: $\mu <$ some specified value

In a **one-tailed test**, the research hypothesis is directional; that is, it specifies that a population mean is either less than (<) or greater than (>) some specified value. We can express our research hypothesis as either

One-tailed test: A type of hypothesis test that involves a directional hypothesis. It specifies that the values of one group are either larger or smaller than some specified population value.

$$H_1: \mu < \text{some specified value}$$

$$H_1: \mu > \text{some specified value}$$

The research hypothesis we've stated for the average price of a gallon of regular gas in California is a one-tailed test.

When a one-tailed test specifies that the population mean is greater than some specified value, we call it a **right-tailed test** because we will evaluate the outcome at the right tail of the sampling distribution. If the research hypothesis specifies that the population mean is less than some specified value, it is called a **left-tailed test** because the outcome will be evaluated at the left tail of the sampling distribution. Our example is a right-tailed test because the research hypothesis states that the mean gas prices in California are higher than \$2.62 (see Figure 7.1).

Right-tailed test: A one-tailed test in which the sample outcome is hypothesized to be at the right tail of the sampling distribution.

Left-tailed test: A one-tailed test in which the sample outcome is hypothesized to be at the left tail of the sampling distribution.

Sometimes, we have some theoretical basis to believe that there is a difference between groups, but we cannot anticipate the direction of that difference. For example, we may have reason to believe that the average price of California gas is different from that of the general population, but we may not have enough

Figure 7.1 Sampling Distribution of Sample Means Assuming H_0 Is True for a Sample N = 100

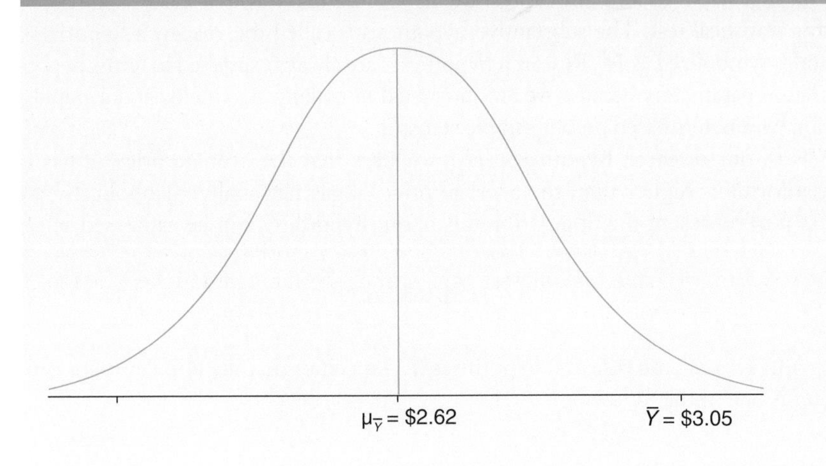

research or support to predict whether it is higher or lower. When we have no theoretical reason for specifying a direction in the research hypothesis, we conduct a **two-tailed test**. The research hypothesis specifies that the population mean is not equal to some specified value. For example, we can express the research hypothesis about the mean price of gas as

> **Two-tailed test:** A type of hypothesis test that involves a nondirectional research hypothesis. We are equally interested in whether the values are less than or greater than one another. The sample outcome may be located at both the low and high ends of the sampling distribution.

$$H_1: \mu \neq \$2.62$$

With both one- and two-tailed tests, our null hypothesis of no difference remains the same. It can be expressed as

$$H_0: \mu = \text{some specified value}$$

The Null Hypothesis (H_0)

> **Null hypothesis (H_0):** A statement of "no difference," which contradicts the research hypothesis and is always expressed in terms of population parameters.

Is it possible that in the population, there is no real difference between the mean price of gas in California and the mean price of gas in the nation and that the observed difference of 0.43 is actually due to the fact that this particular sample happened to contain California gas stations with higher prices? Since statistical inference is based on probability theory, it is not possible to prove or disprove the research hypothesis directly. We can, at best, estimate the likelihood that it is true or false.

To assess this likelihood, statisticians set up a hypothesis that is counter to the research hypothesis. The **null hypothesis**, symbolized as H_0, contradicts the research hypothesis and states that there is no difference between the population mean and some specified value. It is also referred to as the hypothesis of "no difference." Our null hypothesis can be stated symbolically as

$$H_0: \mu = \$2.62$$

Rather than directly testing the substantive hypothesis (H_1) that there is a difference between the mean price of gas in California and the mean price nationally, we test the null hypothesis (H_0) that there is no difference in prices. In hypothesis testing, we hope to reject the null hypothesis to provide indirect support for the research hypothesis. Rejection of the null hypothesis will strengthen our belief in the research hypothesis and increase our confidence in the importance and utility of the broader theory from which the research hypothesis was derived.

PROBABILITY VALUES AND ALPHA

Now let's put all our information together. We're assuming that our null hypothesis ($\mu = \$2.62$) is true, and we want to determine whether our sample evidence casts doubt on that assumption, suggesting that there is evidence for research hypothesis, $\mu > \$2.62$. What are the chances that we would have randomly selected a sample of California gas stations such that the average price per gallon is higher than \$2.62, the average for the nation? We can determine the chances or probability because of what we know about the sampling distribution and its properties. We know, based on the central limit theorem, that if our sample size is larger than 50, the sampling distribution of the mean is approximately normal, with a mean and a standard deviation (standard error) of

$$\sigma_{\bar{Y}} = \frac{\sigma}{\sqrt{N}}$$

We are going to assume that the null hypothesis is true and then see if our sample evidence casts doubt on that assumption. We have a population mean $\mu = \$2.62$ and a standard deviation $\sigma = 0.21$. Our sample size is $N = 100$, and the sample mean is \$3.05. We can assume that the distribution of means of all possible samples of size $N = 100$ drawn from this distribution would be approximately normal, with a mean of \$2.32 and a standard deviation of

$$\sigma_{\bar{Y}} = \frac{0.21}{\sqrt{100}} = 0.02$$

This sampling distribution is shown in Figure 7.1. Also shown in Figure 7.1 is the mean gas price we observed for our sample of California gas stations.

Because this distribution of sample means is normal, we can use Appendix B to determine the probability of drawing a sample mean of $3.05 or higher from this population. We will translate our sample mean into a Z score so that we can determine its location relative to the population mean. In Chapter 4, we learned how to translate a raw score into a Z score by using Formula 4.1:

$$Z = \frac{Y - \bar{Y}}{s}$$

Because we are dealing with a sampling distribution in which our raw score is Y (the mean), and the standard deviation (standard error) is $\sigma / \sqrt{N}$, we need to modify the formula somewhat:

$$Z = \frac{\bar{Y} - \mu_Y}{\sigma / \sqrt{N}} \quad (7.1)$$

Z statistic (obtained): The test statistic computed by converting a sample statistic (such as the mean) to a *Z* score. The formula for obtaining *Z* varies from test to test.

Converting the sample mean to a Z-score equivalent is called computing the test statistic. The Z value we obtain is called the **Z statistic (obtained)**. The obtained Z gives us the number of standard deviations (standard errors) that our sample is from the hypothesized value (μ or μ_Y), assuming the null hypothesis is true. For our example, the obtained Z is

$$Z = \frac{3.05 - 2.62}{0.21 / \sqrt{100}} = \frac{0.43}{0.02} = 21.50$$

Before we determine the probability of our obtained Z statistic, let's determine whether it is consistent with our research hypothesis. Recall that we defined our research hypothesis as a right-tailed test ($\mu >$ $2.62), predicting that the difference would be assessed on the right tail of the sampling distribution. The positive value of our obtained Z statistic confirms that we will be evaluating the difference on the right tail. (If we had a negative obtained Z, it would mean the difference would have to be evaluated at the left tail of the distribution, contrary to our research hypothesis.)

To determine the probability of observing a Z value of 21.50, assuming that the null hypothesis is true, look up the value in Appendix B to find the area to the right of (above) the Z of 21.50. Our calculated Z value is not listed in Appendix B, so we'll need to rely on the last Z value reported in the table, 4.00. Recall from Chapter 4, where we calculated Z scores and their probability, that the Z values are located in column A. The p value is the probability to the right of the obtained Z, or the "area beyond Z" in column C. This area includes the proportion of all sample means that are $3.05 or higher. The proportion is less than .0001 (Figure 7.2). This value is the probability of getting a result as extreme as the sample result if the null hypothesis is true; it is symbolized as p. Thus, for our example, $p \leq .0001$.

Figure 7.2 The Probability (p) Associated With Z = 21.50

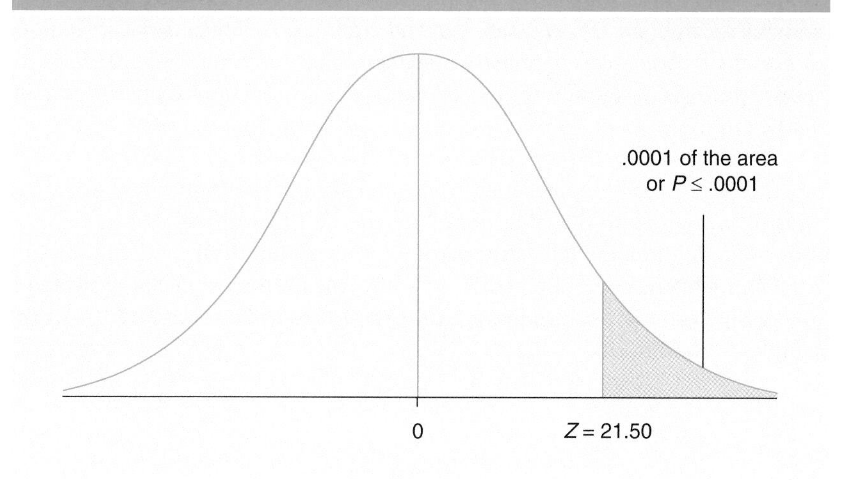

A p value can be defined as the probability associated with the obtained value of Z. It is a measure of how unusual or rare our obtained statistic is compared with what is stated in our null hypothesis. The smaller the p value, the more evidence we have that the null hypothesis should be rejected in favor of the research hypothesis. The larger the p value, we can assume that the null hypothesis is true and fail to reject it. Based on the p value, we can also make a statement regarding the significance of the results. A result is deemed "statistically significant" if the probability is less than or equal to the alpha level.

p value: The probability associated with the obtained value of *Z*.

Researchers usually define in advance what a sufficiently improbable Z value is by specifying a cutoff point below which p must fall to reject the null hypothesis. This cutoff point, called alpha and denoted by the Greek letter α, is customarily set at the .05, .01, or .001 level. Let's say that we decide to reject the null hypothesis if $p \leq .05$. The value .05 is referred to as alpha (α); it defines for us what result is sufficiently improbable to allow us to take the risk and reject the null hypothesis. An alpha (α) of .05 means that even if the obtained Z statistic is due to sampling error, so that the null hypothesis is true, we would allow a 5% risk of rejecting it. Alpha values of .01 and .001 are more cautionary levels of risk. The difference between p and alpha is that p is the actual probability associated with the obtained value of Z, whereas alpha is the level of probability determined in advance at which the null hypothesis is rejected. The null hypothesis is rejected when $p \leq \alpha$.

Alpha (α): The level of probability at which the null hypothesis is rejected. It is customary to set alpha at the .05, .01, or .001 level.

We have already determined that our obtained Z has a probability value less than .0001. Since our observed p is less than .05 ($p = .0001 < \alpha = .05$), we reject the null hypothesis. The value of .0001 means that fewer than 1 out of 10,000 samples drawn from this population are likely to have a mean that is 21.50 Z scores above the hypothesized mean of \$3.05. Another way to say it is as follows:

There is only 1 chance out of 10,000 (or .0001%) that we would draw a random sample with a $Z \geq 21.50$ if the mean price of California gas were equal to the national mean price. We can state that the difference between the average price of gas in California and nationally is statistically significant at the .05 level, or specify the level of significance by saying that the level of significance is less than .0001. For more about significance, refer to A Closer Look 7.1.

Recall that our hypothesis was a one-tailed test ($\mu > \$2.62$). In a two-tailed test, sample outcomes may be located at both the higher and the lower ends of the sampling distribution. Thus, the null hypothesis will be rejected if our sample outcome falls either at the left or right tail of the sampling distribution. For instance, a .05 alpha or p level means that H_0 will be rejected if our sample outcome falls among either the lowest or the highest 5% of the sampling distribution.

Suppose we had expressed our research hypothesis about the mean price of gas as

$$H_1: \mu \neq \$2.62$$

A CLOSER LOOK 7.1

More About Significance

Just because a relationship between two variables is statistically significant does not mean that the relationship is important theoretically or practically. Recall that we are relying on information from a sample to infer characteristics about the population. If you decide to reject the null hypothesis, you must still determine what inferences you can make about the population. Ronald Wasserstein and Nicole Lazar (2016) advise, "Researchers should bring many contextual factors into play to derive scientific inferences, including the design of a study, the quality of the measurements, the external evidence for the phenomenon under study, and the validity of assumptions that underlies the data analysis."[4] Indeed, determining significance is just one part of the research process.

The application of hypothesis testing and significance presented in this text reflects how our discipline currently uses and reports hypothesis testing. Yet scholars and statisticians have expressed concern about reducing scientific inquiry to the pursuit of single measure; that is to say, the only result that matters is when $p < .05$ or some arbitrary level of significance. According to demographer Jan Hoem (2008), "The scientific importance of an empirical finding depends much more on its contribution to the development or falsification of a substantive theory than on the values of indicators of statistical significance."[5]

Many have argued how hypothesis testing is problematic because it fails to provide definitive evidence about the existence of real relationships in the data. Despite these criticisms, hypothesis testing remains the primary model by which we derive statistical inference. Several academic journals have adopted new standards for data (e.g., eliminating p values, reporting nonsignificant findings along with significant ones), in hopes of improving the quality and integrity of research.

The null hypothesis to be directly tested still takes the form $H_0: \mu = \$2.62$ and our obtained Z is calculated using the same formula (Formula 7.1) as was used with a one-tailed test. To find p for a two-tailed test, look up the area in column C of Appendix B that corresponds to your obtained Z (as we did earlier) and then multiply it by 2 to obtain the two-tailed probability. Thus, the two-tailed p value for $Z = 21.50$ is $.0001 \times 2 = .0002$. This probability is less than our stated alpha (.05), and thus, we reject the null hypothesis.

THE FIVE STEPS IN HYPOTHESIS TESTING: A SUMMARY

Statistical hypothesis testing can be organized into five basic steps. Let's summarize these steps:

1. Making assumptions
2. Stating the research and null hypotheses and selecting alpha
3. Selecting the sampling distribution and specifying the test statistic
4. Computing the test statistic
5. Making a decision and interpreting the results

1. *Making Assumptions:* Statistical hypothesis testing involves making several assumptions regarding the level of measurement of the variable, the method of sampling, the shape of the population distribution, and the sample size. In our example, we made the following assumptions:

- A random sample was used.
- The variable *price per gallon* is measured on an interval-ratio level of measurement.
- Because $N > 50$, the assumption of normal population is not required.

2. *Stating the Research and Null Hypotheses and Selecting Alpha:* The substantive hypothesis is called the research hypothesis and is symbolized as H_1. Research hypotheses are always expressed in terms of population parameters because we are interested in making statements about population parameters based on sample statistics. Our research hypothesis was

$$H_1: \mu > \$2.62$$

The null hypothesis, symbolized as H_0, contradicts the research hypothesis in a statement of no difference between the population mean and our hypothesized value. For our example, the null hypothesis was stated symbolically as

$$H_0: \mu = \$2.62$$

We set alpha at .05, meaning that we would reject the null hypothesis if the probability of our obtained Z was less than or equal to .05.

3. *Selecting the Sampling Distribution and Specifying the Test Statistic:* The normal distribution and the Z statistic are used to test the null hypothesis.

4. *Computing the Test Statistic:* Based on Formula 7.1, our Z statistic is 21.50.

5. *Making a Decision and Interpreting the Results:* We confirm that our obtained Z is on the right tail of the distribution, consistent with our research hypothesis. We determine that the p value of 21.50 is less than .0001, lower than our .05 alpha level. We have evidence to reject the null hypothesis of no difference between the mean price of California gas and the mean price of gas nationally. Based on these data, we conclude that the average price of California gas is significantly higher than the national average.

ERRORS IN HYPOTHESIS TESTING

We should emphasize that because our conclusion is based on sample data, we will never really know if the null hypothesis is true or false. In fact, as we have seen, there is a 0.01% chance that the null hypothesis is true and that we are making an error by rejecting it.

The null hypothesis can be either true or false, and in either case, it can be rejected or not rejected. If the null hypothesis is true and we reject it nonetheless, we are making an incorrect decision. This type of error is called a **Type I error**. Conversely, if the null hypothesis is false but we fail to reject it, this incorrect decision is a **Type II error**.

Type I error: The probability associated with rejecting a null hypothesis when it is true.

Type II error: The probability associated with failing to reject a null hypothesis when it is false.

In Table 7.1, we show the relationship between the two types of errors and the decisions we make regarding the null hypothesis. The probability of a Type I error—rejecting a true hypothesis—is equal to the chosen alpha level. For example, when we set alpha at the .05 level, we know that the probability that the null hypothesis is in fact true is .05 (or 5%).

We can control the risk of rejecting a true hypothesis by manipulating alpha. For example, by setting alpha at .01, we are reducing the risk of making a Type I error to 1%. Unfortunately, however, Type I and Type II errors are inversely related; thus, by reducing alpha and lowering the risk of making a Type I error, we are increasing the risk of making a Type II error (Table 7.1).

As long as we base our decisions on sample statistics and not population parameters, we have to accept a degree of uncertainty as part of the process of statistical inference.

Table 7.1 Type I and Type II Errors

	True State of Affairs	
Decision Made	**H_0 Is True**	**H_0 Is False**
Reject H_0	Type I error (α)	Correct decision
Do not reject H_0	Correct decision	Type II error

LEARNING CHECK 7.1

The implications of research findings are not created equal. For example, researchers might hypothesize that eating spinach increases the strength of weightlifters. Little harm will be done if the null hypothesis that eating spinach has no effect on the strength of weightlifters is rejected in error. The researchers would most likely be willing to risk a high probability of a Type I error, and all weightlifters would eat spinach. However, when the implications of research have important consequences (funding of social programs or medical testing), the balancing act between Type I and Type II errors becomes more important. Can you think of some examples where researchers would want to minimize Type I errors? When might they want to minimize Type II errors?

The *t* Statistic and Estimating the Standard Error

The Z statistic we have calculated (Formula 7.1) to test the hypothesis involving a sample of California gas stations assumes that the population standard deviation (σ) is known. The value of σ is required to calculate the standard error:

$$\sigma / \sqrt{N}$$

In most situations, σ will not be known, and we will need to estimate it using the sample standard deviation s. We then use the t statistic instead of the Z statistic to test the null hypothesis. The formula for computing the t statistic is

$$t = \frac{\bar{Y} - \mu}{s / \sqrt{N}} \qquad (7.2)$$

t **statistic (obtained):** The test statistic computed to test the null hypothesis about a population mean when the population standard deviation is unknown and is estimated using the sample standard deviation.

The t value we calculate is called the **t statistic (obtained)**. The obtained t represents the number of standard deviation units (or standard error units) that our sample mean is from the hypothesized value of μ, assuming that the null hypothesis is true.

The *t* Distribution and Degrees of Freedom

t **distribution:** A family of curves, each determined by its degrees of freedom (*df*). It is used when the population standard deviation is unknown and the standard error is estimated from the sample standard deviation.

Degrees of freedom (*df*): The number of scores that are free to vary in calculating a statistic.

To understand the *t* statistic, we should first be familiar with its distribution. The **t distribution** is actually a family of curves, each determined by its degrees of freedom. The concept of degrees of freedom is used in calculating several statistics, including the *t* statistic. The **degrees of freedom (*df*)** represent the number of scores that are free to vary in calculating each statistic.

To calculate the degrees of freedom, we must know the sample size and whether there are any restrictions in calculating that statistic. The number of restrictions is then subtracted from the sample size to determine the degrees of freedom. When calculating the *t* statistic for a one-sample test, we start with the sample size *N* and lose 1 degree of freedom for the population standard deviation we estimate.[6] Note that the degrees of freedom will increase as the sample size increases. In the case of a single-sample mean, the *df* is calculated as follows:

$$df = N - 1 \tag{7.3}$$

Comparing the *t* and *Z* Statistics

Notice the similarities between the formulas for the *t* and *Z* statistics. The only apparent difference is in the denominator. The denominator of *Z* is the standard error based on the population standard deviation σ. For the denominator of *t*, we replace σ / *N* with / *N*, the estimated standard error based on the sample standard deviation.

However, there is another important difference between the *Z* and *t* statistics: Because it is estimated from sample data, the denominator of the *t* statistic is subject to sampling error. The sampling distribution of the test statistic is not normal, and the standard normal distribution cannot be used to determine probabilities associated with it.

In Figure 7.3, we present the *t* distribution for several *df*s. Like the standard normal distribution, the *t* distribution is bell shaped. The *t* statistic, similar to the *Z* statistic, can have positive and negative values. A positive *t* statistic corresponds to the right tail of the distribution; a negative value corresponds to the left tail. Note that when the *df* is small, the *t* distribution is much flatter than the normal curve. But as the degrees of freedom increases, the shape of the *t* distribution gets closer to the normal distribution, until the two are almost identical when *df* is greater than 120.

Appendix C summarizes the *t* distribution. Note that the *t* table differs from the normal (*Z*) table in several ways. First, the column on the left side of the table shows the degrees of freedom. The *t* statistic will vary depending on the degrees of freedom, which must first be computed (*df* = *N* – 1). Second, the probabilities or alpha, denoted as significance levels, are arrayed across the top of the table in two rows, the first for a one-tailed and the second for a two-tailed test. Finally, the values of *t*, listed as the entries of this table, are a function of (a) the degrees of freedom, (b) the level of significance (or probability), and (c) whether the test is a one- or a two-tailed test.

Figure 7.3 The Normal Distribution and t Distributions for 1, 5, 20, and ∞ Degrees of Freedom

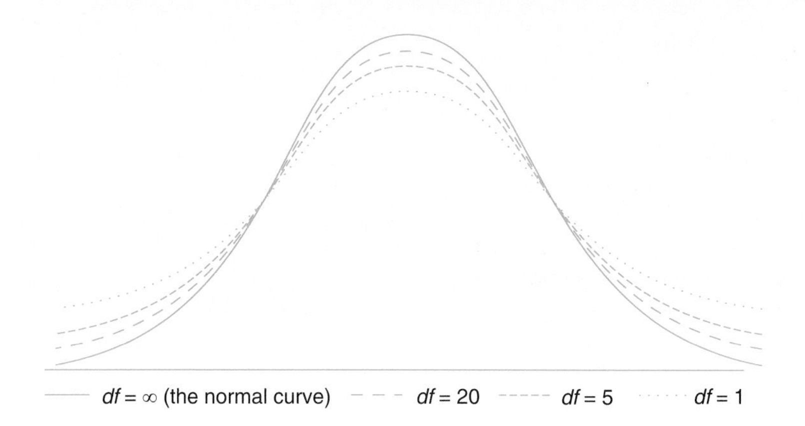

To illustrate the use of this table, let's determine the probability of observing a *t* value of 2.021 with 40 degrees of freedom and a two-tailed test. Locating the proper row ($df = 40$) and column (two-tailed test), we find the *t* statistic of 2.021 corresponding to the .05 level of significance. Restated, we can say that the probability of obtaining a *t* statistic of 2.021 is .05, or that there are fewer than 5 chances out of 100 that we would have drawn a random sample with an obtained *t* of 2.021 if the null hypothesis were correct.

HYPOTHESIS TESTING WITH ONE SAMPLE AND POPULATION VARIANCE UNKNOWN

To illustrate the application of the *t* statistic, let's test a two-tailed hypothesis about a population mean μ. Let's say we drew a random sample of 280 white females who worked full-time in 2017. We found their mean earnings to be $45,785, with a standard deviation, s = $25,563. Based on data from the U.S. Census Bureau,[7] we also know that the 2017 mean earnings for all full-time working women was μ = $41,977. However, we do not know the value of the population standard deviation. We want to determine whether the sample of white women was representative of the population of all full-time women workers in 2017. Although we suspect that white American women experienced a relative advantage in earnings, we are not sure enough to predict that their earnings were indeed higher than the earnings of all women nationally. Therefore, the statistical test is two-tailed.

Let's apply the five-step model to test the hypothesis that the average earnings of white women differed from the average earnings of all women working full-time in the United States in 2017.

1. *Making Assumptions:* Our assumptions are as follows:
   - A random sample is selected.
   - Because $N > 50$, the assumption of normal population is not required.
   - The level of measurement of the variable income is interval ratio.
2. *Stating the Research and the Null Hypotheses and Selecting Alpha:* The research hypothesis is

$$H_1: \mu > \$41{,}977$$

and the null hypothesis is

$$H_0: \mu > \$41{,}977$$

We'll set alpha at .05, meaning that we will reject the null hypothesis if the probability of our obtained statistic is less than or equal to .05.

3. *Selecting the Sampling Distribution and Specifying the Test Statistic:* We use the t distribution and the t statistic to test the null hypothesis.
4. *Computing the Test Statistic:* We first calculate the df associated with our test:

$$df = (N - 1) = (280 - 1) = 279$$

To evaluate the probability of obtaining a sample mean of $45,785, assuming the average earnings of white women were equal to the national average of $41,977, we need to calculate the obtained t statistic by using Formula 7.2:

$$t = \frac{\bar{Y} - \mu}{s / \sqrt{N}} = \frac{45{,}785 - 41{,}977}{25{,}563 / \sqrt{280}} = \frac{3{,}808}{1{,}527.68} = 2.49$$

5. *Making a Decision and Interpreting the Results:* Given our research hypothesis, we will conduct a two-tailed test. To determine the probability of observing a t value of 2.49 with 279 degrees of freedom, let's refer to Appendix C. From the first column, we can see that 279 degrees of freedom is not listed, so we'll have to use the last row, $df = \infty$, to assess the significance of our obtained t statistic.

Our obtained t statistic of 2.49 is not listed in the last row. It is greater than 2.326 (t critical for .01 one-tailed test) but less than 2.576 (t critical for .005 one-tailed test). The probability of obtaining a t of 2.49 can be estimated as being less than .01 but greater than .005 ($.01 > p > .005$), leading to the conclusion that we reject the null hypothesis. The average income for our sample is significantly higher than the national average income.

HYPOTHESIS TESTING WITH TWO SAMPLE MEANS

The two examples that we reviewed at the beginning of this chapter dealt with data from one sample compared with data from the population. In practice, social scientists are often more interested in situations involving two (sample) parameters than those involving one, such as the differences between men and women, Democrats and Republicans, whites and nonwhites, or high school or college graduates. Specifically, we may be interested in finding out whether the average years of education for one racial/ethnic group is the same, lower, or higher than another group.

U.S. data on educational attainment reveal that Asians and Pacific Islanders have more years of education than any other racial/ethnic groups; this includes the percentage of those earning a high school degree or higher or a college degree or higher. Although years of education have steadily increased for blacks and Hispanics since 1990, their numbers remain behind Asians and Pacific Islanders and whites.

Using data from the 2018 General Social Survey (GSS), we examine the difference in white and black educational attainment. From the GSS sample, white respondents reported an average of 14.00 years of education and blacks an average of 13.57 years, as shown in Table 7.2. These sample averages could mean either (a) the average number of years of education for whites is higher than the average for blacks or (b) the average for whites is actually about the same as for blacks, but our sample just happens to indicate a higher average for whites. What we are applying here is a bivariate analysis (for more information, refer to Chapter 2), a method to detect and describe the relationship between two variables—*race/ethnicity* and *educational attainment*.

The statistical procedures discussed in the following sections allow us to test whether the differences that we observe between two samples are large enough for us to conclude that the populations from which these samples are drawn are different as well. We present tests for the significance of the differences between two groups. Primarily, we consider differences between sample means and differences between sample proportions.

Hypothesis testing with two samples follows the same structure as for one-sample tests: The assumptions of the test are stated, the research and null hypotheses are formulated and the alpha level selected, the sampling distribution and the test statistic are specified, the test statistic is computed, and a decision is made whether or not to reject the null hypothesis.

Table 7.2 Years of Education for White and Black Men and Women, GSS 2018

	Whites (Sample 1)	Blacks (Sample 2)
Mean	14.00	13.57
Standard deviation	2.86	2.37
Variance	8.18	5.62
N	830	203

The Assumption of Independent Samples

One important difference between one- and two-sample hypothesis testing involves sampling procedures. With a two-sample case, we assume that the samples are independent of each other. The choice of sample members from one population has no effect on the choice of sample members from the second population. In our comparison of whites and blacks, we are assuming that the selection of whites is independent of the selection of black individuals. (The requirement of independence is also satisfied by selecting one sample randomly, then dividing the sample into appropriate subgroups. For example, we could randomly select a sample and then divide it into groups based on gender, religion, income, or any other attribute that we are interested in.)

Stating the Research and Null Hypotheses

The second difference between one- and two-sample tests is in the form taken by the research and the null hypotheses. In one-sample tests, both the null and the research hypotheses are statements about a single population parameter, μ. In contrast, with two-sample tests, we compare two population parameters.

Our research hypothesis (H_1) is that the average years of education for whites is not equal to the average years of education for black respondents. We are stating a hypothesis about the relationship between race/ethnicity and education in the general population by comparing the mean educational attainment of whites with the mean educational attainment of blacks. Symbolically, we use μ to represent the population mean; the subscript 1 refers to our first sample (whites) and subscript 2 to our second sample (blacks). Our research hypothesis can then be expressed as

$$H_1: \mu_1 \neq \mu_2$$

Because H_1 specifies that the mean education for whites is not equal to the mean education for blacks, it is a nondirectional hypothesis. Thus, our test will be a two-tailed test. Alternatively, if there were sufficient basis for deciding which population mean score is larger (or smaller), the research hypothesis for our test would be a one-tailed test:

$$H_1: \mu_1 < \mu_2 \text{ or } H_1: \mu_1 > \mu_2$$

In either case, the null hypothesis states that there are no differences between the two population means:

$$H_0: \mu_1 = \mu_2$$

We are interested in finding evidence to reject the null hypothesis of no difference so that we have sufficient support for our research hypothesis.

LEARNING CHECK 7.2

For the following research situations, state your research and null hypotheses:

- There is a difference between the mean statistics grades of social science majors and the mean statistics grades of business majors.
- The average number of children in two-parent black families is lower than the average number of children in two-parent nonblack families.
- Grade point averages are higher among girls who participate in organized sports than among girls who do not.

Sampling distribution of the difference between means: A theoretical probability distribution that would be obtained by calculating all the possible mean differences $(\bar{Y}_1 - \bar{Y}_2)$ that would be obtained by drawing all the possible independent random samples of size N_1 and N_2 from two populations where N_1 and N_2 are each greater than 50.

THE SAMPLING DISTRIBUTION OF THE DIFFERENCE BETWEEN MEANS

The sampling distribution allows us to compare our sample results with all possible sample outcomes and estimate the likelihood of their occurrence. Tests about differences between two sample means are based on the sampling distribution of the difference between means. The sampling distribution of the difference between two sample means is a theoretical probability distribution that would be obtained by calculating all the possible mean differences by drawing all possible independent random samples of size N_1 and N_2 from two populations.

The properties of the sampling distribution of the difference between two sample means are determined by a corollary to the central limit theorem. This theorem assumes that our samples are independently drawn from normal populations, but that with sufficient sample size ($N_1 > 50$, $N_2 > 50$), the sampling distribution of the difference between means will be approximately normal, even if the original populations are not normal. This sampling distribution has a mean $\mu_{Y_1} - \mu_{Y_2}$ and a standard deviation (standard error)

$$\sigma_{\bar{Y}_1-\bar{Y}_2} = \sqrt{\frac{\sigma_1^2}{N_1} + \frac{\sigma_2^2}{N_2}} \tag{7.4}$$

which is based on the variances in each of the two populations (σ_1^2 and σ_2^2).

Estimating the Standard Error

Formula 7.4 assumes that the population variances are known and that we can calculate the standard error $\sigma_{\bar{Y}_1-\bar{Y}_2}$ (the standard deviation of the sampling distribution). However, in most situations, the only data we have are based on sample data, and we do not know the true value of the population variances, σ_1^2 and σ_2^2. Thus, we need to estimate the standard error from the sample variances, s_1^2 and s_2^2. The estimated standard error of the difference between means is symbolized as $s_{\bar{Y}_1-\bar{Y}_2}$ (instead of $\sigma_{\bar{Y}_1-\bar{Y}_2}$).

Calculating the Estimated Standard Error

When we can assume that the two population variances are equal, we combine information from the two sample variances to calculate the estimated standard error.

$$S_{\bar{Y}_1-\bar{Y}_2} = \sqrt{\frac{(N_1-1)s_1^2 + (N_2-1)s_2^2}{(N_1+N_2)-2}} \sqrt{\frac{N_1+N_2}{N_1 N_2}} \tag{7.5}$$

where $S_{\bar{Y}_1-\bar{Y}_2}$ is the estimated standard error of the difference between means, and s_1^2 and s_2^2 are the variances of the two samples. As a rule of thumb, when either sample variance is more than twice as large as the other, we can no longer assume that the two population variances are equal and would need to use Formula 7.8.

The *t* Statistic

As with single sample means, we use the *t* distribution and the *t* statistic whenever we estimate the standard error for a difference between means test. The *t* value we calculate is the obtained t. It represents the number of standard deviation units (or standard error units) that our mean difference $\bar{Y}_1 - \bar{Y}_2$ is from the hypothesized value of $\mu_1 - \mu_2$, assuming that the null hypothesis is true.

The formula for computing the t statistic for a difference between means test is

$$t = \frac{\bar{Y}_1 - \bar{Y}_2}{S_{\bar{Y}_1 - \bar{Y}_2}} \tag{7.6}$$

where $S_{\bar{Y}_1 - \bar{Y}_2}$ is the estimated standard error.

Calculating the Degrees of Freedom for a Difference Between Means Test

To use the t distribution for testing the difference between two sample means, we need to calculate the degrees of freedom. As we saw earlier, the degrees of freedom (*df*) represent the number of scores that are free to vary in calculating each statistic. When calculating the t statistic for the two-sample test, we lose 2 degrees of freedom, one for every population variance we estimate. When population variances are assumed to be equal or if the size of both samples is greater than 50, the *df* is calculated as follows:

$$df = (N_1 + N_2) - 2 \tag{7.7}$$

When we cannot assume that the population variances are equal and when the size of one or both samples is equal to or less than 50, we use Formula 7.9 to calculate the degrees of freedom.

THE FIVE STEPS IN HYPOTHESIS TESTING ABOUT DIFFERENCE BETWEEN MEANS: A SUMMARY

As with single-sample tests, statistical hypothesis testing involving two sample means can be organized into five steps.

1. *Making Assumptions:* In our example, we made the following assumptions:
   - Independent random samples are used.
   - The variable *years of education* is measured at an interval-ratio level of measurement.
   - Because $N_1 > 50$ and $N_2 > 50$, the assumption of normal population is not required.
   - The population variances are assumed to be equal.

A CLOSER LOOK 7.2

Calculating the Estimated Standard Error and the Degrees of Freedom (*df*) When the Population Variances Are Assumed to Be Unequal

If the variances of the two samples (s_1^2 and s_2^2) are very different (one variance is twice as large as the other), the formula for the estimated standard error becomes

$$S_{\bar{Y}_1-\bar{Y}_2} = \sqrt{\frac{s_1^2}{N_1} + \frac{s_2^2}{N_2}} \quad (7.8)$$

When the population variances are unequal and the size of one or both samples is equal to or less than 50, we use another formula to calculate the degrees of freedom associated with the t statistic:[8]

$$df = \frac{(s_1^2 / N_1 + s_2^2 / N_2)^2}{(1/N_1 - 1)(s_1^2/N_1)^2 + (1/N_2 - 1)(s_2^2/N_2)^2} \quad (7.9)$$

2. *Stating the Research and Null Hypotheses and Selecting Alpha:* Our research hypothesis is that the mean education of whites is different from the mean education of blacks, indicating a two-tailed test. Symbolically, the research hypothesis is expressed as

$$H_1: \mu_1 \neq \mu_2$$

with μ_1 representing the mean education of whites and μ_2 the mean education of blacks. The null hypothesis states that there are no differences between the two population means, or

$$H_0: \mu_1 = \mu_2$$

We are interested in finding evidence to reject the null hypothesis of no difference so that we have sufficient support for our research hypothesis. We will reject the null hypothesis if the probability of t (obtained) is less than or equal to .05 (our alpha value).

3. *Selecting the Sampling Distribution and Specifying the Test Statistic:* The t distribution and the t statistic are used to test the significance of the difference between the two sample means.

4. *Computing the Test Statistic:* To test the null hypothesis about the differences between the mean education of whites and blacks, we need to translate the ratio of the observed differences to its standard error

into a t statistic (based on data presented in Table 7.2). The obtained t statistic is calculated using Formula 7.6:

$$t = \frac{\bar{Y}_1 - \bar{Y}_2}{S_{\bar{Y}_1 - \bar{Y}_2}}$$

where $S_{\bar{Y}_1 - \bar{Y}_2}$ is the estimated standard error of the sampling distribution. Because the population variances are assumed to be equal, df is $(N_1 + N_2) - 2 = (830 + 203) - 2 = 1{,}031$ and we can combine information from the two sample variances to estimate the standard error (Formula 7.5):

$$S_{\bar{Y}_1 - \bar{Y}_2} = \sqrt{\frac{(830-1)(2.86)^2 + (203-1)(2.37)^2}{830+203-2}}\sqrt{\frac{830+203}{(830)(203)}} = (2.77)(.08) = .22$$

We substitute this value into the denominator for the t statistic (Formula 7.6):

$$t = \frac{14.00 - 13.57}{.22} = 1.95$$

5. *Making a Decision and Interpreting the Results:* We confirm that our obtained t is on the right tail of the distribution. Since our obtained t statistic of 1.95 is less than t critical = 1.96 ($df = \infty$, two tailed; see Appendix C), we can state that its probability is greater than .05. We fail to reject the null hypothesis of no difference between the educational attainment of whites and blacks.

LEARNING CHECK 7.3

Would you change your decision in the previous example if alpha was .01? Why or why not?

STATISTICS IN PRACTICE: VAPE USE AMONG TEENS

Administered annually since 1975, the Monitoring the Future (MTF) survey measures the extent of and beliefs regarding drug use among 8th, 10th, and 12th graders. Data collected from the MTF surveys have revealed decreases or stability in drug use among youths, particularly for cigarettes, alcohol, marijuana, cocaine, and methamphetamine, although vaping (use of e-cigarettes) has been on the rise. According to Nora Volkow, director of the National Institute on Drug Abuse, "Teens are clearly attracted to the marketable technology and flavorings seen in vaping devices."[9]

Let's examine data from the MTF 2017 survey, comparing frequency of lifetime vape use between black and white students. Lifetime vape use is measured on an ordinal scale: 1 = 0 occasions, 2 = 1–2 times, 3 = 3–5 times, 4 = 6–9 times, 5 = 10–19 times, 6 = 20–39 times, and 7 = 40+.

We will rely on SPSS to calculate the *t* obtained for the data. We will not present the complete five-step model and *t*-test calculation because we want to focus here on interpreting the SPSS output. However, we will need a research hypothesis and an alpha level to guide our interpretation. SPSS always estimates a two-tailed test, namely, does the gap of –.83 (1.24 – 2.07) indicate a difference in lifetime vape use between black and white adolescents? We'll set alpha at .05.

The output includes two tables. The Group Statistics table (Figure 7.4) presents descriptive statistics for each group. The survey results indicate that black students are less likely to have vaped nicotine in their lifetime than white students. The mean lifetime use for black students is 1.24 (closer to 0 occasions) and 2.07 (1–2 times) for white students.

In the second table (Figure 7.5), labeled Independent Samples Test, *t* statistics are presented for equal variances assumed (–9.478) and equal variances not assumed (–14.322). Both *t*-obtained statistics are negative, indicating that the average lifetime use for black students is lower than the average lifetime use for white students. To determine which *t* statistic to use, review the results of Levene's test for equality of variances. Levene's test (a calculation that we will not cover in this text) tests the null hypothesis that the population variances are equal. If the significance of the reported *F* statistic is equal to or less than .05 (the baseline alpha for Levene's test), we can reject the null hypothesis that the variances are equal; if the significance is greater than .05, we fail to reject the null hypothesis. (In other words, if the significance for Levene's test is greater than .05, refer to the *t* obtained for equal variances assumed; if the significance is less than .05, refer to the *t* obtained for equal variances not assumed.) Since the significance of *F* is .000 < .05, we reject the null hypothesis and conclude that the variances are not equal. Thus, the *t* obtained that we will use for this model is –14.322 (the one corresponding to equal variances not assumed).

SPSS calculates the exact probability of the *t* obtained for a two-tailed test. There is no need to estimate it based on Appendix C (as we did in our previous example). The significance of –14.322 is .000, which is less than our alpha level of .05. We reject the null hypothesis of no difference for lifetime vaping use between black and white students. On average, black students have vaped less in their lifetime than white students.

Figure 7.4 Group Statistics, MTF 2017

Group Statistics

	171C04(R):Rs RACE B/W/H	N	Mean	Std. Deviation	Std. Error Mean
BY17 34240 #X VAPE NIC/LIFETIME	BLACK:(1)	512	1.24	.897	.040
	WHITE:(2)	2099	2.07	1.925	.042

Figure 7.5 Independent Samples Test, MTF 2017

Independent Samples Test

		Levene's Test for Equality of Variances		t-t		
		F	Sig.	t	df	Sig. (2-tailed)
BY17 34240 #X VAPE NIC/LIFETIME	Equal variances assumed	315.866	.000	-9.478	2609	.000
	Equal variances not assumed			-14.322	1761.290	.000

LEARNING CHECK 7.4

State the null and research hypothesis for this SPSS example. Would you change your decision in the previous example if alpha was .01? Why or why not?

HYPOTHESIS TESTING WITH TWO SAMPLE PROPORTIONS

In the preceding sections, we have learned how to test for the significance of the difference between two population means when the variable is measured at an interval-ratio level. Yet numerous variables in the social sciences are measured at a nominal or an ordinal level. These variables are often described in terms of proportions or percentages. For example, we might be interested in comparing the proportion of those who support immigrant policy reform among Hispanics and non-Hispanics or the proportion of men and women who supported the Democratic candidate during the last presidential election. In this section, we present statistical inference techniques to test for significant differences between two sample proportions.

Hypothesis testing with two sample proportions follows the same structure as the statistical tests presented earlier: The assumptions of the test are stated, the research and null hypotheses are formulated, the sampling distribution and the test statistic are specified, the test statistic is calculated, and a decision is made whether or not to reject the null hypothesis.

In 2013, the Pew Research Center[10] presented a comparison of first-generation Americans (immigrants who were foreign born) and second-generation Americans (adults who have at least one immigrant parent) on several key demographic variables. Based on several measures of success, the Center documented social mobility between the generations, confirming that second-generation Americans were doing better than the first-generation Americans. The statistical question we examine here is whether the difference between the generations is significant.

For example, according to the Center's report, the proportion of first-generation Hispanic Americans who earned a bachelor's degree or higher was 0.11 (p_1); the proportion of second-generation Hispanic Americans with the same response was 0.21 (p_2). A total of 899 first-generation Hispanic Americans (N_1) and 351 second-generation Hispanic Americans (N_2) answered this question. We use the five-step model to determine whether the difference between the two proportions is significant.

1. *Making Assumptions:* Our assumptions are as follows:
   - Independent random samples of $N_1 > 50$ and $N_2 > 50$ are used.
   - The level of measurement of the variable is nominal.

2. *Stating the Research and Null Hypotheses and Selecting Alpha:* We propose a two-tailed test that the population proportions for first-generation and second-generation Hispanic Americans are not equal.

$$H_1: \pi_1 \neq \pi_2$$

$$H_0: \pi_1 = \pi_2$$

We decide to set alpha at .05.

3. *Selecting the Sampling Distribution and Specifying the Test Statistic:* The population distributions of dichotomies are not normal. However, based on the central limit theorem, we know that the sampling distribution of the difference between sample proportions is normally distributed when the sample size is large (when $N_1 > 50$ and $N_2 > 50$), with mean $\mu_{p_1 - p_2}$ and the estimated standard error $S_{p_1 - p_2}$. Therefore, we can use the normal distribution as the sampling distribution, and we can calculate Z as the test statistic.[11]

The formula for computing the Z statistic for a difference between proportions test is

$$Z = \frac{p_1 - p_2}{S_{p_1 - p_2}} \tag{7.10}$$

where p_1 and p_2 are the sample proportions for first- and second-generation Hispanic Americans, and $S_{p_1 - p_2}$ is the estimated standard error of the sampling distribution of the difference between sample proportions.

The estimated standard error is calculated using the following formula:

$$S_{p_1 - p_2} = \sqrt{\frac{p_1(1 - p_1)}{N_1} + \frac{p_2(1 - p_2)}{N_2}} \tag{7.11}$$

4. *Calculating the Test Statistic:* We calculate the standard error using Formula 7.11:

$$S_{p_1 - p_2} = \sqrt{\frac{0.11(1 - 0.11)}{899} + \frac{0.21(1 - 0.21)}{351}} = \sqrt{0.000581547} = 0.02$$

Substituting this value into the denominator of Formula 7.10, we get

$$Z = \frac{0.11 - 0.21}{0.02} = -5.00$$

5. *Making a Decision and Interpreting the Results:* Our obtained Z of –5.00 indicates that the difference between the two proportions will be evaluated at the left tail (the negative side) of the Z distribution. To determine the probability of observing a Z value of –5.00 if the null hypothesis is true, look up the value in Appendix B (column C) to find the area to the right of (above) the obtained Z.

Note that a Z score of 5.00 is not listed in Appendix B; however, the value exceeds the largest Z reported in the table, 4.00. The p value corresponding to a Z score of –5.00 would be less than .0001. For a two-tailed test, we'll have to multiply p by 2 ($.0001 \times 2 = .0002$). If this were a one-tailed test, we would not have to multiply the p value by 2. The probability of –5.00 for a two-tailed test is less than our alpha level of .05 ($.0002 < .05$).

Thus, we reject the null hypothesis of no difference. Based on the Pew Research data, we conclude that there is a significantly higher proportion of college graduates among second-generation Hispanic Americans compared with first-generation Hispanic Americans.

LEARNING CHECK 7.5

If alpha was changed to .01, two-tailed test, would your final decision change? Explain.

We continue our analysis of the 2013 Pew Research Center data, this time examining the difference in educational attainment between first- and second-generation Asian Americans presented in Table 7.3. Our research hypothesis is whether there is a lower proportion of college graduates among first-generation Asian Americans than second-generation Asian Americans, indicating a one-tailed test. We'll set alpha at .05.

The final calculation for Z is

$$Z = \frac{0.50 - 0.55}{0.02} = -2.50$$

The one-tailed probability of –2.50 is .0062. Comparing .0062 to our alpha, we reject the null hypothesis of no difference. We conclude that a significantly higher proportion of second-generation Asian Americans (55%) have a bachelor's degree or higher compared with first-generation Asian Americans (50%). The 5% difference is significant at the .05 level.

Table 7.3 Proportion of College Graduates Among First-Generation and Second-Generation Asian Americans

First-Generation Asian Americans	Second-Generation Asian Americans
$p_1 = .50$	$p_2 = .55$
$N_1 = 2{,}684$	$N_2 = 566$

Source: Pew Research Center, *Second-Generation Americans: A Portrait of the Adult Children of Immigrants* (Washington, D.C.: Pew Research Center, February 7, 2013). http://www.pewsocial-trends.org/2013/02/07/secondgeneration-americans/

LEARNING CHECK 7.6

If alpha was changed to .01, one-tailed test, would your final decision change? Explain.

READING THE RESEARCH LITERATURE: REPORTING THE RESULTS OF HYPOTHESIS TESTING

Let's conclude with an example of how the results of statistical hypothesis testing are presented in the social science research literature. Keep in mind that the research literature does not follow the same format or the degree of detail that we've presented in this chapter. For example, most research articles do not include a formal discussion of the null hypothesis or the sampling distribution. The presentation of statistical analyses and detail will vary according to the journal's editorial policy or the standard format for the discipline.

It is not uncommon for a single research article to include the results of multiple statistical tests. Results have to be presented succinctly and in summary form. An author's findings are usually presented in a summary table that may include the sample statistics (e.g., the sample means), the obtained test statistics (t or Z), the p level, and an indication of whether or not the results are statistically significant.

Robert Emmet Jones and Shirley A. Rainey (2006) examined the relationship between race, environmental attitudes, and perceptions about environmental health and justice.[12] Researchers have documented how people of color and the poor are more likely than whites and more affluent groups to live in areas with poor environmental quality and protection, exposing them to greater health risks. Yet little is known about how this disproportional exposure and risk are perceived by those affected. Jones and Rainey studied black and white residents from the Red River community in Tennessee, collecting data from interviews and a mail survey during 2001 to 2003.

They created a series of index scales measuring residents' attitudes pertaining to environmental problems and issues. The Environmental Concern (EC) Index measures public concern for specific environmental problems in the neighborhood. It includes questions on drinking water quality, landfills, loss of trees, lead paint and poisoning, the condition of green areas, and stream and river conditions. EC-II measures public concern (very unconcerned to very concerned) for the overall environmental quality in the neighborhood. EC-III measures the seriousness (not serious at all to very serious) of environmental problems in the neighborhood. Higher scores on all EC indicators indicate greater concern for environmental problems in their neighborhood. The Environmental Health (EH) Index measures public perceptions of certain physical side effects, such as headaches, nervous disorders, significant weight loss or gain, skin rashes, and breathing problems. The EH Index measures the likelihood (very unlikely to very likely) that the person believes that he or she or a household member experienced health problems due to exposure to environmental contaminants in his or her neighborhood. Higher EH scores reflect a greater likelihood that respondents believe that they have experienced health problems from exposure to environmental contaminants. Finally, the Environmental Justice (EJ) Index measures public perceptions about environmental justice, measuring the extent to which they agreed (or disagreed) that public officials had informed residents about environmental problems, enforced environmental laws, or held meetings to address residents' concerns. A higher mean EJ score indicates a greater likelihood that respondents think public officials failed to deal with environmental problems in their neighborhood.[13] Index score comparisons between black and white respondents are presented in Table 7.4.

Table 7.4 Environmental Concern (EC), Environmental Health (EH), and Environmental Justice (EJ)

Indicator	Group	Mean	Standard Deviation	*t*	Significance (One-Tailed)
EC Index	Blacks	56.2	13.7	6.2	<.001
	Whites	42.6	15.5		
EC-II	Blacks	4.4	1.0	5.6	<.001
	Whites	3.5	1.3		
EC-III	Blacks	3.4	1.1	6.7	<.001
	Whites	2.3	1.0		
EH Index	Blacks	23.0	10.5	5.1	<.001
	Whites	16.0	7.3		
EJ Index	Blacks	31.0	7.3	3.8	<.001
	Whites	27.2	6.3		

Source: Robert E. Jones and Shirley A. Rainey, "Examining Linkages Between Race, Environmental Concern, Health and Justice in a Highly Polluted Community of Color," *Journal of Black Studies* 36, no. 4 (2006): 473–496.

Note: N = 78 blacks, 113 whites.

LEARNING CHECK 7.7

Based on Table 7.4, what would be the t critical at the .05 level for the first indicator, EC Index? Assume a two-tailed test.

Let's examine the table carefully. Each row represents a single index measurement, reporting means and standard deviations separately for black and white residents. Obtained t-test statistics are reported in the second to last column. The probability of each t test is reported in the last column ($p < .001$), indicating a significant difference in responses between the two groups. All index score comparisons are significant at the .001 level. (*Note:* Researchers will use "n.s." to indicate nonsignificant results.)

While not referring to specific differences in index scores or to t-test results, Jones and Rainey use data from this table to summarize the differences between black and white residents on the three environmental index measurements:

> The results presented [in Table 1] suggest that as a group, Blacks are significantly more concerned than Whites about local environmental conditions (EC Index). . . . The results . . . also indicate that as a group, Blacks believe they have suffered more health problems from exposure to poor environmental conditions in their neighborhood than Whites (EH Index). . . . There is greater likelihood that Blacks feel local public agencies and officials failed to deal with environmental problems in their neighborhood in a fair, just, and effective manner (EJ Index).[14]

DATA AT WORK

Stephanie Wood: Campus Visit Coordinator

Photo courtesy of Stephanie Wood

At a Midwest liberal arts university, Stephanie coordinates the campus visit program for the Office of Admission, partnering with other university members (faculty, administrators, coaches, and alumni) to ensure that each prospective student visit is tailored to the student's needs. Stephanie says her work allows her to "think both creatively and strategically in developing innovative and successful events while also providing the opportunity to mentor a group of over fifty college students."

She explains how she uses statistical data and methods to improve the campus visit program. "Emphasis is placed on analyzing the success of each event by tracking the number of campus visitors and event attendees who progress further through the enrollment funnel by later applying, being admitted, and eventually enrolling to the institution. Events that have high yield (a large number of attendees who later enroll) are duplicated while events

(Continued)

(Continued)

with low yield are deconstructed and examined to discover what aspects factored in to the low yield. Variables examined typically include ambassador to student/family ratios, number of students who met with faculty in their major of interest, number of student-athletes able to meet with athletics, university events occurring at competing universities on the same day, weather, and various other factors. After the analysis is complete, conclusions regarding the low yield are made and new strategies are developed to help combat the findings."

Stephanie examines these variables on a daily basis. "I am constantly doing bivariate and multivariate analyses to ensure all events are contributing to increased enrollment across all student profiles."

"Regardless of whether a student *wants* to work with statistics, it is likely they will have to, to some extent. I would advise students to look at statistics in a much simpler and less scary mind-set. Measuring office efficiencies, project successes, and understanding biases is incredibly important in a professional setting."

MAIN POINTS

- Statistical hypothesis testing is a decision-making process that enables us to determine whether a particular sample result falls within a range that can occur by an acceptable level of chance. The process of statistical hypothesis testing consists of five steps: (1) making assumptions, (2) stating the research and null hypotheses and selecting alpha, (3) selecting a sampling distribution and a test statistic, (4) computing the test statistic, and (5) making a decision and interpreting the results.
- Statistical hypothesis testing may involve a comparison between a sample mean and a population mean or a comparison between two sample means. If we know the population variance(s) when testing for differences between means, we can use the Z statistic and the normal distribution. However, in practice, we are unlikely to have this information.
- When testing for differences between means when the population variance(s) are unknown, we use the t statistic and the t distribution.
- Tests involving differences between proportions follow the same procedure as tests for differences between means when population variances are known. The test statistic is Z, and the sampling distribution is approximated by the normal distribution.

KEY TERMS

alpha (α) 207
degrees of freedom (df) 212
left-tailed test 203
null hypothesis (H_0) 204
one-tailed test 203
p value 207
research hypothesis (H_1) 203
right-tailed test 203
sampling distribution of the difference between means 217
statistical hypothesis testing 202
t distribution 212
t statistic (obtained) 211
two-tailed test 204
Type I error 210
Type II error 210
Z statistic (obtained) 206

DIGITAL RESOURCES

Access key study tools at https://edge.sagepub.com/ssdsess4e

- eFlashcards of the glossary terms
- Datasets and codebooks
- SPSS and Excel walk-through videos
- SPSS and Excel demonstrations and problems to accompany each chapter
- Appendix F: Basic Math Review

CHAPTER EXERCISES

1. It is known that, nationally, doctors working for health maintenance organizations (HMOs) average 13.5 years of experience in their specialties, with a standard deviation of 7.6 years. The executive director of an HMO in a western state is interested in determining whether or not its doctors have less experience than the national average. A random sample of 150 doctors from HMOs shows a mean of only 10.9 years of experience.
   a. State the research and the null hypotheses to test whether or not doctors in this HMO have less experience than the national average.
   b. Using an alpha level of .01, calculate this test.
2. In this chapter, we examined the difference in educational attainment between first- and second-generation Hispanic and Asian Americans based on the proportion of each group with a bachelor's degree. We present additional data from the Pew Research Center's 2013 report, measuring the percentage of each group that owns a home.

	Percentage Owning a Home
First-generation Hispanic Americans $N = 899$	43
Second-generation Hispanic Americans $N = 351$	50
First-generation Asian Americans $N = 2,684$	58
Second-generation Asian Americans $N = 566$	51

Source: Pew Research Center, *Second-Generation Americans: A Portrait of the Adult Children of Immigrants* (Washington, D.C.: Pew Research Center, February 7, 2013). http://www.pewsocialtrends.org/2013/02/07/second-generation-americans/

a. Test whether there is a significant difference in the proportion of home owners between first- and second-generation Hispanic Americans. Set alpha at .05.

b. Test whether there is a significant difference in the proportion of home owners between first- and second-generation Asian Americans. Set alpha at .01.

3. For each of the following situations, determine whether a one- or a two-tailed test is appropriate. Also, state the research and the null hypotheses.

a. You are interested in finding out if the average household income of residents in your state is different from the national average household. According to the U.S. Census, for 2017, the national average household income is $57,652.[15]

b. You believe that students in small liberal arts colleges attend more parties per month than students nationwide. It is known that nationally undergraduate students attend an average of 3.2 parties per month. The average number of parties per month will be calculated from a random sample of students from small liberal arts colleges.

c. An epidemiologist believes the risk of death from COVID-19 is lower for children than the elderly.

d. Is there a difference in the amount of study time on-campus and off-campus students devote to their schoolwork during an average week? You prepare a survey to determine the average number of study hours for each group of students.

e. Reading scores for a group of third graders enrolled in an accelerated reading program are predicted to be higher than the scores for nonenrolled third graders.

f. Stress (measured on an ordinal scale) is predicted to be lower for adults who own dogs (or other pets) than for non–pet owners.

4. In 2016, the Pew Research Center[16] surveyed 1,799 white and 1,001 black Americans about their views on race and inequality. Pew researchers found "profound differences between black and white adults in their views on racial discrimination, barriers to black progress and the prospects for change." White and black respondents also disagreed about the best methods to achieve racial equality. For example, 34% of whites and 41% of blacks said that "bringing people of different racial backgrounds together to talk about race" would be a very effective tactic for groups striving to help blacks achieve equality. Test whether the proportion of white respondents who support this tactic is significantly less than the proportion of black respondents.
   a. State the null and research hypotheses.
   b. Calculate the Z statistic and test the hypothesis at the .05 level. What is your Step 5 decision?

5. One way to check on how representative a survey is of the population from which it was drawn is to compare various characteristics of the sample with the population characteristics. A typical variable used for this purpose is age. The GSS 2018 found a mean age of 48.69 and a standard deviation of 17.99 for its sample of 1,495 American adults. Assume that we know from census data that the mean age of all American adults is 37.80. Use this information to answer the following questions:
   a. State the research and the null hypotheses for a two-tailed test of means.
   b. Calculate the t statistic and test the null hypothesis at the .001 significance level. What did you find?
   c. What is your decision about the null hypothesis? What does this tell us about how representative the sample is of the American adult population?

6. According to the GSS 2018, 51% of 218 college graduates reported being interested in environmental issues compared with 43% of 167 high school graduates.
   a. What is the research hypothesis? Should you conduct a one- or a two-tailed test? Why?
   b. Present the five-step model, testing your hypothesis at the .05 level. What do you conclude?

7. GSS 2018 respondents were asked to rate their level of agreement to the statement, "Differences in income in America are too large." Responses were measured on a 5-point scale: 1 = *strongly agree*, 2 = *agree*, 3 = *neutral*, 4 = *disagree*, and 5 = *strongly disagree*. Strong Democrats had an average score of 1.69 ($s = 1.04$, $N = 86$) while strong Republicans had an average score of 2.11 ($s = 1.05$, $N = 67$).
   a. What is the appropriate test statistic? Why?
   b. Test the null hypothesis with a one-tailed test (strong Democrats are more likely to agree with the statement than strong Republicans);

$\alpha = .05$. What do you conclude about the difference in attitudes for these two political groups?

c. If you conducted a two-tailed test with $\alpha = .05$, would your decision have been different?

8. We compare the proportion who indicated speaking a language other than English for two GSS 2018 groups: respondents (1) born in the United States (native born) and (2) not born in the United States (foreign born). Test the research hypothesis that a lower proportion of native-born respondents than foreign-born respondents speak another language. Set alpha at .05.

	Native-Born Sample 1	Foreign-Born Sample 2
Proportion	.25	.72
N	1,294	205

9. In surveys conducted during August 2016 (months before the election), the Pew Research Center reported that among 752 men, 55% indicated that regardless of how they felt about Hillary Clinton, the election of a woman as president would be very important historically. Among 815 women, 65% reported the same. Do these differences reflect a significant gender gap?

a. If you wanted to test the research hypothesis that the proportion of male voters who believe in the historical importance of the election of a woman as president is less than the proportion of female voters who believe the same, would you conduct a one- or a two-tailed test?

b. Test the research hypothesis at the .05 alpha level. What do you conclude?

c. If alpha were changed to .01, would your decision remain the same?

10. In this SPSS output, we examine lifetime use of alcohol among white and black students based on the MTF 2017. Lifetime use is measured on an ordinal scale: 1 = 0 occasions, 2 = 1–2 times, 3 = 3–5 times, 4 = 6–9 times, 5 = 10–19 times, 6 = 20–39 times, and 7 = 40+. Present Step 5 (final decision) for these data. Assume alpha = .05, two-tailed test.

11. Research indicates that charitable giving is more common among older adults, although increased giving by Millennials is part of a growing trend. We examine charitable giving (measured in dollars) for two age groups: (1) 30–39 years and (2) 50–59 years of age, based on data from the GSS2014. Assume alpha = .05 for a two-tailed test. What can you conclude about the difference in giving between the two age groups?

Group Statistics

	recoded Age	N	Mean	Std. Deviation	Std. Error Mean
TOTAL DONATIONS PAST YEAR R AND IMMEDIATE FAMILY	30-39	62	1145.32	2611.793	331.698
	50-59	82	1583.24	3661.567	404.352

Independent Samples Test

		Levene's Test for Equality of Variances		t-		
		F	Sig.	t	df	Sig. (2-tailed)
TOTAL DONATIONS PAST YEAR R AND IMMEDIATE FAMILY	Equal variances assumed	1.318	.253	-.800	142	.425
	Equal variances not assumed			-.837	141.568	.404

12. Based on the 2018 GSS, we compare church attendance (ATTEND) between lower- and upper-class respondents. ATTEND is an ordinal measure: 0 = never, 1 = less than once a year, 2 = once a year, 3 = several times a year, 4 = once a month, 5 = 2–3 times a month, and 6 = nearly every week. A lower score indicates lower church attendance.

Group Statistics

	Subjective class identification	N	Mean	Std. Deviation	Std. Error Mean
How often R attends religious services	LOWER CLASS	137	2.99	2.949	.252
	UPPER CLASS	53	3.51	2.946	.405

Independent Samples Test

		Levene's Test for Equality of Variances		t-		
		F	Sig.	t	df	Sig. (2-tailed)
How often R attends religious services	Equal variances assumed	.161	.689	-1.083	188	.280
	Equal variances not assumed			-1.084	94.704	.281

a. Interpret the group means for lower- and upper-class respondents. Which group attends church more often?

b. Assume alpha = .05 for a two-tailed test. What can you conclude about the difference in church attendance between the two groups?

c. If alpha were changed to .01, would your Step 5 decision change? Explain.

13. We used the Explore command to calculate the confidence intervals for HRSRELAX for men and women. The GSS 2018 asked respondents "after an average work day, about how many hours do you have to relax or pursue the activities you enjoy?" In this exercise, we selected married GSS respondents and calculated the *t* test for HRSRELAX.

a. Is there a significant difference between married men and married women in the number of hours they have to relax during the day? Set alpha at .05.

b. If alpha was changed to .01, would your Step 5 decision change? Explain.

Group Statistics

	Respondents sex	N	Mean	Std. Deviation	Std. Error Mean
Hours per day R have to relax	MALE	136	3.38	3.103	.266
	FEMALE	142	3.35	2.735	.229

Independent Samples Test

		Levene's Test for Equality of Variances		t-		
		F	Sig.	t	df	Sig. (2-tailed)
Hours per day R have to relax	Equal variances assumed	.127	.722	.086	276	.931
	Equal variances not assumed			.086	268.356	.931

8 THE CHI-SQUARE TEST AND MEASURES OF ASSOCIATION

Chapter Learning Objectives

1. Summarize the application of a chi-square test.
2. Calculate and interpret a test for the bivariate relationship between nominal or ordinal variables.
3. Determine the significance of a chi-square test statistic.
4. Explain the concept of proportional reduction of error.
5. Apply and interpret measures of association: lambda, Cramer's V, gamma, and Kendall's tau-*b*.
6. Interpret output for chi-square and measures of association.

Figures collected by the U.S. Census Bureau indicate that educational attainment is increasing in the United States. The percentage of Americans who completed 4 years of high school or more increased from 55.2% in 1970 to 89.8% in 2018. In 1970, only about 11% of Americans completed 4 years or more of college compared with 35% in 2018.[1] Despite this overall increase, educational attainment and one's educational experience still vary by demographic factors such as race/ethnicity, class, or gender.

In the first part of this chapter, we focus on first-generation college students—that is, students whose parents never completed a postsecondary education. The proportion of first-generation students has declined within the total population of first-year, full-time-entering college freshmen, reflecting the overall increase in educational attainment in the U.S. population.[2]

Many first-generation students begin college at 2-year programs or at community colleges. According to W. Elliot Inman and Larry Mayes (1999), since first-generation college students represent a large segment of the community college population, they bring with them a set of distinct goals and constraints. Understanding their experiences and their demographic backgrounds may allow for more intentional recruiting, retention, and graduation efforts. Inman and Mayes set out to examine first-generation college students' experiences, but they began first by determining who was most likely to be a first-generation college student.

Data from Inman and Mayes's study are presented in Table 8.1, a bivariate table, which includes gender and first-generation college status. From the table, we know that a higher percentage of women than men reported being first-generation college students, 46.6% versus 35.4%.

The percentage differences between males and females in first-generation college status, shown in Table 8.1, suggest that

Table 8.1 Percentage of Men and Women Who Are First-Generation College Students

First Generation	Men	Women	Total
Firsts	35.4%	46.6%	41.9%
	(691)	(1,245)	(1,936)
Nonfirsts	64.6%	53.4%	58.1%
	(1,259)	(1,425)	(2,684)
Total (*N*)	100.0%	100.0%	100.0%
	(1,950)	(2,670)	(4,620)

Source: Adapted from W. Elliot Inman and Larry Mayes, "The Importance of Being First: Unique Characteristics of First Generation Community College Students," *Community College Review* 26, no. 3 (1999): 8. Copyright ©North Carolina State University. Published by SAGE.

there is a relationship. In inferential statistics, we base our statements about the larger population on what we observe in our sample. How do we know whether the gender differences in Table 8.1 reflect a real difference in first-generation college status among the larger population? How can we be sure that these differences are not just a quirk of sampling? If we took another sample, would these differences be wiped out or even be reversed?

Let's assume that men and women are equally likely to be first-generation college students—that in the population from which this sample was drawn, there are no real differences between them. What would be the expected percentages of men and women who are first-generation college students versus those who are not?

If gender and first-generation college status were not associated, we would expect the same percentage of men and women to be first-generation college students. Similarly, we would expect to see the same percentage of men and women who are nonfirsts. These percentages should be equal to the percentage of "firsts" and "nonfirsts" respondents in the sample as a whole (categories used by Inman and Mayes). The last column of Table 8.1—the row marginals—displays these percentages: 41.9% of all respondents were first-generation students, whereas 58.1% were nonfirsts. Therefore, if there were no association between gender and first-generation college status, we would expect to see 41.9% of the men and 41.9% of the women in the sample as first-generation students. Similarly, 58.1% of the men and 58.1% of the women would not be.

Table 8.2 shows these hypothetical expected percentages. Because the percentage distributions of the variable *first-generation college status* are identical for men and women, we can say that Table 8.2 demonstrates a perfect model of "no association" between the variable *first-generation college status* and the variable *gender.*

If there is an association between gender and first-generation college status, then at least some of the observed percentages in Table 8.1 should differ from the hypothetical expected percentages shown in Table 8.2. Conversely, if gender and

Table 8.2 Percentage of Men and Women Who Are First-Generation College Students: Hypothetical Data Showing No Association

First Generation	Men	Women	Total
Firsts	41.9%	41.9%	41.9%
			(1,936)
Nonfirsts	58.1%	58.1%	58.1%
			(2,684)
Total (*N*)	100.0%	100.0%	100.0%
	(1,950)	(2,670)	(4,620)

first-generation college status are not associated, the observed percentages should approximate the expected percentages shown in Table 8.2. In a cell-by-cell comparison of Tables 8.1 and 8.2, you can see that there is quite a disparity between the observed percentages and the hypothetical percentages. For example, in Table 8.1, 35.4% of the men reported that they were first-generation college students, whereas the corresponding cell for Table 8.2 shows that 41.9% of the men reported the same. The remaining three cells reveal similar discrepancies.

Are the disparities between the observed and expected percentages large enough to convince us that there is a genuine pattern in the population? The chi-square statistic helps us answer this question. It is obtained by comparing the actual observed frequencies in a bivariate table with the frequencies that are generated under an assumption that the two variables in the cross-tabulation are not associated with each other. If the observed and expected values are very close, the chi-square statistic will be small. If the disparities between the observed and expected values are large, the chi-square statistic will be large. In the following sections, we will learn how to compute the chi-square statistic to determine whether the differences between men's and women's first-generation college status could have occurred simply by chance.

THE CONCEPT OF CHI-SQUARE AS A STATISTICAL TEST

The chi-square test (pronounced kai-square and written as χ^2) is an inferential statistical technique designed to test for significant relationships between two variables organized in a bivariate table. The test has a variety of research applications and is one of the most widely used tests in the social sciences. Chi-square requires no assumptions about the shape of the population distribution from which a sample is drawn. It can be applied to nominal or ordinal data (including grouped interval-level data).

Chi-square test: An inferential technique designed to test for a significant relationship between two variables organized in a bivariate table.

The chi-square test can also be applied to the distribution of scores for a single variable. Also referred to as the goodness-of-fit test, the chi-square can compare the actual distribution of a variable with a set of expected frequencies. This application is not presented in this chapter.

THE CONCEPT OF STATISTICAL INDEPENDENCE

Statistical independence: The absence of association between two cross-tabulated variables. The percentage distributions of the dependent variable within each category of the independent variable are identical.

When two variables are not associated (as in Table 8.2), one can say that they are **statistically independent**. That is, an individual's score on one variable is independent of his or her score on the second variable. We identify statistical independence in a bivariate table by comparing the distribution of the dependent variable in each category of the independent variable. When two variables are statistically independent, the percentage distributions of the dependent variable within each category of the independent variable are identical. The hypothetical data presented in Table 8.2 illustrate the notion of statistical independence. Based on Table 8.2, we would say that first-generation college status is independent of one's gender.[3]

LEARNING CHECK 8.1

The data we will use to practice calculating chi-square are also from Inman and Mayes's research. We will examine the relationship between *age* (independent variable) and *first-generation college status* (the dependent variable), as shown in the following bivariate table:

Age and First-Generation College Status

First-Generation Status	Years of Age		
	19 Years or Younger	20 Years or Older	Total
Firsts	916 (33.7%)	1,018 (53.6%)	1,934 (41.9%)
Nonfirsts	1,802 (66.3%)	881 (46.4%)	2,683 (58.1%)
Total (*N*)	2,718 (100.0%)	1,899 (100.0%)	4,617 (100.0%)

Source: Adapted from W. Elliot Inman and Larry Mayes, "The Importance of Being First: Unique Characteristics of First Generation Community College Students," *Community College Review* 26, no. 3 (1999): 8.

Construct a bivariate table (in percentages) showing no association between age and first-generation college status.

THE STRUCTURE OF HYPOTHESIS TESTING WITH CHI-SQUARE

The chi-square test follows the same five basic steps as the statistical tests presented in Chapter 7: (1) making assumptions, (2) stating the research and null hypotheses and selecting alpha, (3) selecting the sampling distribution and specifying the test statistic, (4) computing the test statistic, and (5) making a decision and interpreting the results. Before we apply the five-step model to a specific example, let's discuss some of the elements that are specific to the chi-square test.

The Assumptions

The chi-square test requires no assumptions about the shape of the population distribution from which the sample was drawn. However, like all inferential techniques, it assumes random sampling. It can be applied to variables measured at a nominal and/or an ordinal level of measurement.

Stating the Research and the Null Hypotheses

The research hypothesis (H_1) proposes that the two variables are related in the population.

H_1: The two variables are related in the population. (*Gender* and *first-generation college status* are statistically dependent.)

Like all other tests of statistical significance, the chi-square is a test of the null hypothesis. The null hypothesis (H_0) states that no association exists between two cross-tabulated variables in the population, and therefore, the variables are statistically independent.

H_0: There is no association between the two variables in the population. (*Gender* and *first-generation college status* are statistically independent.)

LEARNING CHECK 8.2

Refer to the data in the previous Learning Check. Are the variables *age* and *first-generation college status* statistically independent? Write out the research and the null hypotheses for your practice data.

The Concept of Expected Frequencies

Assuming that the null hypothesis is true, we compute the cell frequencies that we would expect to find if the variables are statistically independent. These

Expected frequencies (f_e): The cell frequencies that would be expected in a bivariate table if the two variables were statistically independent.

Observed frequencies (f_o): The cell frequencies actually observed in a bivariate table.

frequencies are called **expected frequencies** (and are symbolized as f_e). The chi-square test is based on cell-by-cell comparisons between the expected frequencies (f_e) and the frequencies actually observed (**observed frequencies** are symbolized as f_o).

Calculating the Expected Frequencies

The difference between f_o and f_e will determine the likelihood that the null hypothesis is true and that the variables are, in fact, statistically independent. When there is a large difference between f_o and f_e, it is unlikely that the two variables are independent, and we will probably reject the null hypothesis. On the other hand, if there is little difference between f_o and f_e, the variables are probably independent of each other, as stated by the null hypothesis (and therefore, we will not reject the null hypothesis).

The most important element in using chi-square to test for the statistical significance of cross-tabulated data is the determination of the expected frequencies. Because chi-square is computed on actual frequencies instead of on percentages, we need to calculate the expected frequencies based on the null hypothesis.

In practice, the expected frequencies are more easily computed directly from the row and column frequencies than from the percentages. We can calculate the expected frequencies using this formula:

$$f_e = \frac{\text{(Column Marginal)(Row Marginal)}}{N} \tag{8.1}$$

To obtain the expected frequencies for any cell in any cross-tabulation in which the two variables are assumed independent, multiply the row and column totals for that cell and divide the product by the total number of cases in the table.

Let's use this formula to recalculate the expected frequencies for our data on gender and first-generation college status as displayed in Table 8.1. Consider the men who were first-generation college students (the upper left cell). The expected frequency for this cell is the product of the column total (1,950) and the row total (1,936) divided by all the cases in the table (4,620):

$$f_e = \frac{(1{,}950)(1{,}936)}{4{,}620} = 817.14$$

For men who are nonfirsts (the lower left cell), the expected frequency is

$$f_e = \frac{(1{,}950)(2{,}684)}{4{,}620} = 1{,}132.86$$

Next, let's compute the expected frequencies for women who are first-generation college students (the upper right cell):

$$f_e = \frac{(2{,}670)(1{,}936)}{4{,}620} = 1{,}118.86$$

Finally, the expected frequency for women who are nonfirsts (the lower right cell) is

$$f_e = \frac{(2{,}670)(2{,}684)}{4{,}620} = 1{,}551.14$$

These expected frequencies are displayed in Table 8.3.

Note that the table of expected frequencies contains identical row and column marginals as the original table (Table 8.1). Although the expected frequencies usually differ from the observed frequencies (depending on the degree of relationship between the variables), the row and column marginals must always be identical with the marginals in the original table.

LEARNING CHECK 8.3

Refer to the data in Learning Check 8.1. Calculate the expected frequencies for age and first-generation college status and construct a bivariate table. Are your column and row marginals the same as in the original table?

Table 8.3 Expected Frequencies of Men and Women and First-Generation College Status

First Generation	Men	Women	Total
Firsts	817.14	1,118.86	1,936
Nonfirsts	1,132.86	1,551.14	2,684
Total (N)	1,950	2,670	4,620

Calculating the Obtained Chi-Square

The next step in calculating chi-square is to compare the differences between the expected and observed frequencies across all cells in the table. In Table 8.4, the expected frequencies are shown next to the corresponding observed frequencies. Note that the difference between the observed and expected frequencies in each cell is quite large. Is it large enough to be significant? The way we decide is by calculating the **obtained chi-square** statistic:

Chi-square (obtained): The test statistic that summarizes the differences between the observed (f_o) and the expected (f_e) frequencies in a bivariate table.

$$\chi^2 = \sum \frac{(f_o - f_e)^2}{f_e} \tag{8.2}$$

where

f_o = observed frequencies

f_e = observed frequencies

According to this formula, for each cell, subtract the expected frequency from the observed frequency, square the difference, and divide by the expected frequency. After performing this operation for every cell, sum the results to obtain the chi-square statistic.

Let's follow these procedures using the observed and expected frequencies from Table 8.4. Our calculations are displayed in Table 8.5. The obtained chi-square statistic, 58.00, summarizes the differences between the observed frequencies and the frequencies that we would expect to see if the null hypothesis were true and the variables—*gender* and *first-generation college status*—were not associated. Next, we need to interpret our obtained chi-square statistic and decide whether it is large enough to allow us to reject the null hypothesis.

Table 8.4 Observed and Expected Frequencies of Men and Women Who Are First-Generation College Students

First Generation	Men		Women		Total
	f_o	f_e	f_o	f_e	
Firsts	691	817.14	1,245	1,118.86	1,936
Nonfirsts	1,259	1,132.86	1,425	1,551.14	2,684
Total (N)	1,950		2,670		4,620

Table 8.5 Calculating Chi-Square					
Gender and First-Generation College Status	f_o	f_e	$f_o - f_e$	$(f_o - f_e)^2$	$\frac{(f_o - f_e)^2}{f_e}$
Men/firsts	691	817.14	−126.14	15,911.2996	19.47
Men/nonfirsts	1,259	1,132.86	126.14	15,911.2996	14.05
Women/firsts	1,245	1,118.86	126.14	15,911.2996	14.22
Women/nonfirsts	1,425	1,551.14	−126.14	15,911.2996	10.26

$$\chi^2 = \Sigma \frac{(f_o - f_e)^2}{f_e} = 58.00$$

LEARNING CHECK 8.4

Using the format of Table 8.5, construct a table to calculate chi-square for age and educational attainment.

The Sampling Distribution of Chi-Square

In Chapter 7, we learned that test statistics such as Z and t have characteristic sampling distributions that tell us the probability of obtaining a statistic, assuming that the null hypothesis is true. In the same way, the sampling distribution of chi-square tells the probability of getting values of chi-square, assuming no relationship exists in the population.

Like other sampling distributions, the chi-square sampling distributions depend on the degrees of freedom. In fact, the chi-square sampling distribution is not one distribution but—like the t distribution—is a family of distributions. The shape of a particular chi-square distribution depends on the number of degrees of freedom. This is illustrated in Figure 8.1, which shows chi-square distributions for 1, 5, and 9 degrees of freedom. Here are some of the main properties of the chi-square distributions that can be observed in this figure:

- The distributions are positively skewed.
- Chi-square values are always positive. The minimum possible value is zero, with no upper limit to its maximum value. A chi-square of zero means that the variables are completely independent and the observed

Figure 8.1 Chi-Square Distributions for 1, 5, and 9 Degrees of Freedom

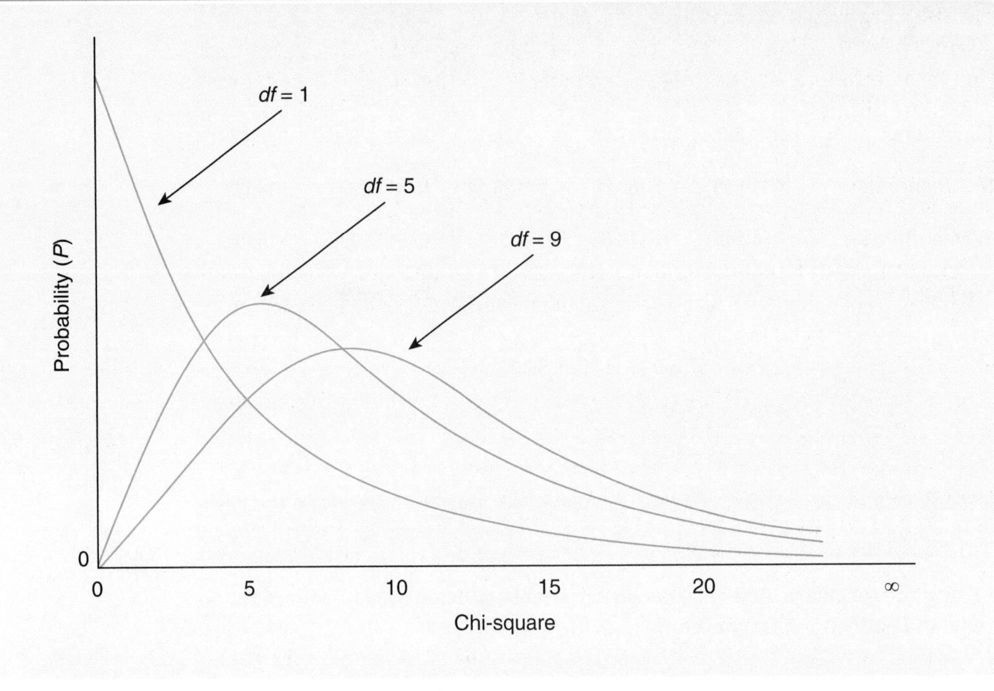

frequencies in every cell are equal to the corresponding expected frequencies.

- As the number of degrees of freedom increases, the chi-square distribution becomes more symmetrical and, with degrees of freedom greater than 30, begins to resemble the normal curve.

Determining the Degrees of Freedom

In Chapter 7, we defined degrees of freedom (*df*) as the number of values that are free to vary. With cross-tabulation data, we find the degrees of freedom by using Formula 8.3:

$$df = (r - 1)(c - 1) \quad (8.3)$$

where

r = the number of rows

c = the number of columns

Thus, Table 8.1 with two rows and two columns has (2 – 1)(2 – 1) or 1 degree of freedom. If the table had three rows and two columns, it would have (3 – 1)(2 – 1) or 2 degrees of freedom.

Appendix D shows values of the chi-square distribution for various degrees of freedom. Notice how the table is arranged with the degrees of freedom listed down the first column and the level of significance (or *p* values) arrayed across the top. For example, with 5 degrees of freedom, the probability associated with a chi-square as large as 15.086 is .01. An obtained chi-square as large as 15.086 would occur only once in 100 samples.

The degrees of freedom in a bivariate table can be interpreted as the number of cells in the table for which the expected frequencies are free to vary, given that the marginal totals are already set. Based on our data in Table 8.3, suppose we first calculate the expected frequencies for men who are first-generation college students ($f_e = 817.14$). Because the sum of the expected frequencies in the first column is set at 1,950, the expected frequency of men who are nonfirsts has to be 1,132.86 (1,950 – 817.14). Similarly, all other cells are predetermined by the marginal totals and are not free to vary. Therefore, this table has only 1 degree of freedom.

Data in a bivariate table can be distorted if by chance one cell is over- or undersampled and may therefore influence the chi-square calculation. Calculation of the degrees of freedom compensates for this, but in the case of a 2 × 2 table with just 1 degree of freedom, the value of chi-square should be adjusted by applying the Yates's correction for continuity. Formula 8.4 reduces the absolute value of each ($f_o - f_e$) by .5, and then the difference is squared and then divided by the expected frequency for each cell. The formula for the Yates's correction for continuity is as follows:

$$\chi_c^2 = \Sigma \frac{(|f_o - f_e| - 0.5)^2}{f_e} \qquad (8.4)$$

LEARNING CHECK 8.5

Based on Appendix D, identify the probability for each chi-square value (*df* in parentheses):

- 12.307 (15)
- 20.337 (21)
- 54.052 (24)

Making a Final Decision

With the Yates's correction, the corrected chi-square is 57.54. Refer to Table 8.6 for calculations.

We can see that 57.54 does not appear on the first row ($df = 1$); in fact, it exceeds the largest chi-square value of 10.827 ($p = .001$). We can establish that the probability of obtaining a chi-square of 57.54 is less than .001 if the null hypothesis were true. If our alpha was preset at .05, the probability of 10.827 would be well below this. Therefore, we can reject the null hypothesis that gender and first-generation college status are not associated in the population from which our sample was drawn. Remember, the larger the chi-square statistic, the smaller the p value providing us with more evidence to reject the null hypothesis. We can be very confident of our conclusion that there is a relationship between gender and first-generation college status in the population because the probability of this result occurring owing to sampling error is less than .001, a very rare occurrence.

Review

To summarize our discussion, let's apply the five-step process of hypothesis testing.

1. *Making Assumptions:*
   - A random sample of $N = 4,620$ was selected.
   - The level of measurement of the variable gender is nominal.

Table 8.6 Calculating Yates's Correction

Gender and First-Generation College Status	$\lvert f_o - f_e \rvert$	$(\lvert f_o - f_e \rvert - 0.50)^2$	f_e	$\chi_c^2 = \sum \frac{(\lvert f_o - f_e \rvert - 0.5)^2}{f_e}$
Men firsts	126.14	$(125.64)^2 = 15,785.41$	817.14	19.32
Men nonfirsts	126.14	$(125.64)^2 = 15,785.41$	1,132.86	13.93
Women firsts	126.14	$(125.64)^2 = 15,785.41$	1,118.86	14.11
Women nonfirsts	126.14	$(125.64)^2 = 15,785.41$	1,551.14	10.18
Total				57.54

- The level of measurement of the variable *first-generation college status* is nominal.

2. *Stating the Research and Null Hypotheses and Selecting Alpha:* The research hypothesis, H_1, is that there is a relationship between gender and first-generation college status (i.e., gender and first-generation college status are statistically dependent). The null hypothesis, H_0, is that there is no relationship between gender and first-generation college status in the population (i.e., gender and first-generation college status are statistically independent). Alpha is set at .05.

3. *Selecting the Sampling Distribution and Specifying the Test Statistic:* Both the sampling distribution and the test statistic are chi-square.

4. *Computing the Test Statistic:* We should first determine the degrees of freedom associated with our test statistic:

$$df = (r - 1)(c - 1) = (2 - 1)(2 - 1) = (1)(1) = 1$$

Next, to calculate chi-square, we calculate the expected frequencies under the assumption of statistical independence. To obtain the expected frequencies for each cell, we multiply its row and column marginal totals and divide the product by N. The expected frequencies are displayed in Table 8.3.

Are these expected frequencies different enough from the observed frequencies presented in Table 8.1 to justify rejection of the null hypothesis? To find out, we calculate the chi-square statistic of 57.54 (with the Yates's correction). The calculations are shown in Table 8.6.

5. *Making a Decision and Interpreting the Results:* To determine the probability of obtaining our chi-square of 57.54, we refer to Appendix D. With 1 degree of freedom, the probability of obtaining 57.54 is less than .001 (less than our alpha of .05). We reject the null hypothesis that there is no difference in first-generation college status among men and women. Thus, we can conclude that in the population from which our sample was drawn, first-generation college status does vary by gender. Based on our sample data, we know that women are more likely to report being first-generation college students than men.

LEARNING CHECK 8.6

What decision can you make about the association between age and first-generation college status? Should you reject the null hypothesis at the .05 alpha level or at the .01 level?

STATISTICS IN PRACTICE: RESPONDENT AND MOTHER EDUCATION

Each year, the General Social Survey (GSS) collects data on individual degree attainment along with the highest degrees attained by both parents. We know that parental education is an important predictor of a child's educational attainment. For this Statistics in Practice, we will use GSS 2018 data and take a look at mother and individual degree attainment as shown in Table 8.7. Based on our sample of 1,390 men and women, the bivariate table shows a pattern of positive association between mother's education (the independent variable) and respondent's education (the dependent variable)—as mother's education increases, so does the respondent's education. Among those with a mother who earned less than a high school degree, only 24.7% reported some college education. In contrast, among those with a mother with some college education or more, 63.7% reported having some college education.

It is not clear whether these differences are owing to chance or to sampling fluctuations, or whether they reflect a real pattern of association in the population. In the following discussion, we will not review our calculations (although they are presented in Table 8.8). Rather, our focus will be on the five-step model and drawing conclusions about the relationship between mother's and respondent's educational degrees.

1. *Making Assumptions:*
   - A random sample of N = 1,390 is selected.

Table 8.7 Mother's Degree by Respondent's Degree, GSS 2018

Respondent's Degree	Mother's Degree: Less Than High School	Mother's Degree: High School Degree	Mother's Degree: Some College or More	Total
Less than high school	103	37	10	150
	(27.3%)	(5.6%)	(2.9%)	(10.8%)
High school degree	181	384	117	682
	(48.0%)	(57.9%)	(33.4%)	(49.1%)
Some college or more	93	242	223	558
	(24.7%)	(36.5%)	(63.7%)	(40.1%)
Total	377	663	350	1390
	(100%)	(100%)	(100%)	(100%)

- The level of measurement of the variable respondent's degree is ordinal.
- The level of measurement of the variable mother's degree is ordinal.

2. *Stating the Research and Null Hypotheses and Selecting Alpha:*

   H_1: There is a relationship between mother's educational attainment and respondent's educational attainment in the population. (Mother's degree and respondent's degree are statistically dependent.)

   H_0: There is no relationship between mother's educational attainment and respondent's educational attainment in the population. (Mother's degree and respondent's degree are statistically independent.)

For this test, we'll select an alpha of .01.

3. *Selecting the Sampling Distribution and Specifying the Test Statistic:* The sampling distribution is chi-square; the test statistic is also chi-square.
4. *Computing the Test Statistic:* The degrees of freedom for Table 8.7 is

$$df = (r - 1)(c - 1) = (3 - 1)(3 - 1) = (2)(2) = 4$$

Table 8.8 Calculating Chi-Square for Respondent's and Mother's Educational Degree

Respondent's Degree and Mother's Degree	f_o	f_e	$f_o - f_e$	$(f_o - f_e)^2$	$\frac{f_o - f_e}{f_e}$
Less than high school/less than high school	103	40.7	62.3	3,881.29	95.36
Less than high school/high school	37	71.5	−34.5	1,190.25	16.65
Less than high school/some college or more	10	37.8	−27.8	772.84	20.45
High school/less than high school	181	185	−4	16	.09
High school/high school	384	325.3	58.7	3,445.69	10.59
High school/some college or more	117	171.7	−54.7	2,992.09	17.43
Some college or more/less than high school	93	151.3	−58.3	3,398.89	22.46
Some college or more/high school	242	266.2	−24.2	585.64	2.2
Some college or more/some college or more	223	140.5	82.5	6,806.25	48.44

$$\chi^2 = \Sigma \frac{(f_o - f_e)^2}{f_e} = 233.67$$

The chi-square obtained is 233.67. The detailed calculations are shown in Table 8.8.

5. *Making a Decision and Interpreting the Results:* To determine if the observed frequencies are significantly different from the expected frequencies, we compare our calculated chi-square with Appendix D. With 4 degrees of freedom, our chi-square of 233.67 exceeds the largest listed chi-square value of 18.465 ($p = .001$). We determine that the probability of observing our obtained chi-square of 233.67 is less than .001 and less than our alpha of .01. We can reject the null hypothesis that there is no relationship between mother's educational attainment and respondent's educational attainment. Thus, we conclude that in the population from which our sample was drawn, respondent's degree is related to mother's degree. The positive relationship between the two variables is significant.

A CLOSER LOOK 8.1

A Cautionary Note: Sample Size and Statistical Significance for Chi-Square

Although we found the relationship between gender and first-generation college status to be statistically significant, this in itself does not give us much information about the strength of the relationship or its substantive significance in the population. Statistical significance only helps us evaluate whether the argument (the null hypothesis) that the observed relationship occurred by chance is reasonable. It does not tell us anything about the relationship's theoretical importance or even if it is worth further investigation.

The distinction between statistical and substantive significance is important in applying any of the statistical tests discussed in Chapter 7. However, this distinction is of particular relevance for the chi-square test because of its sensitivity to sample size. The size of the calculated chi-square is directly proportional to the size of the sample, independent of the strength of the relationship between the variables.

For instance, suppose that we cut the observed frequencies for every cell in Table 8.1 exactly into half—which is equivalent to reducing the sample size by one half. This change will not affect the percentage distribution of firsts among men and women; therefore, the size of the percentage difference and the strength of the association between gender and first-generation college status will remain the same. However, reducing the observed frequencies by half will cut down our calculated chi-square by exactly half, from 57.54 to 28.77. (Can you verify this calculation?) Conversely, had we doubled the frequencies in each cell, the size of the calculated chi-square would have doubled, thereby making it easier to reject the null hypothesis.

This sensitivity of the chi-square test to the size of the sample means that a relatively strong association between the variables may not be significant when the sample size is small. Similarly, even when the association between variables is very weak, a large sample may result in a statistically significant relationship. However, just because the calculated chi-square is large and we are able to reject the null hypothesis by a large margin does not imply that the relationship between the variables is strong and substantively important.

Another limitation of the chi-square test is that it is sensitive to small expected frequencies in one or more of the cells in the table. Generally, when the expected frequency in one or more of the cells is below 5, the chi-square statistic may be unstable and lead to erroneous conclusions. There is no hard-and-fast rule regarding the size of the expected frequencies. Most researchers limit the use of chi-square to tables that either have no f_e values below 5 or have no more than 20% of the f_e values below 5.

Testing the statistical significance of a bivariate relationship is only a small step, although an important one, in examining a relationship between two variables. A significant chi-square suggests that a relationship, weak or strong, probably exists in the population and is not due to sampling fluctuation. However, to establish the strength of the association, we need to employ measures of association such as gamma, lambda (both discussed later in this chapter), or Pearson's *r* (refer to Chapter 10). Used in conjunction, statistical tests of significance and measures of association can help determine the importance of the relationship and whether it is worth additional investigation.

LEARNING CHECK 8.7

For the bivariate table with age and first-generation college status (first presented in Learning Check 8.1), the value of the obtained chi-square is 181.15 with 1 degree of freedom. Based on Appendix D, we determine that its probability is less than .001. This probability is less than our alpha level of .05. We reject the null hypothesis of no relationship between age and first-generation college status. If we reduce our sample size by half, the obtained chi-square is 90.58. Determine the *p* value for 90.58. What decision can you make about the null hypothesis?

PROPORTIONAL REDUCTION OF ERROR

In this section, we review special **measures of association** for nominal and ordinal variables. These measures enable us to use a single summarizing measure or number for analyzing the pattern of relationship between two variables. Unlike chi-square, measures of association reflect the strength of the relationship and, at times, its direction (whether it is positive or negative). They also indicate the usefulness of predicting the dependent variable from the independent variable.

Measure of association: A single summarizing number that reflects the strength of a relationship, indicates the usefulness of predicting the dependent variable from the independent variable, and often shows the direction of the relationship.

We discuss four measures of association: (1) lambda (measures of association for nominal variables), (2) gamma and (3) Kendall's tau-*b* (measures of association between ordinal variables), and (4) Cramer's *V* (a chi-square–related measure of association). In Chapter 10, we introduce Pearson's correlation coefficient, which is used for measuring bivariate association between interval-ratio variables.

Proportional reduction of error (PRE): A measure that tells us how much we can improve predicting the value of a dependent variable based on information about an independent variable.

All the measures of association discussed here and in Chapter 10 are based on the concept of the **proportional reduction of error**, often abbreviated as **PRE**. According to the concept of PRE, two variables are associated when information about one variable (an independent variable) can help us improve our prediction of the other variable (a dependent variable).

Table 8.9 may help us grasp intuitively the general concept of PRE. Using General Social Survey (GSS) 2018 data, Table 8.9 shows a moderate relationship between the independent variable, *educational attainment*, and the dependent variable, *support for abortion if the woman is poor and can't afford any more children*. The table shows that 56.6% of the respondents who did not receive a bachelor's degree were antiabortion, compared with only 41.3% of the respondents who had a bachelor's degree or more.

The conceptual formula for all[4] PRE measures of association is

$$PRE = \frac{E_1 - E_2}{E_1} \tag{8.5}$$

A CLOSER LOOK 8.2

What Is Strong? What Is Weak? A Guide to Interpretation

The more you work with various measures of association, the better feel you will have for what particular values mean. Until you develop this skill, here are some guidelines regarding what is generally considered a strong relationship and what is considered a weak relationship. Keep in mind that these are only rough guidelines. Often, the interpretation for a measure of association will depend on the research context. A +0.30 in one research field will mean something a little different from a +0.30 in another research field. Zero, however, always means the same thing: no relationship.

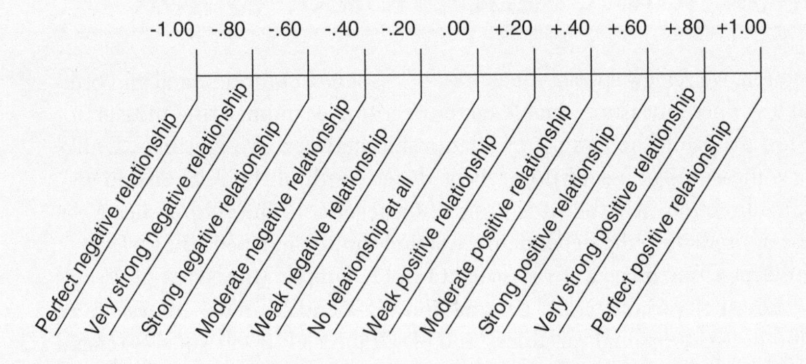

Table 8.9 Support for Abortion by Degree, GSS 2018

Support for Abortion	Degree: Less Than Bachelor's	Degree: Bachelor's or More	Total
No	371	129	500
	56.6%	41.3%	51.7%
Yes	284	183	467
	43.4%	58.7%	48.3%
Total	655	312	967
	100.0%	100.0%	100.0%

where

*E*1 = errors of prediction made when the independent variable is ignored (Prediction 1)

*E*2 = errors of prediction made when the prediction is based on the independent variable (Prediction 2)

All PRE measures are based on comparing predictive error levels that result from each of the two methods of prediction. Let's say that we want to predict a respondent's position on abortion, but we do not know anything about the degree one has. Based on the row totals in Table 8.9, we could predict that every respondent in the sample is antiabortion because this is the modal category of the variable *abortion position*. With this prediction, we would make 467 errors because in fact 500 respondents in this group are antiabortion but 467 respondents are prochoice. Thus,

$$E_1 = 967 - 500 = 467$$

How can we improve this prediction by using the information we have on each respondent's educational attainment? For our new prediction, we will use the following rule: If a respondent has less than a bachelor's degree, we predict that they will be antiabortion; if a respondent has a bachelor's degree or more, we predict that they are prochoice. It makes sense to use this rule because we know, based on Table 8.9, that respondents with a lower educational attainment are more likely to be antiabortion, while respondents who have a bachelor's degree or more are more likely to be prochoice. Using this prediction rule, we will make

413 errors (instead of 467) because 129 of the respondents who have a bachelor's degree or more are actually antiabortion, whereas 284 of the respondents who have less than a bachelor's degree are prochoice (129 + 284 = 413). Thus,

$$E_2 = 129 + 284 = 413$$

Our first prediction method, ignoring the independent variable (*educational attainment*), resulted in 467 errors. Our second prediction method, using information we have about the independent variable (*educational attainment*), resulted in 413 errors. If the variables are associated, the second method will result in fewer errors of prediction than the first method. The stronger the relationship is between the variables, the larger will be the reduction in the number of errors of prediction.

Let's calculate the PRE for Table 8.9 using Formula 8.5. The PRE resulting from using educational attainment to predict position on abortion is

$$\text{PRE} = \frac{467 - 413}{467} = 0.12$$

PRE measures of association can range from 0.0 to ±1.0. A PRE of zero indicates that the two variables are not associated; information about the independent variable will not improve predictions about the dependent variable. A PRE of ±1.0 indicates a perfect positive or negative association between the variables; we can predict the dependent variable without error using information about the independent variable. Intermediate values of PRE will reflect the strength of the association between the two variables and therefore the utility of using one to predict the other. The more the measure of association departs from 0.00 in either direction, the stronger the association. PRE measures of association can be multiplied by 100 to indicate the percentage improvement in prediction.

A PRE of 0.12 indicates that there is a weak relationship between respondents' educational attainment and their position on abortion. (Refer to A Closer Look 8.2 for a discussion of the strength of a relationship.) A PRE of 0.12 means that we have improved our prediction of respondents' position on abortion by just 12% (0.12 × 100 = 12%) by using information on their educational attainment.

Lambda: An asymmetrical measure of association, lambda is suitable for use with nominal variables and may range from 0.0 to 1.0. It provides us with an indication of the strength of an association between the independent and dependent variables.

Asymmetrical measure of association: A measure of association whose value may vary depending on which variable is considered the independent variable and which the dependent variable.

LAMBDA: A MEASURE OF ASSOCIATION FOR NOMINAL VARIABLES

Lambda is an asymmetrical measure used to determine the strength of the relationship between two nominal variables. An **asymmetrical measure** will vary depending on which variable is considered the independent variable and which the dependent variable.

In U.S. colleges, there is a difference in the rate of first-generation college students by race and ethnicity. Specifically, Latinx people have a higher percentage of first-generation college students (47.8%) than any other racial/ethnic group.[5]

In Table 8.10, we present fictional data on the relationship between Latinx ethnic identity and first-generation college student status.

Examine the row totals, which show the subtotals by first-generation student status, firsts versus nonfirsts. If we had to predict one first-generation category, our best bet would be nonfirsts (132), the largest category or mode. This prediction will result in the smallest possible error. The number of wrong predictions or errors we would make is 109 (241 – 132).

Now take another look at Table 8.10, but this time let's consider student Latinx ethnic identity when we predict first-generation college student status. We can use the mode, but this time we will identify it separately for Latinx and non-Latinx respondents. The mode for Latinx is "first generation"; therefore, we can predict that all Latinx students would be non-first-generation college students. With this method of prediction, we make 32 errors, since 55 out of 87 Latinx students were first-generation college students. Next let's examine the group of non-Latinx students. The mode for non-Latinx students is "non–first generation"; this will be our prediction, our best guess. This method of prediction results in 54 errors (154 – 100 = 54). The total number of errors is 86 (32 + 54).

Let's put it all together and calculate lambda.

1. Find E_1, the errors of prediction made when the independent variable is ignored. To find E_1, find the mode of the dependent variable and subtract its frequency from N. For Table 8.10,

$$E_1 = N - \text{Modal frequency}$$
$$E_1 = 241 - 132 = 109$$

2. Find E_2, the errors made when the prediction is based on the independent variable. To find E_2, find the modal frequency for each category of the independent variable, subtract it from the category total to find the number of errors, and then add up all the errors. For Table 8.10,

Table 8.10 Latinx Ethnic Identity and First-Generation College Status

	Latinx	Non-Latinx	Row Total
First generation	55	54	109
Non–first generation	32	100	132
Column total	87	154	241

$$E_2 = N_k - \text{Mode}_k$$

$$\text{Latinx} = 87 - 55 = 32$$

$$\text{Non-Latinx} = 154 - 100 = 54$$

$$E_2 = 32 + 54 = 86$$

3. Calculate lambda (denoted by the Greek symbol λ) using Formula 8.5:

$$\lambda = \frac{E_1 - E_2}{E_1} = \frac{109 - 86}{109} = 0.21$$

Lambda may range in value from 0.0 to 1.0. Zero indicates that there is nothing to be gained by using the independent variable to predict the dependent variable. A lambda of 1.0 indicates that by using the independent variable as a predictor, we are able to predict the dependent variable without any error. In our case, a lambda of 0.21 indicates that for this sample of respondents, there is a weak association between Latinx identity and first-generation college status.

The PRE when multiplied by 100 can be interpreted as follows: By using information on respondent's Latinx ethnic identity to predict first-generation college status, we reduced our error of prediction by 21%. In other words, if we rely on students' Latinx identity to predict their first-generation status, we would reduce the error by 21%.

A reminder that lambda is an asymmetrical measure. If we considered *Latinx identity* as the dependent variable and *first-generation status* as the independent variable, we would have obtained a slightly different lambda value.

Lambda is always zero in situations in which the mode for each category of the independent variable falls into the same category of the dependent variable. A problem with interpreting lambda arises in situations in which lambda is zero, but other measures of association indicate that the variables are associated. To avoid this potential problem, examine the percentage differences in the table whenever lambda is exactly equal to zero. If the percentage differences are very small (usually 5% or less), lambda is an appropriate measure of association for the table. However, if the percentage differences are larger, indicating that the two variables may be associated, lambda will be a poor choice as a measure of association. In such cases, we may want to discuss the association in terms of the percentage differences or select an alternative measure of association.

We've presented the lambda calculation based on the dependent variable as the row variable. The method of calculation is the same when the variables are switched (the independent variable presented in columns and the dependent variable presented in rows). Although lambda can be calculated either way, ultimately what guides the decision of which variables to consider as independent or dependent is the researcher's theory.

CRAMER'S *V*: A CHI-SQUARE-RELATED MEASURE OF ASSOCIATION FOR NOMINAL VARIABLES

Cramer's V is an alternative measure of association that can be used for nominal variables. It is based on the value of chi-square and ranges between 0 to 1, with 0 indicating no association and 1 indicating perfect association. Because it cannot take negative values, it is considered a nondirectional measure. Unfortunately, Cramer's V is somewhat limited because the results cannot be interpreted using the PRE framework. It is calculated using Formula 8.6:

Cramer's V: A chi-square-related measure of association for nominal variables. Cramer's V is based on the value of chi-square and ranges between 0 and 1.

$$V = \sqrt{\frac{\chi^2}{N(m)}} \tag{8.6}$$

where m = smaller of $(r - 1)$ or $(c - 1)$

Earlier, we tested the hypothesis that first-generation college enrollment was related to student gender. The analysis yielded a chi-square value of 57.54, leading us to reject the null hypothesis that there are no differences between men and women. We concluded that in the population from which our sample was drawn, first-generation college student status does vary by student sex.

We can use Cramer's V to measure the relative strength of the association between student gender and first-generation college student status using Formula 8.6:

$$V = \sqrt{\frac{\chi^2}{N(m)}} = \sqrt{\frac{57.54}{4{,}620(1)}} = .11$$

A Cramer's V of 0.11 tells us that there is a weak association between these two variables.

GAMMA AND KENDALL'S TAU-*b*: SYMMETRICAL MEASURES OF ASSOCIATION FOR ORDINAL VARIABLES

In this section, we discuss a way to measure and interpret an association between two ordinal variables. When both the independent and dependent variables in a bivariate table are measured at the ordinal level, we can talk about the direction of the relationship, the last property of a bivariate table we'll consider.

Positive relationship: A bivariate relationship between two variables measured at the ordinal level or higher in which the variables vary in the same direction.

Negative relationship: A bivariate relationship between two variables measured at the ordinal level or higher in which the variables vary in opposite directions.

Gamma: A symmetrical measure of association suitable for use with ordinal variables or with dichotomous nominal variables. It can vary from 0.0 to ±1.0 and provides us with an indication of the strength and direction of the association between the variables. Gamma is also referred to as Goodman and Kruskal's gamma.

Kendall's tau-*b*: A symmetrical measure of association suitable for use with ordinal variables. Unlike gamma, it accounts for pairs tied on the independent and dependent variable. It can vary from 0.0 to ±1.0. It provides an indication of the strength and direction of the association between the variables.

A **positive** bivariate relationship is said to exist when the variables vary in the same direction. Higher values of one variable go together with higher values of the other variable. In a **negative** bivariate relationship, the variables vary in opposite directions: Higher values of one variable go together with lower values of the other variable (and lower values of one go together with the higher values of the other).

Gamma and **Kendall's tau-*b*** are **symmetrical measures of association** suitable for use with ordinal variables or with dichotomous nominal variables. This means that their value will be the same regardless of which variable is the independent variable or the dependent variable. Both gamma and Kendall's tau-*b* can vary from 0.0 to ±1.0 and provide us with an indication of the strength and direction of the association between the variables. Gamma and Kendall's tau-*b* can be positive or negative. A gamma or Kendall's tau-*b* of 1.0 indicates that the relationship between the variables is positive and that the dependent variable can be predicted without any errors based on the independent variable. A gamma of −1.0 indicates a perfect, negative association between the variables. A gamma or a Kendall's tau-*b* of zero reflects no association between the two variables; hence, there is nothing to be gained by using the independent variable to predict the dependent variable.

Ordinal variables are very common in social science research. The GSS contains many questions that ask people to indicate their responses on an ordinal scale—for example, strongly agree, agree, neutral, disagree, and strongly disagree.

Let's consider the association between two ordinal variables. We hypothesize that as one's educational attainment increases, the lack of confidence in the executive branch of the federal government will also increase. Our analysis is based on two variables: (1) education (EDUC) with three degree categories and (2) confidence in the executive branch of the federal government (CONFED) measured in three categories—(1) hardly any, (2) only some, and (3) a great deal.

Table 8.11 displays the cross-tabulation of these two variables, with *education* as the independent variable and *confidence in the executive branch of the federal government* as the dependent variable. We find that 36.6% of those with less than a high school degree report having hardly any confidence in the executive branch of the federal government, as compared with 52.5% of those with at least some college. The percentage difference (36.6% − 52.5% = −15.9%) suggests that the variables are related, albeit weakly. We can examine the percentages for those who report a great deal of confidence by their educational attainment (12.5% − 11.1% = 1.4%) and reach the same conclusion.

Since both variables are ordinal measures, we select gamma to assess the strength and the direction of the relationship between education and political discussions. We will not demonstrate how to calculate gamma; rather, we will rely on SPSS output to interpret gamma.

The SPSS output reports that the gamma for the table is 0.168, indicating a weak positive relationship between education and confidence in the

Table 8.11 Confidence in the Executive Branch of the Federal Government by Education, GSS 2018

	Less Than High School	High School	Some College or More	Total
Hardly any	41	197	199	437
	36.6%	41.6%	52.5%	45.3%
Only some	57	211	138	406
	50.9%	44.5%	36.4%	42.1%
A great deal	14	66	42	122
	12.5%	13.9%	11.1%	12.6%
Total	112	474	379	965
	100%	100%	100%	101%

Figure 8.2

Symmetric Measures

		Value	Asymptotic Standard Error[a]	Approximate T[b]	Approximate Significance
Ordinal by Ordinal	Gamma	.168	.048	3.454	.001
N of Valid Cases		965			

a. Not assuming the null hypothesis.

b. Using the asymptotic standard error assuming the null hypothesis.

executive branch of the federal government. The positive sign of gamma indicates that as education increases, so does the lack of confidence in the executive branch of the federal government. By using education to predict frequency of political discussion, we've reduced our prediction error by 16.8% (Figure 8.2).

Measures of association for ordinal data are not influenced by the modal category as is lambda. Consequently, an ordinal measure of association might be preferable for tables when an association cannot be detected by lambda. We can use an ordinal measure for some tables where one or both variables would appear to be measured on a nominal scale. Dichotomous variables (those with

Symmetrical measure of association: A measure of association whose value will be the same when either variable is considered the independent variable or the dependent variable.

Elaboration: A process designed to further explore a bivariate relationship; it involves the introduction of control variables.

Control variable: An additional variable considered in a bivariate relationship. The variable is controlled for when we take into account its effect on the variables in the bivariate relationship.

only two categories) can be treated as ordinal variables for most purposes. In this chapter, we calculated lambda to examine the association between abortion attitudes and educational attainment (Table 8.9). Although both variables might be considered as nominal variables—because both are dichotomized (yes/no; less than a bachelor's degree/more than a bachelor's degree)—they could also be treated as ordinal variables. Thus, the association might also be examined using gamma, an ordinal measure of association.

Elaboration is a process designed to further explore a bivariate relationship, involving the introduction of additional variables, called **control variables**. By adding a control variable to our analysis, we are considering or "controlling" for the variable's effect on the bivariate relationship. Each potential control variable represents an alternative explanation for the bivariate relationship under consideration.

READING THE RESEARCH LITERATURE: INDIA'S INTERNET-USING POPULATION

Most social science research considers multivariate causal relationships, much more complex than the bivariate relationships we considered in this chapter. Thus, gamma or Kendall's tau-*b* is not often presented in scholarly research. These statistics are most appropriate for descriptive, rather than inferential, analyses.

For example, Bharti Varshney, Prashant Kumar, Vivek Sapre, and Sanjeev Vanrshey (2014) present a demographic summary of Internet usage habits among the online population of India. India has more than 74 million Internet users out of a total population of 1.2 billion. The researchers test the hypothesis that usage varies by age, gender, occupation, and city/town tier.[6]

Age was measured as an ordinal variable. *Tier* is also an ordinal measure, identifying city or town of residence by population and technology infrastructure (Internet connection at home). Categories for both independent variables and for all dependent variables were not reported in the article.[7]

The researchers summarize Table 8.12 by first describing why they used cross-tabulations, then explaining the relationship between *age*, *tier*, and the dependent variables.

> In [the] present study, cross tabulations were performed to find up to what extent usage and activity variables are related to the demographic variables. The cross tabulation table is the basic technique for examining the relationship between two categorical (nominal or ordinal) variables. The purpose of a cross tabulation is to show the relationship (or lack thereof) between two variables . . . with age group and tier being the ordinal measure, cross tab

was carried out with [Gamma and Kendall's tau-*b*]. For statistically significant and strong relationship, significance value of each measure should be less than or equal to .05 and the value of each measure should be greater than 0.150.

Based on the above outputs, for cross tab on age group to Internet access days in a week and for age group to weekly actual time spent on the Net, statistically significant and strong relationship is obtained.

For cross tab with independent variable city/town tier, for all of the variables, that is, city/town tier to Internet access days in a week, city/town tier to average session length and city/town tier to weekly actual time spent on the Net, statistically significant and strong relationship is obtained. The possible reason being the usage rate goes down with the decreasing size of the city. Most likely because the infrastructure and surfing speed get poorer while the surfing charges increase with the decreasing size of the city. (pp. 431–432)[8]

Table 8.12 Results for Age And City/Town Tier With Internet Use and Duration

Independent Variable	Dependent Variable	Kendall's tau-*b*	Sig.	Gamma	Sig.
Age	Internet access in a week	0.257	.000	0.356	.000
	Average session length	0.174	.002	0.225	.002
	Weekly time spent on net	0.166	.004	0.230	.004
Tier	Internet access in a week	0.187	.002	0.288	.002
	Average session length	0.115	.008	0.226	.008
	Weekly time spent on net	0.153	.009	0.234	.009

Source: Adapted from Bharti Varshney, Prashant Kumar, Vivek Sapre, and Sanjeev Varshney, "Demographic Profile of the Internet-Using Population of India," *Management and Labour Studies* 39, no. 4 (2014): 431.

DATA AT WORK

Patricio Cumsille: Professor

Photo courtesy of Patricio Cumsille

Dr. Cumsille became a clinical psychologist because he wanted to understand and help people. Along with his PhD in human development and family studies, he has a minor in statistics. When asked about using statistics in his work, he explains, "As a developmental researcher it is essential to collect longitudinal data and use appropriate techniques to uncover how developmental processes unfold over time. So, statistics is an essential part of what I do, both in my teaching and research activities." Based in Chile, he applies advanced statistical techniques and longitudinal methods in his research.

Cumsille offers four recommendations to undergraduates interested in pursuing a career in quantitative research or statistics. "This is a fascinating field, with new research methods developing and expanding. The first thing you need to do is to get a solid training in the basics of statistics. Second, you need to get involved in research projects from very early in your career, so you get exposed to the multiple details that are part of a research project. Starting as a research assistant doing the most basic tasks required in a research project is essential to become a seasoned researcher later on. Third, make sure to get exposed to as many perspectives on how to develop a research project. Finally, it is important to have a solid theoretical training in order for your research to be informed by theory."

MAIN POINTS

- Bivariate tables are interpreted by comparing percentages across different categories of the independent variable. A relationship is said to exist if the percentage distributions vary across the categories of the independent variable.
- The chi-square test is an inferential statistical technique designed to test for a significant relationship between nominal and ordinal variables organized in a bivariate table. This is conducted by testing the null hypothesis that no association exists between two cross-tabulated variables in the population, and therefore, the variables are statistically independent.
- The obtained chi-square (χ^2) statistic summarizes the differences between the observed frequencies (f_o) and the expected frequencies (f_e)—the frequencies we would have expected to see if the null hypothesis were true and the variables were not associated. The Yates's correction for continuity is applied to all 2 × 2 tables.

- The sampling distribution of chi-square tells the probability of getting values of chi-square, assuming no relationship exists in the population. The shape of a particular chi-square sampling distribution depends on the number of degrees of freedom.
- Measures of association are single summarizing numbers that reflect the strength of the relationship between variables, indicate the usefulness of predicting the dependent from the independent variable, and often show the direction of the relationship.
- Proportional reduction of error (PRE) underlies the definition and interpretation of several measures of association. PRE measures are derived by comparing the errors made in predicting the dependent variable while ignoring the independent variable with errors made when making predictions that use information about the independent variable.
- Measures of association may be symmetrical or asymmetrical. When the measure is symmetrical, its value will be the same regardless of which of the two variables is considered the independent or dependent variable. In contrast, the value of asymmetrical measures of association may vary depending on which variable is considered the independent variable and which the dependent variable.
- Lambda is an asymmetrical measure of association suitable for use with nominal variables. It can range from 0.0 to 1.0 and gives an indication of the strength of an association between the independent and the dependent variables.
- Variables measured at the ordinal or interval-ratio levels may be positively or negatively associated. With a positive association, higher values of one variable correspond to higher values of the other variable. When there is a negative association between variables, higher values of one variable correspond to lower values of the other variable.
- Gamma is a symmetrical measure of association suitable for ordinal variables or for dichotomous nominal variables. It can vary from 0.0 to ±1.0 and reflects both the strength and direction of the association between two variables.
- Kendall's tau-*b* is a symmetrical measure of association suitable for use with ordinal variables. Unlike gamma, it accounts for pairs tied on the independent and dependent variable. It can vary from 0.0 to ±1.0. It provides an indication of the strength and direction of the association between two variables.
- Cramer's *V* is a measure of association for nominal variables. It is based on the value of chi-square and ranges between 0.0 to 1.0. Because it cannot take negative values, it is considered a nondirectional measure.

KEY TERMS

asymmetrical measure of association 256
chi-square (obtained) 244
chi-square test 239
control variable 262
Cramer's *V* 259
elaboration 262
expected frequencies (f_e) 242
gamma 260
Kendall's tau-*b* 260
lambda 256
measure of association 253
negative relationship 260
observed frequencies (f_o) 242
positive relationship 260
proportional reduction of error (PRE) 254
statistical independence 240
symmetrical measure of association 261

DIGITAL RESOURCES

Access key study tools at https://edge.sagepub.com/ssdsess4e

- eFlashcards of the glossary terms
- Datasets and codebooks
- SPSS and Excel walk-through videos
- SPSS and Excel demonstrations and problems to accompany each chapter
- Appendix F: Basic Math Review

CHAPTER EXERCISES

1. We examine the relationship between sex and one's opinion of whether or not people can be trusted based on GSS 2018 data.

Can People Be Trusted?	Men	Women	Total
Can trust	158	156	314
Cannot trust	279	345	624
Depends	16	24	40
Total	453	525	978

a. What is the number of degrees of freedom for this table?

b. Test the null hypothesis that sex and opinion of whether or not people can be trusted are independent (alpha = .01). What do you conclude?

c. If alpha were set at .05, would your decision change? Explain.

d. Calculate lambda for the table. Interpret this measure of association.

2. Below we present a cross-tabulation of views on restoring the voting rights of felons in prison by gender based on a March 2018 randomly selected sample of U.S. citizens 18 years of age and older.

Views on Restoring Voting Rights of Felons in Prison	Gender		Total
	Men	Women	
Support	110	130	240
Oppose	249	318	567
Total	359	448	807

Source: HuffPost, "Restoration of Voting Rights, 2018." Retrieved and modified from http://big.assets.huffingtonpost.com/tabsHPRestorationofvotingrights20180316.pdf

a. Percentage the cross-tabulation and describe whether or not there appears to be a relationship between gender and views on restoring the voting rights of felons in prison.

b. The obtained chi-square is .2512. Based on alpha of .05, what can you conclude about the relationship between gender and views on restoring the voting rights of felons in prison?

3. We extend our analysis of sex and trust, from Exercise 1, with the addition of a control variable, race. Bivariate tables for whites and blacks are presented.

For Whites

Can People Be Trusted?	Men	Women	Total
Can trust	136	129	265
Cannot trust	186	221	407
Depends	10	14	24
Total	332	364	696

For Blacks			
Can People Be Trusted?	**Men**	**Women**	**Total**
Can trust	11	12	23
Cannot trust	59	80	139
Depends	3	4	7
Total	73	96	169

a. Which racial group has a higher percentage of respondents indicating that they can trust people?

b. For black respondents, are women more likely to report than men that they can trust people?

c. For each table, test the hypothesis that sex and trust are independent (alpha = .01). What do you conclude?

4. Suppose we used General Social Survey data to examine the relationship between educational attainment (RDEGREE: 0 = *less than high school*, 1 = *high school degree*, 2 = *some college or more*) and the variable HELPPOOR, which measures support of the statement that the government in Washington should do everything possible to improve the standard of living of all poor Americans. The obtained chi-square is 40.43. Based on an alpha of .05, do you reject the null hypothesis? Explain.

helppoor SHOULD GOVT IMPROVE STANDARD OF LIVING? * RDegree Recoded Degree Crosstabulation

Count					
		RDegree Recoded Degree			
		.00	1.00	2.00	Total
helppoor SHOULD GOVT IMPROVE STANDARD OF LIVING?	1 GOVT ACTION	30	65	32	127
	2	6	32	41	79
	3 AGREE WITH BOTH	32	153	126	311
	4	6	56	58	120
	5 PEOPLE HELP SELVES	14	69	35	118
Total		88	375	292	755

5. Is there a relationship between the race of violent offenders and the race of their victims? Data from the U.S. Department of Justice for 2011 are presented below.

Characteristics of Victim	Characteristics of Offender		
	White	Black	Other
White	2,630	448	33
Black	193	2,447	9
Other	180	45	99

Source: U.S. Department of Justice, Expanded Homicide Data, Table 6, 2011.

a. Let's treat race of offenders as the independent variable and race of victims as the dependent variable. If we first ignore the independent variable and try to predict race of victim, how many errors will we make?

b. If we now take into account the independent variable, how many errors of prediction will we make for those offenders who are white? Black offenders? Other offenders?

c. Combine the answers in (a) and (b) to calculate the proportional reduction in error for this table based on the independent variable. How does this statistic improve our understanding of the relationship between the two variables?

6. Let's continue our analysis of offenders and victims of violent crime. In the following table, U.S. Department of Justice 2011 data for the sex of offenders and the sex of victims are reported.

Sex of Victim	Sex of Offender	
	Male	Female
Male	3,760	450
Female	1,590	140

Source: U.S. Department of Justice, Expanded Homicide Data, Table 6, 2011.

a. Treating sex of offender as the independent variable, how many errors of prediction will be made if the independent variable is ignored?

b. How many fewer errors will be made if the independent variable is taken into account?

c. Combine your answers in (a) and (b) to calculate lambda. Discuss the relationship between these two variables.

d. Which lambda is stronger, the one for sex of offenders/victims or race of offenders/victims (Exercise 5)?

7. Earlier in this chapter, we reviewed research from Inman and Mayes on first-generation college students. They also examined the relationship between student race and first-generation college status. Based on their data, test whether race is independent of first-generation college status (alpha = .01).

First-Generation College Status	Student Race					Total
	White	Black	Native American	Hispanic	Asian American	
Firsts	1,742	102	41	19	6	1,910
Nonfirsts	2,392	119	45	25	22	2,603
Total	4,134	221	86	44	28	4,513

Source: Adapted from W. Elliot Inman and Larry Mayes, "The Importance of Being First: Unique Characteristics of First Generation Community College Students," *Community College Review* 26, no. 3 (1999): 8.

8. GSS 2018 respondents were asked their views on teens (between the ages of 14 and 16) engaging in sex before marriage (TEENSEX). Responses are cross-tabulated by educational attainment (DEGREE).
   a. What percentage of those surveyed felt teen sex was always wrong?
   b. What percentage of those with less than a high school diploma felt teen sex was always wrong?
   c. What percentage of those with a graduate degree felt teen sex was always wrong?
   d. Interpret the gamma statistic.
   e. Using chi-square, test the null hypothesis that DEGREE and TEENSEX are statistically independent. Set alpha at .05.

Sex before marriage - - teens 14–16 * Rs highest degree Crosstabulation									
			Rs highest degree						
			Lt high school	High school	Junior college	Bachelor	Graduate	Total
Sex before marriage - - teens 14–16	Always wrong	Count	78	305	47	95	54	579
		% within Rs highest degree	65.5%	59.8%	58.8%	53.1%	49.1%	58.0%

		Rs highest degree					
		Lt high school	High school	Junior college	Bachelor	Graduate	Total
Almst always wrg	Count	19	81	18	36	27	181
	% within Rs highest degree	16.0%	15.9%	22.5%	20.1%	24.5%	18.1%
Sometimes wrong	Count	16	77	6	30	20	149
	% within Rs highest degree	13.4%	15.1%	7.5%	16.8%	18.2%	14.9%
Not wrong at all	Count	6	47	9	18	9	89
	% within Rs highest degree	5.0%	9.2%	11.3%	10.1%	8.2%	8.9%
Total	Count	119	510	80	179	110	998
	% within Rs highest degree	100.0%	100.0%	100.0%	100.0%	100.0%	100.0%

Chi-Square Tests			
	Value	df	Asymptotic Significance (2-sided)
Pearson Chi-Square	16.364[a]	12	.175
Likelihood Ratio	17.110	12	.146
Linear-by-Linear Association	4.747	1	.029
N of Valid Cases	998		

a. 0 cells (0.0%) have expected count less than 5. The minimum expected count is 7.13.

Symmetric Measures					
		Value	Asymptotic Standard Error[a]	Approximate T[b]	Approximate Significance
Ordinal by Ordinal	Gamma	.110	.041	2.643	.008
N of Valid Cases		998			

a. Not assuming the null hypothesis.

b. Using the asymptotic standard error assuming the null hypothesis.

9. Data from the ISSP 2014 are presented below for 975 French respondents, cross-tabulating highest completed educational level (DEGREE) with religious service attendance (ATTEND). Notice the international categories for degree. Test the null hypothesis that the variables are not related. Set alpha at .05.

ATTEND Attendance of religious services * DEGREE Highest completed education level: Categories for international comparison Crosstabulation

Count								
		DEGREE Highest completed education level: Categories for international comparison						
		No Formal Education	Primary School	Lower Secondary	Upper Secondary	Lower Level Tertiary	Upper Level Tertiary	Total
ATTEND Attendance of religious services	1 Several times a week or more often (incl. every day, several	1	0	10	4	4	4	23
	2 Once a week	1	2	24	4	5	12	48
	3 2 or 3 times a month	1	3	12	4	3	4	27
	4 Once a month	0	3	14	1	1	6	25
	5 Several times a year	3	3	80	21	23	34	164
	6 Once a year	0	6	44	16	22	22	110
	7 Less Frequently than once a year	1	6	54	18	26	14	119
	8 Never	3	9	175	70	91	111	459
Total		10	32	413	138	175	207	975

Chi-Square Tests	Value	df	Asymptotic Significance (2-sided)
Pearson Chi-Square	52.047[a]	35	.032
Likelihood Ratio	51.313	35	.037
Linear-by-Linear Association	8.439	1	.004
N of Valid Cases	975		

a. 21 cells (43.8%) have expected count less than 5. The minimum expected count is .24.

10. Examine the relationship between DEGREE and CONEDUC (confidence in education) based on the following GSS 2018 sample of 971 respondents. Test the null hypothesis that there is no relationship between the two variables. Set alpha at .05.

Confidence in education * Rs highest degree Crosstabulation			Rs highest degree					
			Lt high school	High school	Junior college	Bachelor	Graduate	Total
Confidence in education	A great deal	Count	46	128	18	32	27	251
		% within Rs highest degree	40.4%	26.8%	22.5%	17.1%	23.9%	25.8%
	Only some	Count	47	254	46	118	65	530
		% within Rs highest degree	41.2%	53.2%	57.5%	63.1%	57.5%	54.6%
	Hardly any	Count	21	95	16	37	21	190
		% within Rs highest degree	18.4%	19.9%	20.0%	19.8%	18.6%	19.6%
Total		Count	114	477	80	187	113	971
		% within Rs highest degree	100.0%	100.0%	100.0%	100.0%	100.0%	100.0%

Chi-Square Tests			
	Value	df	Asymptotic Significance (2-sided)
Pearson Chi-Square	22.334[a]	8	.004
Likelihood Ratio	21.966	8	.005
Linear-by-Linear Association	4.970	1	.026
N of Valid Cases	971		

a. 0 cells (0.0%) have expected count less than 5. The minimum expected count is 15.65.

11. Using the output from GSS 2018 data that is presented below, describe the association between marital status and a respondent's preferred candidate in the 2016 U.S. presidential election. While we present multiple measures of associations, you should only report and describe the appropriate measure of association.

Vote Clinton or Trump * Marital status Crosstabulation									
			Marital status						
			Married	Widowed	Divorced	Separated	Never Married	Total
Vote Clinton or Trump	Clinton	Count	185	61	88	20	147	501
		% within Marital status	46.0%	64.9%	56.8%	66.7%	81.2%	58.1%
	Trump	Count	217	33	67	10	34	361
		% within Marital status	54.0%	35.1%	43.2%	33.3%	18.8%	41.9%
Total		Count	402	94	155	30	181	862
		% within Marital status	100.0%	100.0%	100.0%	100.0%	100.0%	100.0%

Directional Measures						
			Value	Asymptotic Standard Error[a]	Approximate T[b]	Approximate Significance
Nominal by Nominal	Lambda	Symmetric	.039	.024	1.598	.110
		Vote Clinton or Trump Dependent	.089	.053	1.598	.110
		Marital status Dependent	.000	.000	.[c]	.[c]
	Goodman and Kruskal tau	Vote Clinton or Trump Dependent	.077	.017		.000[d]
		Marital status Dependent	.033	.008		.000[d]

a. Not assuming the null hypothesis.

b. Using the asymptotic standard error assuming the null hypothesis.

c. Cannot be computed because the asymptotic standard error equals zero.

d. Based on chi-square approximation.

Symmetric Measures					
		Value	Asymptotic Standard Error[a]	Approximate T[b]	Approximate Significance
Ordinal by Ordinal	Gamma	–.405	.047	–8.168	.000
N of Valid Cases		862			

a. Not assuming the null hypothesis.

b. Using the asymptotic standard error assuming the null hypothesis.

12. We take another look at the confidence in the executive branch of the federal government (CONFED), first presented in Table 8.11 (based on GSS 2018). CONFED is presented with two independent variables—(1) *respondent race* and (2) *age* (measured in categories as reported in corresponding tables). Interpret each measure of association. Which variable—race or age—has a higher proportional reduction of error?

Confid. in exec branch of fed govt * Race of respondent Crosstabulation						
			Race of respondent			
			White	Black	Other	Total
Confid. in exec branch of fed govt	A great deal	Count	87	19	16	122
		% within Race of respondent	12.6%	11.4%	14.8%	12.6%
	Only some	Count	294	63	49	406
		% within Race of respondent	42.6%	37.7%	45.4%	42.1%
	Hardly any	Count	309	85	43	437
		% within Race of respondent	44.8%	50.9%	39.8%	45.3%
Total		Count	690	167	108	965
		% within Race of respondent	100.0%	100.0%	100.0%	100.0%

Directional Measures						
			Value	Asymptotic Standard Error[a]	Approximate T[b]	Approximate Significance
Nominal by Nominal	Lambda	Symmetric	.007	.012	.626	.532
		Confid. in exec branch of fed govt Dependent	.011	.018	.626	.532
		Race of respondent Dependent	.000	.000	.[c]	.[c]
	Goodman and Kruskal tau	Confid. in exec branch of fed govt Dependent	.002	.003		.326[d]
		Race of respondent Dependent	.001	.002		.623[d]

a. Not assuming the null hypothesis.

b. Using the asymptotic standard error assuming the null hypothesis.

c. Cannot be computed because the asymptotic standard error equals zero.

d. Based on chi-square approximation.

Confid. in exec branch of fed govt * Recoded Age, 5 Categories Crosstabulation								
			Recoded Age, 5 Categories					
			18–29	30–39	40–49	50–59	60 and Older	Total
Confid. in exec branch of fed govt	A great deal	Count	21	20	18	22	41	122
		% within Recoded Age, 5 Categories	12.6%	10.6%	12.5%	12.7%	14.0%	12.6%
	Only some	Count	81	80	56	78	111	406
		% within Recoded Age, 5 Categories	48.5%	42.6%	38.9%	45.1%	37.9%	42.1%
	Hardly any	Count	65	88	70	73	141	437
		% within Recoded Age, 5 Categories	38.9%	46.8%	48.6%	42.2%	48.1%	45.3%
Total		Count	167	188	144	173	293	965
		% within Recoded Age, 5 Categories	100.0%	100.0%	100.0%	100.0%	100.0%	100.0%

Symmetric Measures					
		Value	Asymptotic Standard Error[a]	Approximate T[b]	Approximate Significance
Ordinal by Ordinal	Gamma	.033	.040	.826	.409
N of Valid Cases		965			

a. Not assuming the null hypothesis.

b. Using the asymptotic standard error assuming the null hypothesis.

13. In 2010, Paul Mazerolle, Alex Piquero, and Robert Brame examined whether violent onset offenders have distinct career dimensions from

offenders whose initial offending involves nonviolence.[9] In this table, the researchers investigate the relationship between gender, race, and age, as well as nonviolent versus violent onset using chi-square analysis. Their data are based on 1,503 juvenile offenders in Queensland, Australia. The independent variables are reported in rows.

The chi-square models for gender and age at first offense are significant at the .01 level. Interpret the relationship between gender and age at first offense with a nonviolent or violent initial offense.

Violent Offense Onset by Gender, Race, and Age

	Nonviolent Onset, *N* (%)	Violent Onset, *N* (%)	Total, *N* (%)
Gender			
Male	1,146 (88.29)	152 (11.71)	1,298 (100)
Female	156 (81.68)	35 (18.32)	191 (100)
Total	1,302	187	1,489
			$\chi^2 = 6.331$**
Indigenous status			
Nonindigenous	815 (87.17)	120 (12.83)	935 (100)
Indigenous	477 (88.17)	64 (11.83)	541 (100)
			$\chi^2 = 0.317$**
Age at first offense			
Less than 14 years	579 (90.33)	62 (9.67)	641 (100)
14 years and older	723 (85.26)	125 (14.74)	848 (100)
			$\chi^2 = 8.539$**

Source: Paul Mazerolle, Alex Piquero, and Robert Brame, "Violent Onset Offenders: Do Initial Experiences Shape Criminal Career Dimensions?" *International Criminal Justice Review* 20, no. 2 (2010): 132–146.

**$p < .01$.

9

ANALYSIS OF VARIANCE

Many research questions require us to look at multiple samples or groups, at least more than two at a time. We may be interested in studying the relationship between ethnic identity (white, African American, Asian American, Latinx) and church attendance, the influence of one's social class (lower, working, middle, and upper) on President Donald Trump's job approval ratings, or the effect of educational attainment (less than high school, high school graduate, some college, and college graduate) on household income. Note that each of these examples requires a comparison between multiple demographic or ethnic groups, more than the two-group comparisons we reviewed in Chapter 7. While it would be easy to confine our analyses between two groups, our social world is much more complex and diverse.

Let's say that we're interested in examining educational attainment—on average, how many years of education do Americans achieve? For 2018, the U.S. Census reported that 89.8% of adults (25 years and older) completed at least a high school degree, and 35% of all adults attained at least a bachelor's degree.[1] Special attention has been paid to the educational achievement of Latinx students. Data from the U.S. Census, as well as from the U.S. Department of Education, confirm that Latinx students continue to have lower levels of educational achievement than other racial or ethnic groups.

In Chapter 7, we introduced statistical techniques to assess the difference between two sample means or proportions. In Table 7.2, we compared the difference in educational attainment for blacks and whites. But what if we wanted to examine more than two racial or ethnic groups?

Suppose we collect a random sample of 23 men and women, grouped them into four demographic categories, and included their educational attainment in Table 9.1. With the t-test statistic we discussed in Chapter 7, we could analyze only two samples at a time. We would have to analyze the mean educational attainment

Chapter Learning Objectives

1. Explain the application of a one-way analysis of variance (ANOVA) model.
2. Define the concepts of between and within total variance.
3. Calculate and interpret a test for two or more sample cases with means.
4. Determine the significance of an F-ratio test statistic.
5. Interpret output for ANOVA.

Table 9.1 Educational Attainment (Measured in Years) for Four Racial/Ethnic Groups

White	Black	Asian	Latino
$n_1 = 6$	$n_2 = 5$	$n_3 = 6$	$n_4 = 6$
12	16	16	11
12	12	18	12
14	13	16	12
12	13	14	13
12	14	16	16
16		20	11

Source: U.S. Census Bureau, Educational Attainment in the United States: 2018, Table 1.

of whites and blacks, whites and Asians, whites and Latinos, blacks and Asians, and so on. (Confirm that we would have to analyze six different pairs.) In the end, we would have a series of *t*-test statistic calculations, and we still wouldn't be able to answer our original question: Is there a difference in educational attainment among all *four* demographic groups?

LEARNING CHECK 9.1

Identify the independent and dependent variables in Table 9.1.

Analysis of variance (ANOVA): An inferential statistics technique designed to test for the significant relationship between two variables in two or more groups or samples.

There is a statistical technique that will allow us to examine all the four groups or samples simultaneously. This technique is called **analysis of variance (ANOVA)**. ANOVA follows the same five-step model of hypothesis testing that we used with *t* test for means and *Z* test for proportions (in Chapter 7) and chi-square (in Chapter 8). In this chapter, we review the calculations for ANOVA and discuss two applications of ANOVA from the research literature.

UNDERSTANDING ANALYSIS OF VARIANCE

Recall that the *t* test examines the difference between two means, $\bar{Y}_1 - \bar{Y}_2$, while the null hypothesis assumed that there was no difference between them: $\mu_1 = \mu_2$. Rejecting the null hypothesis meant that there was a significant difference between the two mean scores (or the populations from which the samples were drawn).

The logic of ANOVA is the same but extending to two or more groups. For the data presented in Table 9.1, ANOVA will allow us to examine the variation among four means $(\bar{Y}_1, \bar{Y}_2, \bar{Y}_3, \bar{Y}_4)$, and the null hypothesis can be stated as follows: $\mu_1 = \mu_2 = \mu_3 = \mu_4$. Rejecting the null hypothesis for ANOVA indicates that there is a significant variation among the four samples (or the four populations from which the samples were drawn) and that at least one of the sample means is significantly different from the others. In our example, it suggests that years of education (dependent variable) do vary by group membership (independent variable).

When ANOVA procedures are applied to data with one dependent and one independent variable, it is called a **one-way ANOVA**.

One-way ANOVA: Analysis of variance application with one dependent variable and one independent variable.

The means, standard deviations, and variances for the samples have been calculated and are shown in Table 9.2. Note that the four mean educational years are not identical, with Asians having the highest educational attainment. Also, based on the standard deviations, we can tell that the samples are relatively homogeneous with deviations within 1.51 to 2.07 years of the mean. We already know that there is a difference between the samples, but the question remains: Is this difference significant? Do the samples reflect a relationship between demographic group membership and educational attainment in the general population?

LEARNING CHECK 9.2

We've calculated the mean and standard deviation scores for each group in Table 9.2. Compute each mean (Chapter 3) and standard deviation (Chapter 3) and confirm that our statistics are correct.

Table 9.2 Years of Education, Means, Standard Deviations, and Variances for Four Racial/Ethnic Groups

White	Black	Asian	Latino
$n_1 = 6$	$n_2 = 5$	$n_3 = 6$	$n_4 = 6$
12	16	16	11
12	12	18	12
14	13	16	12
12	13	14	13
12	14	16	16
16	—	20	11

(Continued)

(Continued)

White	Black	Asian	Latino
$n_1 = 6$	$n_2 = 5$	$n_3 = 6$	$n_4 = 6$
$\bar{Y}_1 = 13.00$	$\bar{Y}_2 = 13.60$	$\bar{Y}_3 = 16.67$	$\bar{Y}_4 = 12.50$
$s_1 = 1.67$	$s_2 = 1.51$	$s_3 = 2.07$	$s_4 = 1.87$
$s_1^2 = 2.79$	$s_2^2 = 2.28$	$s_3^2 = 4.28$	$s_4^2 = 3.50$
$\bar{Y}$ (for all groups) = 13.96			

To determine whether the differences are significant, ANOVA examines the differences *between* our four samples, as well as the differences *within* a single sample. The differences can also be referred to as variance or variation, which is why ANOVA is the analysis of *variance*. What is the difference between one sample's mean score and the overall mean? What is the variation of individual scores within one sample? Are all the scores alike (no variation), or is there a broad variation in scores? ANOVA allows us to determine whether the variance between samples is larger than the variance within the samples. If the variance is larger between samples than the variance within samples, we know that educational attainment varies significantly across the samples. It would support the notion that group membership explains the variation in educational attainment.

THE STRUCTURE OF HYPOTHESIS TESTING WITH ANOVA

The Assumptions

ANOVA requires several assumptions regarding the method of sampling, the level of measurement, the shape of the population distribution, and the homogeneity of variance.

Independent random samples are used. Our choice of sample members from one population has no effect on the choice of sample members from the second, third, or fourth population. For example, the selection of Asians has no effect on the selection of any other sample.

The dependent variable, years of education, is an interval-ratio level of measurement. Some researchers also apply ANOVA to ordinal-level measurements.

The population is normally distributed. Although we cannot confirm whether the populations are normal, given that our N is so small, we must assume that the population is normally distributed to proceed with our analysis.

The population variances are equal. Based on our calculations in Table 9.2, we see that the sample variances, although not identical, are relatively homogeneous.[2]

Stating the Research and the Null Hypotheses and Setting Alpha

The research hypothesis (H_1) proposes that at least one of the means is different. We do not identify which one(s) will be different, or larger or smaller; we only predict that a difference does exist.

H_1: At least one mean is different from the others.

ANOVA is a test of the null hypothesis of no difference between any of the means. Since we're working with four samples, we include four μ_s in our null hypothesis.

$$H_0: \mu_1 = \mu_2 = \mu_3 = \mu_4$$

As we did in other models of hypothesis testing, we'll have to set our alpha. Alpha is the level of probability at which we'll reject our null hypothesis. For this example, we'll set alpha at .05.

The Concepts of Between and Within Total Variance

A word of caution before we proceed: Since we're working with four different samples and a total of 23 respondents, we'll have a lot of calculations. It's important to be consistent with your notations (don't mix up numbers for the different samples) and be careful with your calculations.

Our primary set of calculations has to do with the two types of variance: (1) between-group variance and (2) within-group variance. The estimate of each variance has two parts, the sum of squares and degrees of freedom (*df*).

The **between-group sum of squares (*SSB*)** measures the difference in average years of education between our four groups. Sum of squares is the abbreviation for "sum of squared deviations." For *SSB*, what we're measuring is the sum of squared deviations between each sample mean to the overall mean score. The formula for the *SSB* can be presented as follows:

Between-group sum of squares (*SSB*): The sum of squared deviations between each sample mean to the overall mean score.

$$SSB = \sum n_k(\bar{Y}_k - \bar{Y})^2 \quad (9.1)$$

where

n_k = the number of cases in a sample (k represents the number of different samples)

$\bar{Y}_k$ = the mean of a sample

$\bar{Y}$ = the overall mean

SSB can also be understood as the amount of variation in the dependent variable (years of education) that can be attributed to or explained by the independent variable (the four demographic groups).

Within-group sum of squares (*SSW*): Sum of squared deviations within each group, calculated between each individual score and the sample mean.

Within-group sum of squares (*SSW*) measures the variation of scores within a single sample or, as in our example, the variation in years of education within one group. *SSW* is also referred to as the amount of unexplained variance, since this is what remains after we consider the effect of the specified independent variable. The formula for *SSW* measures the sum of squared deviations within each group, between each individual score with its sample mean.

$$SSW = \sum(Y_i - \bar{Y}_k)^2 \quad (9.2)$$

where

Y_i = each individual score in a sample

$\bar{Y}_k$ = the mean of a sample

Even with our small sample size, if we were to use Formula 9.2, we'd have a tedious and cumbersome set of calculations. Instead, we suggest using the following computational formula for within-group variation or *SSW*:

$$SSW = \sum Y_i^2 - \sum\frac{(\sum Y_k)^2}{n_k} \quad (9.3)$$

where

Y_i^2 = the squared scores from each sample

$\sum Y_k$ = the sum of the scores of each sample

n_k = the number of cases in a sample

Total sum of squares (*SST*): The total variation in scores, calculated by adding *SSB* (between-group sum of squares) and *SSW* (within-group sum of squares).

Together, the explained (*SSB*) and unexplained (*SSW*) variances compose the amount of total variation in scores. The **total sum of squares (*SST*)** can be represented by

$$SST = \sum(Y_i - \bar{Y})^2 = SSB + SSW \quad (9.4)$$

where

Y_i = each individual score

$\bar{Y}$ = the overall mean

The second part of estimating the between-group and within-group variances is calculating the degrees of freedom. Degrees of freedom are also discussed in Chapters 7 and 8. For ANOVA, we have to calculate two degrees of freedom. For *SSB*, the degrees of freedom are determined by

$$df_b = k - 1 \tag{9.5}$$

where k is the number of samples.

For *SSW*, the degrees of freedom are determined by

$$df_w = N - k \tag{9.6}$$

where

N = total number of cases

k = number of samples

Finally, we can estimate the between-group variance by calculating **mean square between**. Simply stated, mean squares are averages computed by dividing each sum of squares by its corresponding degrees of freedom. Mean square between can be represented by

Mean square between: Sum of squares between divided by its corresponding degrees of freedom.

$$\text{Mean square between} = SSB\ /\ df_b \tag{9.7}$$

and the within-group variance or **mean square within** can be represented by

Mean square within: Sum of squares within divided by its corresponding degrees of freedom.

$$\text{Mean square within} = SSW\ /\ df_w \tag{9.8}$$

The *F* Statistic

Together, the mean square between (Formula 9.7) and mean square within (Formula 9.8) compose the ***F* ratio** obtained or ***F* statistic**. Developed by R. A. Fisher, the *F* statistic is the ratio of between-group variance to within-group variance and is determined by Formula 9.9:

F ratio or *F* statistic: The test statistic for ANOVA, calculated by the ratio of mean square to mean square within.

$$F = \frac{\text{Mean square between}}{\text{Mean square within}} = \frac{SSB\ /\ df_b}{SSW\ /\ df_w} \tag{9.9}$$

We know that a larger F-obtained statistic means that there is more between-group variance than within-group variance, increasing the chances of rejecting our null hypothesis. In Table 9.3, we present additional calculations to compute F.

Let's calculate between-group sum of squares and degrees of freedom based on Formulas 9.1 and 9.5. The calculation for SSB is

$$\sum n_k(Y - Y)^2 = 6(13 - 13.96)^2 + 5(13.6 - 13.96)^2 + 6(16.67 - 13.96)^2 + 6(12.5 - 13.96)^2$$

$$= 5.53 + .65 + 44.06 + 12.79$$

$$= 63.03$$

The degrees of freedom for SSB is $k - 1$ or $4 - 1 = 3$. Based on Formula 9.7, the mean square between is

$$\text{Mean square between} = 63.03 / 3 = 21.01$$

Table 9.3 Computational Worksheet for ANOVA

White	Black	Asian	Latino
$n_1 = 6$	$n_2 = 5$	$n_3 = 6$	$n_4 = 6$
12	16	16	11
12	12	18	12
14	13	16	12
12	13	14	13
12	14	16	16
16	—	20	11
$\bar{Y}_1 = 13.00$	$\bar{Y}_2 = 13.60$	$\bar{Y}_3 = 16.67$	$\bar{Y}_4 = 12.50$
$s_1 = 1.67$	$s_2 = 1.51$	$s_3 = 2.07$	$s_4 = 1.87$
$s_1 = 2.79$	$s_2^2 = 2.28$	$s_3^2 = 4.28$	$s_4^2 = 3.50$
$\sum Y_1 = 78$	$\sum Y_2 = 68$	$\sum Y_3 = 100$	$\sum Y_4 = 75$
$\sum Y_1^2 = 1028$	$\sum Y_2^2 = 934$	$\sum Y_3^2 = 1{,}688$	$\sum Y_4^2 = 955$
$\bar{Y}$ (for all groups) = 13.96			

The within-group sum of squares and degrees of freedom are based on Formulas 9.3 and 9.6. The calculation for SSW is

$$\sum Y_i^2 - \sum \frac{(\Sigma Y_k)^2}{n_k} = (1{,}028 + 934 + 1{,}688 + 955) - \left(\frac{78^2}{6} + \frac{68^2}{5} + \frac{100^2}{6} + \frac{75^2}{6}\right)$$

$$= 4{,}605 - (1{,}014 + 924.8 + 1{,}666.67 + 937.50)$$

$$= 4{,}605 - 4{,}542.97$$

$$= 62.03$$

The degrees of freedom for SSW is $N - k = 23 - 4 = 19$. Based on Formula 9.8, the mean square within is

$$\text{Mean square within} = 62.03 \,/\, 19 = 3.26$$

Finally, our calculation of F is based on Formula 9.9:

$$F \text{ ratio} = 21.01 \,/\, 3.26 = 6.44$$

A CLOSER LOOK 9.1

Decomposition of SST

According to Formula 9.4, sum of squares total (SST) is equal to

$$\text{SST} = \Sigma\left(Y_i - \bar{Y}\right)^2 = \text{SSB} + \text{SSW}$$

You can see that the between sum of squares (explained variance) and within sum of squares (unexplained variance) account for the total variance (SST) in a particular dependent variable. How does that apply to a single case in our educational attainment example? Let's take the first black respondent in Table 9.1 with 16 years of education.

This respondent's total deviation (corresponding to SST) is based on the difference between the years of education from the overall mean (Formula 9.4). The individual mean is quite a bit higher than the overall mean education of 8 years. The difference of the individual from the overall mean is 2.04 years (16 – 13.96). Between-group deviation (corresponding to SSB) can be determined by measuring the difference between the group average from the overall mean (Formula 9.1). The deviation between the group average and overall average for blacks is –.36 years (13.60 – 13.96). Finally, the within-group deviation (corresponding to SSW, Formula 9.2) is based on the difference between the first black person's years of education and the group average for blacks: 2.40 years (16 – 13.60). So for the first black person in our sample, SSB + SSW = SST or 2.40 + –.36 = 2.04. In a complete ANOVA problem, we're computing these two sources of deviation (SSB and SSW) to obtain SST (Formula 9.4) for everyone in the sample.

Making a Decision

To determine the probability of calculating an F statistic of 6.44, we rely on Appendix E, the distribution of the F statistic. Appendix E lists the corresponding values of the F distribution for various degrees of freedom and two levels of significance, .05 and .01.

Since we set alpha at .05, we'll refer to the table marked "p = .05." Note that Appendix E includes two *df*s. These refer to our degrees of freedom, $df_1 = df_b$ and $df_2 = df_w$.

Because of the two degrees of freedom, we'll have to determine the probability of our F obtained differently than we did with t test or chi-square. For this ANOVA example, we'll have to determine the corresponding F, also called the F critical, when $df_b = 3$ and $df_w = 19$, and $\alpha = .05$.

F critical: *F*-test statistic that corresponds to the alpha level, df_w, and df_b.

F obtained: The *F*-test statistic that is calculated.

Based on Appendix E, the **F critical** (as in Appendix E) is 3.13, while our **F obtained** (the one that we calculated) is 6.44. Since our F obtained is greater than the F critical (6.44 > 3.13), we know that its probability is <.05, extending into the shaded area. (If our F obtained was <3.13, we could determine that its probability was greater than our alpha of .05, in the unshaded area of the F-distribution curve. Refer to Figure 9.1.) We can reject the null hypothesis of no difference and conclude that there is a significant difference in educational attainment between the four groups.

THE FIVE STEPS IN HYPOTHESIS TESTING: A SUMMARY

To summarize, we've calculated an ANOVA test examining the difference between four demographic groups and their average years of education.

1. *Making Assumptions:*
    - Independent random samples are used.
    - The dependent variable, *years of education*, is an interval-ratio level of measurement.
    - The population is normally distributed.
    - The population variances are equal.
2. *Stating the Research and Null Hypothesis and Selecting Alpha:*

    H_1: At least one mean is different from the others.

$$H_0: \mu_1 = \mu_2 = \mu_3 = \mu_4$$

$$\alpha = .05$$

Figure 9.1 Comparing F Obtained Versus F Critical

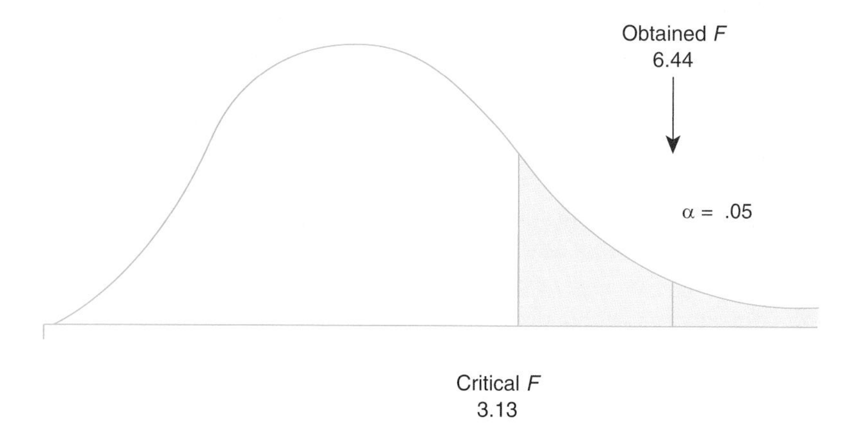

3. *Selecting the Sampling Distribution and Specifying the Test Statistic:* The F distribution and F statistic are used to test the significance of the difference between the four sample means.

4. *Computing the Test Statistic:* We need to calculate the between-group and within-group variation (sum of squares and degrees of freedom). We estimate $SSB = 21.01$ ($df_b = 3$) and $SSW = 3.26$ ($df_w = 19$).

Based on Formula 9.9,

$$F = 21.01 / 3.26 = 6.44$$

5. *Making a Decision and Interpreting the Results:* We reject the null hypothesis of no difference. Our F obtained of 6.44 is greater than the F critical of 3.13. The probability of 6.44 is $<.05$. F doesn't advise us about which groups are different, only that educational attainment does differ significantly by demographic group members. Based on the sample data, we know that the only group to achieve a college education average was Asians (16.67 years). If we were to rank the other group means, second highest educational attainment was among blacks (13.60), followed by whites (13.00) and Latinos (12.50).

LEARNING CHECK 9.3

If alpha were changed to .01, would our final decision change?

STATISTICS IN PRACTICE: THE ETHICAL CONSUMER

Consumers increasingly avoid products that are perceived as having a negative social or environmental impact. By purchasing fair labor, organic, or environmentally safe products, consumers express their politics through their purchases, a practice that has been referred to as ethical consumerism. The GSS 2014 included a series of questions on what it takes to be a good citizen. Respondents were asked how important it was to "choose products for political, ethical or environmental reasons, even if they cost a bit more." Answers were ranked on a 7-point scale: 1 = *not at all important* to 7 = *very important*. We use SPSS and the GSS 2014 data to examine the relationship between responses to this question and political party identification (Democrat, Independent, or Republican). We will set alpha at .05 to assess our results. SPSS output is presented as Figure 9.2.

The ANOVA output includes two tables: (1) Descriptives and (2) ANOVA. In the Descriptives table, the *N*, mean, and standard deviation are reported for each group and the entire sample, along with the 95% confidence interval for each mean.

The *F* obtained is reported in the ANOVA table, along with its level of significance (or probability). For these data, *F* obtained is 1.766 with a significance of .172. Since the level of significance is greater than our alpha (.172 > .05), we fail to reject the null hypothesis. Although the means are different (Democrats had the highest mean of 5.11), these differences are not significant.

Figure 9.2 SPSS ANOVA Output: Political Party Identification and Choosing Products for Political Reasons

Descriptives

buypol HOW IMPORTANT TO CHOOSE PRODUCTS FOR POL REASONS

	N	Mean	Std. Deviation	Std. Error	95% Confidence Interval for Mean		Minimum	Maximum
					Lower Bound	Upper Bound		
1.00 Democrat	208	5.11	1.648	.114	4.89	5.34	1	7
2.00 Independent	98	4.85	1.852	.187	4.48	5.22	1	7
3.00 Republican	135	4.78	1.718	.148	4.49	5.07	1	7
Total	441	4.95	1.719	.082	4.79	5.11	1	7

ANOVA

buypol HOW IMPORTANT TO CHOOSE PRODUCTS FOR POL REASONS

	Sum of Squares	df	Mean Square	F	Sig.
Between Groups	10.408	2	5.204	1.766	.172
Within Groups	1290.494	438	2.946		
Total	1300.902	440			

LEARNING CHECK 9.4

Calculate eta square for the model presented in Figure 9.2.

A CLOSER LOOK 9.2

Assessing the Relationship Between Variables

Based on our five-step model of *F*, we've determined that there is a significant difference between the four demographic groups in their educational attainment. We rejected the null hypothesis and concluded that the years of education (our dependent variable) do vary by group membership (our independent variable). But can we say anything about how strong the relationship is between the variables?

The correlation ratio or eta square (η^2) allows us to make a statement about the strength of the relationship or the effect size. Eta square is determined by the following:

$$\eta^2 = \frac{SSB}{SST} \tag{9.10}$$

The ratio of *SSB* to *SST* (*SSB* + *SSW*) represents the proportion of variance that is explained by the group (or independent) variable. Eta square indicates the strength of the relationship between the independent and dependent variables, ranging in value from 0 to 1.0. As eta square approaches 0, the relationship between the variables is weaker, and as eta square approaches 1, the relationship between the variables is stronger.

Based on our ANOVA example,

$$\eta^2 = \frac{63.03}{63.03 + 62.03} = .50$$

We can state that 50% of the variation in educational attainment can be attributed to demographic group membership. Or we can say that 50% of the variation in the dependent variable (educational attainment) can be explained by the independent variable (group membership). So how strong is this relationship? We can base our determination of the strength on A Closer Look 8.2 from Chapter 8, the same scale that we used to assess gamma. We conclude that there is a very strong relationship between group membership and educational attainment.

READING THE RESEARCH LITERATURE: COLLEGE SATISFACTION AMONG LATINO STUDENTS

Like bivariate, *t* test, or chi-square analyses, ANOVA can help us understand how the categories of experience—race, age, class, and/or gender—shape our social lives. ANOVA allows us to investigate a variety of social categories by comparing the differences between them. We conclude this chapter with an example of how ANOVA is presented and interpreted in the social science literature.

Young Kim, Liz Rennick, and Marla Franco (2014)[3] investigated the levels of college engagement and satisfaction among Latino undergraduate students attending highly selective colleges and universities in comparison with other racial/ethnic groups. The Latino student population is often characterized as academically unprepared for college; however, recent research has indicated that Latino students demonstrate a stronger drive to success and a higher level of academic effort compared to their peers. The 2014 research by Kim et al. is among the few studies to consider the experience of Latino students at selective colleges and universities.

Their research is based on the 2010 University of California Undergraduate Experience Survey. Kim and their colleagues limited their analyses to the outcomes of 33,415 junior and senior students. All factors were measured on a 6-point Likert-type scale: 1 = *very poor* to 6 = *excellent*. Selected results are presented in Table 9.4.[4]

The table presents ANOVA results for four models regarding the student satisfaction with the educational experience. The mean and standard deviation for each racial/ethnic group are reported in the table along with the obtained *F* ratio. The sample sizes for each racial/ethnic student group are not reported. All models are significant at the .001 level. The researchers briefly summarize their findings:

> In terms of satisfaction with the educational experience, Latino students rated their satisfaction with quality of instruction and courses in the major, advising and out-of-class contact, and library support as equal to or higher than all other racial/ethnic groups. Sense of belonging and overall college satisfaction was the only area where White students reported higher levels of satisfaction than Latino students. (p. 253)[5]

They state, "We may argue that the college experiences of Latino students at highly selective research institutions are favorable and meaningful,

Table 9.4 Patterns of College Experiences by Student Race/Ethnicity

	Latino	African American	Asian	White	F
Quality of instruction and courses in the major	$M = 5.38$, $SD = 1.75$	$M = 5.20$, $SD = 1.80$	$M = 4.78$, $SD = 1.76$	$M = 5.37$, $SD = 1.77$	225.28***
Sense of belonging and satisfaction	$M = 4.88$, $SD = 1.80$	$M = 4.51$, $SD = 1.84$	$M = 4.45$, $SD = 1.74$	$M = 5.12$, $SD = 1.81$	287.07***
Satisfaction with advising and out of class contact	$M = 5.21$, $SD = 1.80$	$M = 5.16$, $SD = 1.87$	$M = 4.92$, $SD = 1.74$	$M = 5.21$, $SD = 1.74$	66.26***
Satisfaction with library support	$M = 5.30$, $SD = 1.80$	$M = 5.23$, $SD = 1.85$	$M = 4.76$, SD = 1.76	$M = 5.14$, $SD = 1.75$	142.00***

Source: Adapted from Young Kim, Liz Rennick, and Marla Franco, "Latino College Students at Highly Selective Institutions: A Comparison of Their College Experiences and Outcomes to Other Racial/Ethnic Groups," *Journal of Hispanic Higher Education* 13, no. 4 (2014): 254.

***$p < .001$.

which may contribute to development in desired educational outcomes at these types of institutions" (p. 258).

LEARNING CHECK 9.5

For the "Sense of Belonging and Satisfaction" ANOVA model, what additional information would you need to determine the *F* critical?

DATA AT WORK

Kevin Hemminger: Sales Support Manager/ Graduate Program in Research Methods and Statistics

Photo courtesy of Kevin Hemminger

While working as a sales support manager for a travel management company, Kevin decided to return to school to pursue a degree in criminal justice. Although originally he intended to become a probation officer for the juvenile court system, his career trajectory went into a different direction after taking a statistics course. "Due in great part to an incredible professor who made the subject enjoyable and interesting, I decided to pursue a PhD in Research Methods and Statistics in order to teach research methods and statistics. . . . My hope is to teach on either the graduate level, or to gifted high school students preparing for college."

Kevin has incorporated what he's learned about methods and statistics into his sales support work. "Being a very customer-focused company in a very customer-centric field, the idea of surveying customers is critical and being able to incorporate a knowledgeable administration of surveys has been integral to my team's success. Knowing research best practices and understanding what constitutes good, valid research has made a tremendous impact on the value of the data I provide to my leaders. My knowledge in the field has also given me a confidence level in speaking about quantitative data that I never had before. I work with enormous data sets and spreadsheets on a daily basis and understanding even the most basic concepts such as mean, median, mode, standard deviation, and variance has allowed me to lead projects and present findings in a much more intelligent way."

He recently completed his master's degree and is currently working on his doctorate. Kevin offers these fine words of encouragement for anyone considering data and statistical work: "If you're thinking about pursuing a career involving statistics, my recommendation is to dive full throttle into a study or project you're incredibly passionate about. Watching your study move from an idea, to a literature review, to a hypothesis, data collection, and then data analysis is one of the best ways to confirm your interest—and, by far, the best way to experience the power behind statistics as a discipline. If you get excited watching your data come in, and even more excited wondering [what] you're going to do with that data, I think you've confirmed there is no question you'd enjoy the field! If you're learning stats concepts and all you can think about is how you'd teach that concept to others—it's over! You're perfect for the field."

MAIN POINTS

- Analysis of variance (ANOVA) procedures allows us to examine the variation in means in more than two samples. To determine whether the difference in mean scores is significant, ANOVA examines the differences between multiple samples, as well as the differences within a single sample.
- One-way ANOVA is a procedure using one dependent variable and one independent variable. The five-step hypothesis testing model is applied to one-way ANOVA.
- The test statistic for ANOVA is *F*. The *F* statistic is the ratio of between-group variance to within-group variance.

KEY TERMS

analysis of variance (ANOVA) 280
between-group sum of squares (*SSB*) 283
F critical 288
F obtained 288
F ratio or *F* statistic 285
mean square between 285
mean square within 285
one-way ANOVA 281
total sum of squares (*SST*) 284
within-group sum of squares (*SSW*) 284

DIGITAL RESOURCES

Access key study tools at https://edge.sagepub.com/ssdsess4e

- eFlashcards of the glossary terms
- Datasets and codebooks
- SPSS and Excel walk-through videos
- SPSS and Excel demonstrations and problems to accompany each chapter
- Appendix F: Basic Math Review

CHAPTER EXERCISES

1. For a random sample of 32 GSS cases, health is measured according to a 4-point scale: 1 = *excellent*, 2 = *good*, 3 = *fair*, and 4 = *poor*. Four social classes are reported here: 1 = *lower*, 2 = *working*, 3 = *middle*, and 4 = *upper*. Present the five-step model for these data, using alpha = .05.

Lower Class	Working Class	Middle Class	Upper Class
3	2	2	2
2	1	3	1
2	3	1	1
2	2	1	2
3	2	2	1
3	2	3	1
4	3	3	1
4	3	1	2

2. We take another look at health, this time examining the relationship between educational attainment and perceived quality of health care. Data for three groups are presented based on the Health Information National Trends Survey (HINTS) 2012 data set. HINTS is an annual survey measuring the use of cancer-related information for adults 18 years and older. Present the five-step model for these data, using alpha = .01. QUALITYCARE is measured on a 5-point scale: 1 = *excellent*, 2 = *very good*, 3 = *good*, 4 = *fair*, and 5 = *poor*. Note how a lower score indicates a higher quality of care.

Present the five-step model for these data, using alpha = .05.

Less Than High School	Some College	College Graduate
1	2	1
4	3	1
2	2	1
2	2	2
3	4	1
3	2	2

3. We selected a sample of 30 International Social Science Programme respondents, noting their educational status (no degree, secondary degree, and university degree) and their level of church attendance (0 = *never*, 1 = *infrequently*, and 2 = *two to three times per month or more*). Is there a relationship between educational attainment and church attendance?

Complete the five-step model for these data, using alpha = .01.

No Degree	Secondary Degree	University Degree
2	2	0
1	2	0
1	2	0
2	1	0
2	1	1
2	0	1
0	2	0
2	1	1
2	1	2
2	2	1

4. Based on a sample of 21 Monitoring the Future respondents, we present their racial/ethnic background and the numbers of school days missed in the past 4 weeks.
   a. Complete the five-step model for these data, and set alpha at .05.
   b. If alpha were set at .01, would your decision change? Explain.

White	Black	Hispanic
4	1	4
5	2	3
3	2	5
4	1	1
4	3	5
4	4	2
6	3	2

5. We selected a sample of 14 Monitoring the Future respondents. We present their number of moving (traffic) violations in the past 12 months along with their residential area (residential area is the independent variable). Complete the five-step model for these data, using alpha = .05.

Small Town	Medium-Sized City	Large City
0	2	3
0	3	4
1	1	4
2	1	3
1		2

6. The GSS 2014 asked respondents to identify what was important for "truly being American." When asked "How important to have American ancestry?" answers were measured on a 4-point scale: 1 = *very important*, 2 = *fairly important*, 3 = *not very important*, and 4 = *not important at all*. We selected a sample of 20 GSS respondents and present their individual responses. Complete the five-step model for these data, using alpha = .01 to assess the significance of the model.

White	Black	Other
3	1	4
2	2	2
2	2	2
2	2	2
3	1	3
4	1	2
4	2	—

7. Nan Sook Park and her colleagues (2012) investigated racial/ethnic differences in predictors of self-rating health and the use of sociocultural resources. Their data are based on the Survey of Older Floridians, a statewide sample of white, African American, Cuban, and non-Cuban Hispanic seniors.

We present ANOVA results for two sociocultural resources. Social support was measured with the question: In times of trouble, can you count on at least some of your family and friends? (1 = *hardly ever*, 2 = *some of the time*, and 3 = *most of the time*). Religious attendance was measured according to the following scale: 1 = *never or almost never* to 5 = *more than once a week*. Mean scores are presented for each racial/ethnic group, along with the standard deviation in parentheses.

Review each measure of sociocultural resource and determine whether the null hypothesis would be rejected. Set alpha at .05 for each.

	Racial/Ethnic Group Mean (Standard Deviation)				
Variable	Whites (n = 503)	African Americans (n = 360)	Cubans (n = 328)	Non-Cuban Hispanics (n = 241)	F
Social support	2.85 (0.47)	2.75 (0.57)	2.73 (0.60)	2.58 (0.70)	12.17***
Religious attendance	2.79 (1.57)	3.94 (1.21)	2.74 (1.49)	3.37 (1.43)	56.43***

Source: Nan Sook Park, Yuri Jan, Beom Lee, and David Chiriboga, "Racial/Ethnic Differences in Predictors of Self-Rated Health: Findings From the Survey of Older Floridians," *Research on Aging* 35, no. 1 (2012): 207.

***$p < .001$.

8. The GSS 2014 included a series of questions about what it takes to be a good citizen. Respondents were asked, to be a good citizen, how important was it "to help people in America who are worse off than yourself" (HELPUSA) and "to help people in the rest of the world who are worse off than yourself" (HELPWRLD). Answers were measured on a 7-point scale: 1 = *not important at all* to 7 = *very important*. We ran ANOVA models for each with respondent's political party (RPartyId) as the independent variable. Set alpha at .05 to assess the significance of each model.

Descriptives

		N	Mean	Std. Deviation	Std. Error	95% Confidence Interval for Mean Lower Bound	95% Confidence Interval for Mean Upper Bound	Minimum	Maximum
helpusa HOW IMPORTANT TO HELP WORSE OFF PPL IN AMERICA	1.00 Democrat	167	6.15	1.117	.086	5.98	6.32	1	7
	2.00 Independent	87	5.94	1.214	.130	5.68	6.20	2	7
	3.00 Republican	101	5.76	1.184	.118	5.53	6.00	3	7
	Total	355	5.99	1.169	.062	5.87	6.11	1	7
helpwrld HOW IMPORTANT TO HELP WORSE OFF PPL IN REST OF WORLD	1.00 Democrat	166	4.91	1.712	.133	4.65	5.17	1	7
	2.00 Independent	83	4.82	1.829	.201	4.42	5.22	1	7
	3.00 Republican	102	4.45	1.614	.160	4.13	4.77	1	7
	Total	351	4.75	1.720	.092	4.57	4.94	1	7

ANOVA

		Sum of Squares	df	Mean Square	F	Sig.
helpusa HOW IMPORTANT TO HELP WORSE OFF PPL IN AMERICA	Between Groups	9.688	2	4.844	3.595	.028
	Within Groups	474.267	352	1.347		
	Total	483.955	354			
helpwrld HOW IMPORTANT TO HELP WORSE OFF PPL IN REST OF WORLD	Between Groups	13.740	2	6.870	2.341	.098
	Within Groups	1021.189	348	2.934		
	Total	1034.929	350			

9. The following ANOVA model examines the relationship between DEGREE (respondent's educational degree) and VALGIVEN (total dollar value of all donations made in the past year). Data are from the GSS 2014. Is there a significant difference in donation amount by educational degree? Set alpha at .01.

Descriptives

VALGIVEN TOTAL DONATIONS PAST YEAR R AND IMMEDIATE FAMILY

	N	Mean	Std. Deviation	Std. Error	95% Confidence Interval for Mean		Minimum	Maximum
					Lower Bound	Upper Bound		
0 LT HIGH SCHOOL	54	593.85	1229.945	167.374	258.14	929.56	0	7000
1 HIGH SCHOOL	177	1164.01	2904.570	218.321	733.15	1594.88	0	23000
2 JUNIOR COLLEGE	40	857.50	1826.298	288.763	273.42	1441.58	0	10000
3 BACHELOR	52	3397.40	9054.431	1255.624	876.63	5918.17	0	59300
4 GRADUATE	41	5590.61	13340.635	2083.457	1379.79	9801.43	0	70000
Total	364	1863.40	6188.306	324.355	1225.55	2501.25	0	70000

ANOVA

VALGIVEN TOTAL DONATIONS PAST YEAR R AND IMMEDIATE FAMILY

	Sum of Squares	df	Mean Square	F	Sig.
Between Groups	906027134	4	226506784	6.257	.000
Within Groups	1.300E+10	359	36198062.6		
Total	1.390E+10	363			

10. Is there a significant difference in e-mail hours per week among the same educational groups? Using data from the GSS 2018, we ran an ANOVA model using DEGREE (educational attainment) as the independent variable and EMAILHR (e-mail hours per week) as the dependent variable. Based on an alpha of .05, what do you conclude?

Descriptives

Email hours per week

	N	Mean	Std. Deviation	Std. Error	95% Confidence Interval for Mean		Minimum	Maximum
					Lower Bound	Upper Bound		
LT HIGH SCHOOL	68	3.63	10.117	1.227	1.18	6.08	0	70
HIGH SCHOOL	328	5.05	10.378	.573	3.92	6.18	0	100
JUNIOR COLLEGE	68	5.93	10.722	1.300	3.33	8.52	0	70
BACHELOR	124	9.62	12.399	1.113	7.42	11.82	0	60
GRADUATE	76	10.42	9.840	1.129	8.17	12.67	0	40
Total	664	6.46	10.957	.425	5.63	7.30	0	100

ANOVA

Email hours per week

	Sum of Squares	df	Mean Square	F	Sig.
Between Groups	3647.686	4	911.922	7.912	.000
Within Groups	75953.372	659	115.255		
Total	79601.059	663			

11. We examine the relationship between educational attainment (DEGREE) and agreement to the statement, "Scientific research is necessary and should be supported by the federal government" as measured in the GSS 2018.

Responses to the scientific research statement (ADVFRONT) are measured on an ordinal scale: 1 = *strongly agree*, 2 = *agree*, 3 = *neither*, 4 = *disagree*, and 5 = *strongly disagree*. Does agreement vary by one's educational degree?

Descriptives

Sci Rsch is necessary and should be supported by federal govt

	N	Mean	Std. Deviation	Std. Error	95% Confidence Interval for Mean Lower Bound	95% Confidence Interval for Mean Upper Bound	Minimum	Maximum
LT HIGH SCHOOL	70	2.11	.713	.085	1.94	2.28	1	4
HIGH SCHOOL	277	1.90	.700	.042	1.82	1.98	1	4
JUNIOR COLLEGE	47	1.85	.551	.080	1.69	2.01	1	3
BACHELOR	98	1.79	.596	.060	1.67	1.91	1	4
GRADUATE	56	1.63	.676	.090	1.44	1.81	1	3
Total	548	1.87	.679	.029	1.82	1.93	1	4

ANOVA

Sci Rsch is necessary and should be supported by federal govt

	Sum of Squares	df	Mean Square	F	Sig.
Between Groups	8.474	4	2.119	4.718	.001
Within Groups	243.838	543	.449		
Total	252.312	547			

a. Set alpha at .01, and test the null hypothesis of equal means.

b. What is the eta square for this model?

12. Using Monitoring the Future 2017 data, we examine whether a relationship exists between the race/ethnicity of a student and how the student rates the importance of being a community leader. IMPLDRCOMUNTY is measured on an ordinal scale: 1 = *not important*, 2 = *somewhat important*, 3 = *quite important*, and 4 = *extra important*. Analysis of variance results are presented.

Descriptives

171A007H:IMP LDR COMUNTY

	N	Mean	Std. Deviation	Std. Error	95% Confidence Interval for Mean Lower Bound	95% Confidence Interval for Mean Upper Bound	Minimum	Maximum
BLACK:(1)	276	2.91	.996	.060	2.79	3.03	1	4
WHITE:(2)	1041	2.51	.969	.030	2.45	2.57	1	4
HISPANIC:(3)	380	2.53	1.005	.052	2.43	2.64	1	4
Total	1697	2.58	.991	.024	2.53	2.63	1	4

ANOVA

171A007H:IMP LDR COMUNTY

	Sum of Squares	df	Mean Square	F	Sig.
Between Groups	35.836	2	17.918	18.605	.000
Within Groups	1631.435	1694	.963		
Total	1667.270	1696			

a. Set alpha at .05. What do you conclude about the relationship between student race/ethnicity and the importance of being a community leader?

b. What is the eta square for this model?

c. If alpha were set at .01, would your decision change? Explain.

13. In Chapter 3's Statistics in Practice, we reviewed Myron Pope's 2002 research on community college mentoring. Pope compared responses from four groups of minority students, measuring their perception of multilevel mentoring at their school. Student responses were based on a 5-point scale: 1 = *no agreement* to 5 = *strong agreement*. Pope's results are presented again with an additional column of *F* test and significance.

a. What are the degrees of freedom for the models?

b. Based on an alpha of .05, assess each model. What do you conclude?

	African American (n = 178), M (SD)	Asian (n = 12), M (SD)	Hispanic (n = 28), M (SD)	Native American (n = 22), M (SD)	Multiethnic (n = 14), M (SD)	F Test Sig.
There are persons of color in administrative roles from whom I would seek mentoring at this institution.	3.76 (1.10)	3.50 (1.31)	3.14 (1.08)	4.09 (.68)	4.14 (.66)	3.508 .008
There are peer mentors who can advise me.	3.48 (1.11)	2.17 (1.40)	3.14 (1.20)	3.91 (.68)	3.29 (1.54)	5.245 .000
I mentor other students.	3.30 (1.25)	2.00 (1.21)	3.00 (1.09)	3.46 (1.10)	3.29 (1.07)	3.702 .006

Source: Adapted from Myron Pope, "Community College Mentoring Minority Student Perception," *Community College Review* 30, no. 3 (2002): 37.

10

REGRESSION AND CORRELATION

The differential access to the Internet is real and persistent. Take for example Celeste Campos-Castillo's (2015) research which confirmed the impact of gender and race on the digital divide. Pew researchers Andrew Perrin and Maeve Duggan (2015) documented other sources of the divide. For example, Americans with college degrees continue to have higher rates of Internet use than Americans with less than a college degree. Although less-educated adults have increased their Internet use since 2000, the percentage who use the Internet is still lower than the percentage of college graduates.[1]

In this chapter, we apply regression and correlation techniques to examine the relationship between interval-ratio variables. **Correlation** is a measure of association used to determine the existence and strength of the relationship between variables and is similar to the proportional reduction of error (PRE) measures reviewed in Chapter 8. **Regression** is a linear prediction model, using one or more independent variables to predict the values of a dependent variable. We will present two basic models: (1) **Bivariate regression** examines how changes in one independent variable affect the value of a dependent variable, while (2) **multiple regression** estimates how several independent variables affect one dependent variable.

We begin with exploring the relationship between educational attainment and Internet hours per week with the bivariate regression model. We will use *years of educational attainment* as our independent variable (X) to predict *Internet hours per week* (our dependent variable or Y). Fictional data are presented for a sample of 10 individuals in Table 10.1.

Chapter Learning Objectives

1. Describe linear relationships and prediction rules for bivariate and multiple regression models.
2. Construct and interpret straight-line graphs and best-fitting lines.
3. Calculate and interpret a and b.
4. Calculate and interpret the coefficient of determination (r^2) and Pearson's correlation coefficient (r).
5. Interpret multiple regression output.
6. Test the significance of r^2 and R^2 using ANOVA.

Correlation: A measure of association used to determine the existence and strength of the relationship between interval-ratio variables.

Regression: A linear prediction model using one or more independent variables to predict the values of a dependent variable.

Bivariate regression: A regression model that examines the effect of one independent variable on the values of a dependent variable.

Multiple regression: A regression model that examines the effects of several independent variables on the values of one dependent variable.

Table 10.1 Educational Attainment and Internet Hours per Week, N = 10

Educational Attainment (X)	Internet Hours per Week (Y)
10	1
9	0
12	3
13	4
19	7
11	2
16	6
23	9
14	5
21	8
$\bar{X} = 14.80$	$\bar{Y} = 4.50$
$s_x^2 = 9.17$	$s_y^2 = 9.17$
Range = 23 − 9 = 14	Range = 9 − 0 = 9

Source: Andrew Perrin and Maeve Duggan, *Americans' Internet Access: 2000–2015*, Pew Research Center, June 6, 2015.

THE SCATTER DIAGRAM

Scatter diagram (scatterplot): A visual method used to display a relationship between two interval-ratio variables.

One quick visual method used to display the relationship between two interval-ratio variables is the **scatter diagram** (or **scatterplot**). Often used as a first exploratory step in regression analysis, a scatter diagram can suggest whether two variables are associated.

The scatter diagram showing the relationship between *educational attainment* and *Internet hours per week* is shown in Figure 10.1. In a scatter diagram, the scales for the two variables form the vertical and horizontal axes of a graph. Usually, the independent variable, X, is arrayed along the horizontal axis and the dependent variable, Y, along the vertical axis. In Figure 10.1, each dot represents a person; its location lies at the exact intersection of that person's years of education and Internet hours per week. Note that individuals with lower educational attainment have fewer hours of Internet use, while individuals with higher educational attainment spend more time on the Internet per week. Educational attainment and Internet hours are positively associated.

Figure 10.1 Scatter Diagram of Educational Attainment and Internet Hours per Week

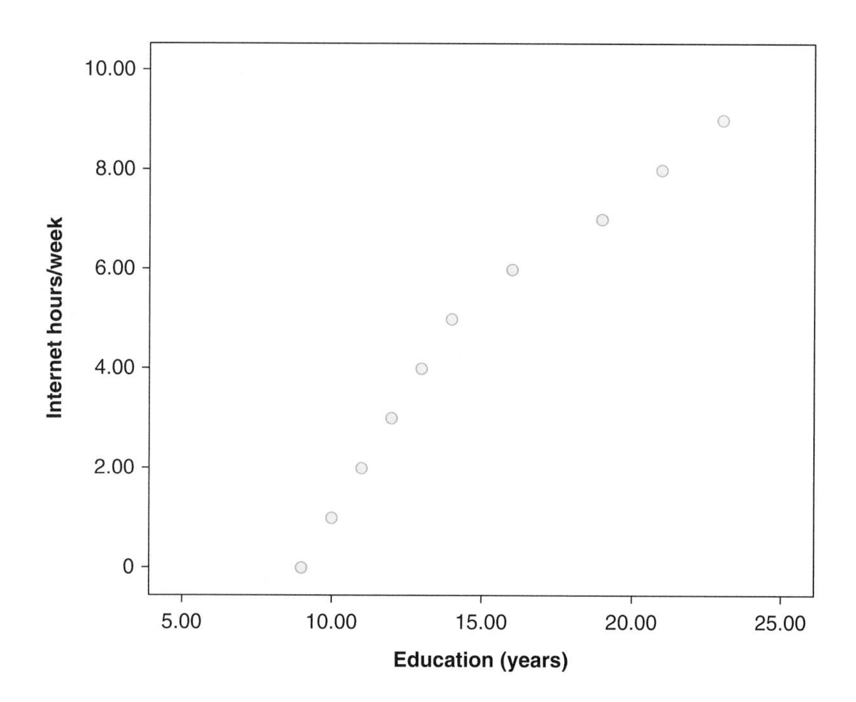

Scatter diagrams may also reveal a negative association between two variables or no relationship at all. We will review a negative relationship between two variables later in this chapter. Nonlinear relationships are explained in A Closer Look 10.1.

Linear relationship: A relationship between two interval-ratio variables in which the observations displayed in a scatter diagram can be approximated with a straight line.

Deterministic (perfect) linear relationship: A relationship between two interval-ratio variables in which all the observations (the dots) fall along a straight line. The line provides a predicted value of Y (the vertical axis) for any value of X (the horizontal axis).

LINEAR RELATIONSHIPS AND PREDICTION RULES

Although we can use a scatterplot as a first step to explore a relationship between two interval-ratio variables, we need a more systematic way to express the relationship between two interval-ratio variables. One way to express them is as a linear relationship. A **linear relationship** allows us to approximate the observations displayed in a scatter diagram with a straight line. In a perfectly linear relationship, all the observations (the dots) fall along a straight line. A perfect relationship is sometimes called a **deterministic relationship**, and the line itself provides a predicted value of Y (the vertical axis) for any value of X (the horizontal axis). For example, in Figure 10.2, we have superimposed a straight line on the scatterplot originally displayed in Figure 10.1. Using this line, we can obtain a predicted value of Internet hours per week for any individual by starting with a value from the education axis and then moving up to the Internet hours per week axis (indicated by the dotted lines). For example, the predicted value of Internet hours per week for someone with 12 years of education is approximately 3 hours.

Finding the Best-Fitting Line

As indicated in Figure 10.2, the actual relationship between years of education and Internet hours is not perfectly linear; that is, although some individual points lie very close to the line, none fall exactly on the line. Most relationships we study in the social sciences are not deterministic, and we are not able to come up with a linear equation that allows us to predict Y from X with perfect accuracy. We are much more likely to find relationships approximating linearity but in which numerous cases don't follow this trend perfectly.

Figure 10.2 A Straight-Line Graph for Educational Attainment and Internet Hours per Week

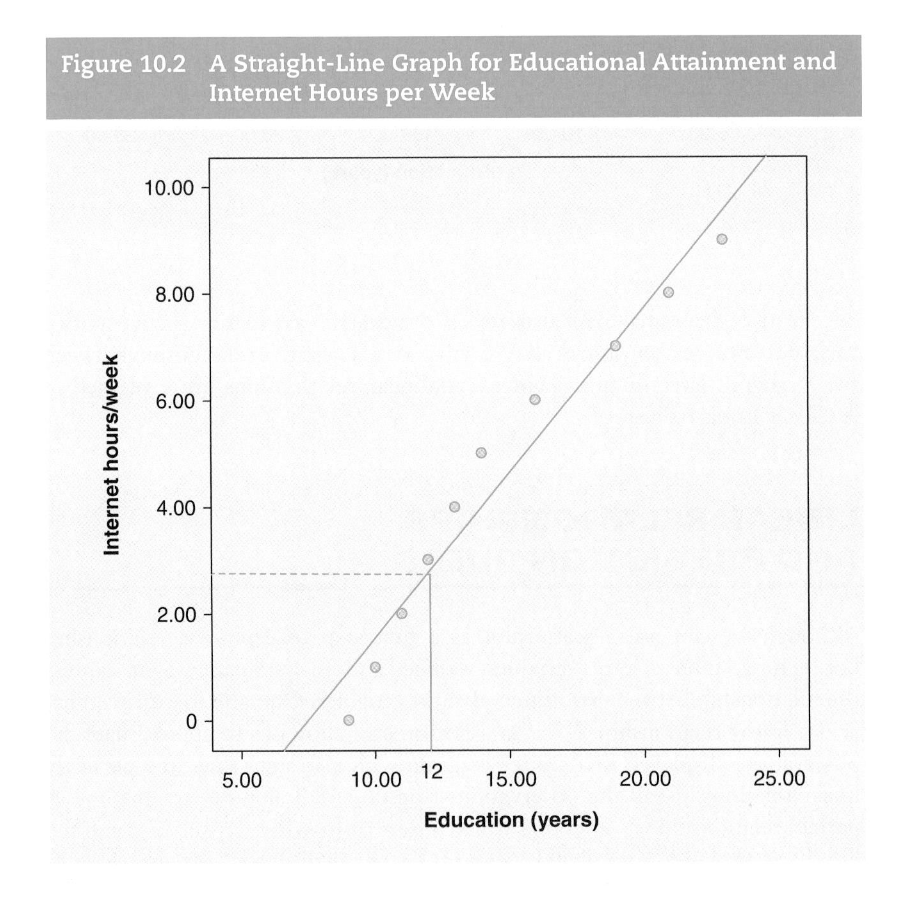

The relationship between educational attainment and Internet hours, as depicted in Figure 10.2, can also be described with the following algebraic equation, an equation for a straight line:

$$\hat{Y} = a + b(X) \quad (10.1)$$

where

$\hat{Y}$ = the predicted score on the dependent variable

X = the score on the independent variable

a = the Y-intercept, or the point where the line crosses the Y-axis; therefore, a is the value of Y when X is 0

b = the slope of the regression line, or the change in Y with a unit change in X

Y-intercept (a): The point where the regression line crosses the Y-axis and where $X = 0$.

Slope (b): The amount of change in a dependent variable per unit change in an independent variable.

LEARNING CHECK 10.1

For each of these four lines, as X goes up by 1 unit, what does Y do? Be sure you can answer this question using both the equation and the line.

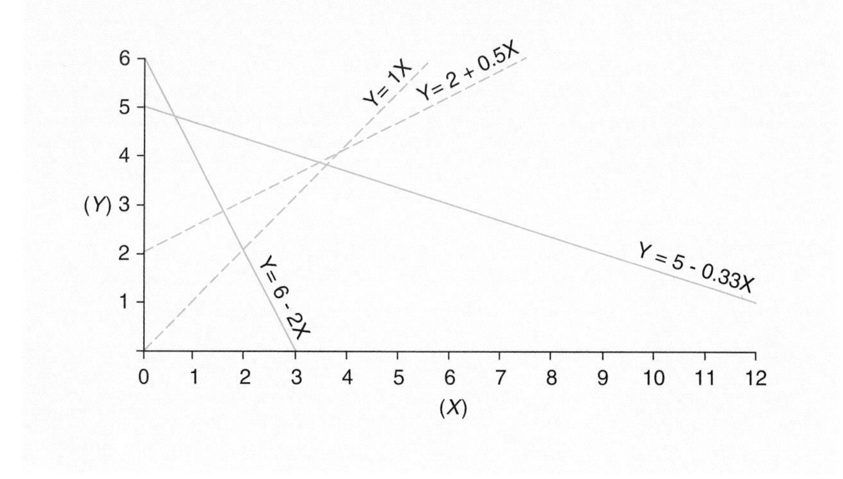

Defining Error

The best-fitting line is the one that generates the least amount of error, also referred to as the residual. Look again at Figure 10.2. For each education level, the line (or the equation that this line represents) predicts a value of Internet

A CLOSER LOOK 10.1

Other Regression Techniques

The regression examples we present in this chapter reflect two assumptions.

The first assumption is that the dependent and independent variables are interval-ratio measurements. In fact, regression models often include ordinal measures such as social class, income, and attitudinal scales. (Later in the chapter, we feature a regression model based on ordinal attitudinal scales.) Dummy variable techniques (creating a dichotomous variable, coded 1 or 0) permit the use of nominal variables, such as sex, race, religion, or political party affiliation. For example, in measuring gender, males could be coded as 0 and females coded as 1. Dummy variable techniques will not be elaborated here.

Our second assumption is that the variables have a linear or straight-line relationship. For the most part, social science relationships can be approximated using a linear equation. It is important to note, however, that sometimes a relationship cannot be approximated by a straight line and is better described by some other, nonlinear function. For example, Figure 10.3 shows a nonlinear relationship between age and hours of reading (hypothetical data). Hours of reading increase with age until the twenties, remain stable until the forties, and then tend to decrease with age.

Figure 10.3 A Nonlinear Relationship Between Age and Hours of Reading per Week

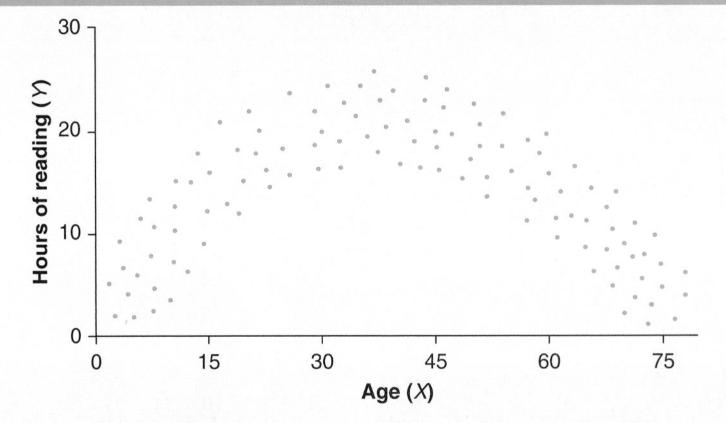

There are regression models for many nonlinear relationships, for nominal or dichotomous dependent variables, or even when there are multiple dependent variables. These advanced regression techniques will not be covered in this text.

hours. For example, with 21 years of education, the predicted value for Y is 8.34 hours. But we know from Table 10.1 that the actual value for 21 years of education is 8.0 hours. Thus, we have two values for Y: (1) a predicted Y, which we symbolize as $\hat{Y}$ and is generated by the prediction equation, also called the linear regression equation $\hat{Y} = a + b(X)$, and (2) the observed Y, symbolized simply as Y. Thus, for someone with 21 years of education, $\hat{Y} = 8.34$, whereas $Y = 8.0$.

We can think of the residual as the difference between the observed Y and the predicted $\hat{Y}$. If we symbolize the residual as e, then

$$e = Y - \hat{Y}$$

The residual is $8.34 - 8.0 = 0.34$ hours.

The Residual Sum of Squares (Σe^2)

Our goal is to identify a line or a prediction equation that minimizes the error for each individual observation. However, any line we choose will minimize the residual for some observations but may maximize it for others. We want to find a prediction equation that minimizes the residuals over all observations.

There are many mathematical ways of defining the residuals. For example, we may take the algebraic sum of residuals $\Sigma(Y - \hat{Y})$, the sum of the absolute residuals $\Sigma(|Y - \hat{Y}|)$, or the sum of the squared residuals $\Sigma(Y - \hat{Y})^2$. For mathematical reasons, statisticians prefer to work with the third method—squaring and summing the residuals over all observations. The result is the residual sum of squares, or Σe^2. Symbolically, Σe^2 is expressed as

$$\sum e^2 = \sum \left(Y - \hat{Y}\right)^2$$

The Least Squares Line

The best-fitting regression line is that line where the sum of the squared residuals, or Σe^2, is at a minimum. Such a line is called the **least squares line** (or **best-fitting line**), and the technique that produces this line is called the **least squares method**. The technique involves choosing a and b for the equation such that Σe^2 will have the smallest possible value. In the next section, we use the data from the 10 individuals to find the least squares equation.

> **Least squares line (best-fitting line):** A line where the residual sum of squares, or Σe^2, is at a minimum.
>
> **Least squares method:** The technique that produces the least squares line.

Computing *a* and *b*

Using calculus, it can be shown that to figure out the values of a and b in a way that minimizes Σe^2, we need to apply the following formulas:

$$b = \frac{s_{XY}}{s_X^2} \qquad (10.2)$$

$$a = \bar{Y} - b\left(\bar{X}\right) \qquad (10.3)$$

where

s_{XY} = the covariance of X and Y

s_X^2 = the variance of X

$\bar{Y}$ = the mean of Y

$\bar{X}$ = the mean of X

a = the Y-intercept

b = the slope of the line

These formulas assume that X is the independent variable and Y is the dependent variable.

Before we compute a and b, let's examine these formulas. The denominator for b is the variance of the variable X. It is defined as follows:

$$\text{Variance}\left(X\right) = s_X^2 = \frac{\sum\left(X - \bar{X}\right)^2}{N - 1}$$

This formula should be familiar to you from Chapter 3. The numerator (s_{XY}), however, is a new term. It is the covariance of X and Y and is defined as

$$\text{Covariance}\left(X,Y\right) = s_{XY} = \frac{\sum\left(X - \bar{X}\right)\left(Y - \bar{Y}\right)}{N - 1} \quad (10.4)$$

The covariance is a measure of how X and Y vary together. Essentially, the covariance tells us to what extent higher values of one variable are associated with higher values of the second variable (in which case we have a positive covariation) or with lower values of the second variable (which is a negative covariation). Based on the formula, we subtract the mean of X from each X score and the mean of Y from each Y score and then take the product of the two deviations. The results are then summed for all the cases and divided by $N - 1$.

In Table 10.2, we show the computations necessary to calculate the values of a and b for our 10 individuals. The means for educational attainment and Internet hours per week are obtained by summing column 1 and column 2, respectively, and dividing each sum by N. To calculate the covariance, we first subtract from each X score (column 3) and from each Y score (column 5) to obtain the mean deviations. We then multiply these deviations for every observation. The products of the mean deviations are shown in column 7.

The covariance is a measure of the linear relationship between two variables, and its value reflects both the strength and the direction of the relationship. The covariance will be close to zero when X and Y are unrelated; it will be larger than zero when the relationship is positive and smaller than zero when the relationship is negative.

Table 10.2 Worksheet for Calculating a and b for the Regression Equation

(1)	(2)	(3)	(4)	(5)	(6)	(7)
Educational Attainment	Internet Hours per Week					
X	Y	$(X-\bar{X})$	$(X-\bar{X})^2$	$(Y-\bar{Y})$	$(Y-\bar{Y})^2$	$(X-\bar{X})(Y-\bar{Y})$
10	1	–4.8	23.04	–3.5	12.25	16.80
9	0	–5.8	33.64	–4.5	20.25	26.10
12	3	–2.8	7.84	–1.5	2.25	2.25
13	4	–1.8	3.24	–0.5	.25	0.90
19	7	4.2	17.64	2.5	6.25	10.50
11	2	–3.8	14.44	–2.25	6.25	9.50
16	6	1.2	1.44	1.5	2.25	1.80
23	9	8.2	67.24	4.5	20.25	36.90
14	5	–0.8	.64	0.5	0.25	–0.40
21	8	6.2	38.44	3.5	12.25	21.70
$\Sigma X = 148$	$\Sigma Y = 45$	0[a]	207.60	0[a]	82.50	128

$$\bar{X} = \frac{\Sigma X}{N} = \frac{148}{10} = 14.8$$

$$\bar{Y} = \frac{\Sigma Y}{N} = \frac{45}{10} = 4.5$$

$$s_X^2 = \frac{\Sigma(X-\bar{X})^2}{N-1} = \frac{207.60}{9} = 23.07$$

$$s_X = \sqrt{23.07} = 4.80$$

$$s_Y^2 = \frac{\Sigma(Y-\bar{Y})^2}{N-1} = \frac{82.50}{9} = 9.17$$

$$s_Y = \sqrt{9.17} = 3.03$$

$$s_{XY} = \frac{\Sigma(X-\bar{X})(Y-\bar{Y})}{N-1} = \frac{128}{9} = 14.22$$

[a]Answers may differ due to rounding; however, the exact value of these column totals, properly calculated, will always be equal to zero.

A CLOSER LOOK 10.2

Understanding the Covariance

Let's say we have a set of eight data points for which the mean of X is 6 and the mean of Y is 3.

So the covariance in this case will be positive, giving us a positive b and a positive r.

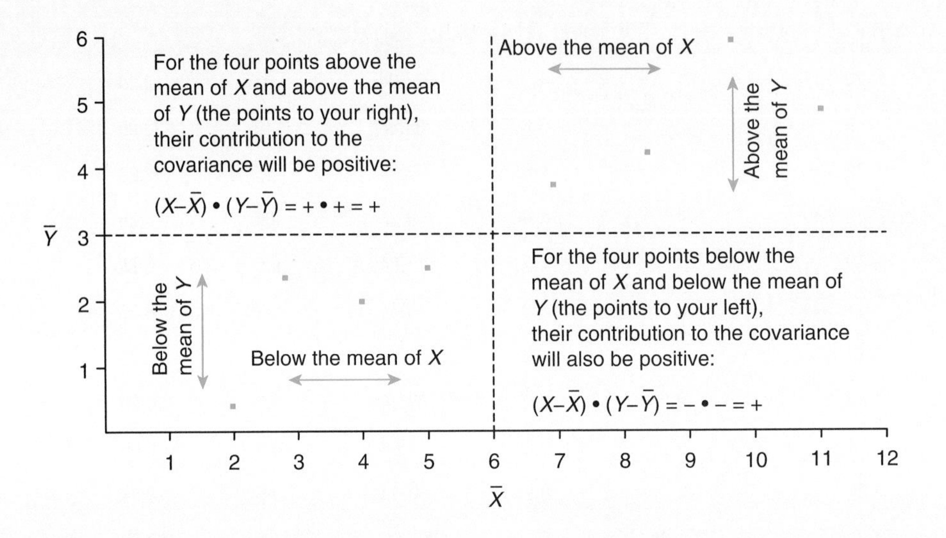

Now let's say we have a set of eight points that look like this:

So the covariance in this case will be negative, giving us a negative b.

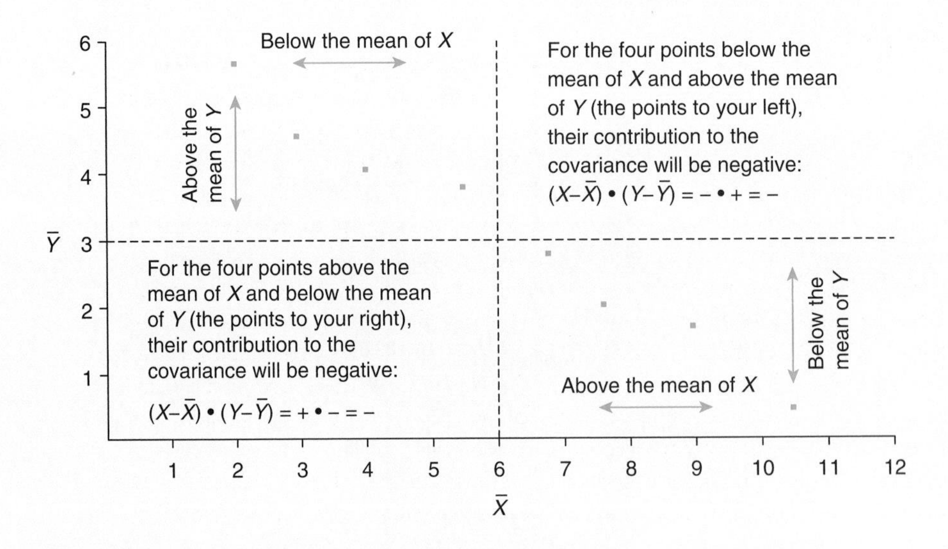

Now, let's substitute the values for the covariance and the variance from Table 10.2 to calculate b:

$$b = \frac{s_{XY}}{s_X^2} = \frac{14.22}{23.07} = 0.62$$

Once b has been calculated, we can solve for a, the intercept:

$$a = \bar{Y} - b(\bar{X}) = 4.5 - .62(14.8) = -4.68$$

The prediction equation is therefore

$$\hat{Y} = -4.68 + .62(X)$$

This equation can be used to obtain a predicted value for Internet hours per week given an individual's years of education. For example, for a person with 15 years of education, the predicted Internet hours is

$$\hat{Y} = -4.68 + 0.62(15) = 4.62$$

We can plot the straight-line graph corresponding to the regression equation. To plot a straight line, we need only two points, where each point corresponds to an X, Y value predicted.

Interpreting *a* and *b*

The b coefficient is equal to 0.62. This tells us that with each additional year of educational attainment, Internet hours per week are predicted to increase by 0.62 hours.

Note that because the relationships between variables in the social sciences are inexact, we don't expect our regression equation to make perfect predictions for every individual case. However, even though the pattern suggested by the regression equation may not hold for every individual, it gives us a tool by which to make the best possible guess about how Internet usage is associated, on average, with educational attainment. We can say that the slope of 0.62 is the estimate of this relationship.

The Y intercept a is the predicted value of Y, when $X = 0$. Thus, it is the point at which the regression line and the Y-axis intersect. The Y intercept can have positive or negative values. In this instance, it is unusual to consider someone with 0 years of education. As a general rule, be cautious when making predictions for Y based on values of X that are outside the range of the data, such as the −4.68 intercept calculated for our model. The intercept may not have a clear substantive interpretation.

We can plot the regression equation with two points: (1) the mean of X and the mean of Y and (2) 0 and the value of a. We've displayed this regression line in Figure 10.4.

Figure 10.4 The Best-Fitting Line for Educational Attainment and Internet Hours per Week

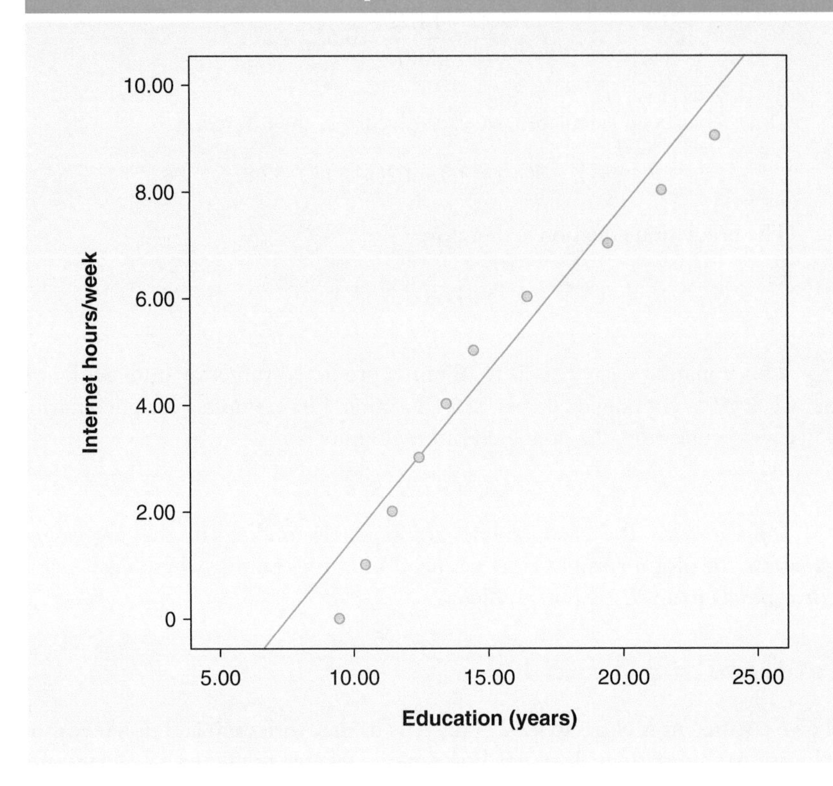

LEARNING CHECK 10.2

Use the prediction equation to calculate the predicted values of Y if X equals 9, 11, or 14. Verify that the regression line in Figure 10.2 passes through these points.

A NEGATIVE RELATIONSHIP: AGE AND INTERNET HOURS PER WEEK

Pew researchers Perrin and Duggan (2015) also documented how older adults have lagged behind younger adults in their Internet adoption. The majority of seniors, about 58%, currently use the Internet.[2] In this section, we'll examine the relationship between respondent age and Internet hours per week, defining *Internet hours* as the dependent variable (Y) and *age* as the independent variable (X). The fictional data are presented in Table 10.3 and the corresponding scatter diagram in Figure 10.5.

The scatter diagram reveals that age and Internet hours per week are

Table 10.3 Age and Internet Hours per Week, $N = 10$; Worksheet for Calculating a and b for the Regression Equation

(1)	(2)	(3)	(4)	(5)	(6)	(7)
Age	Internet Hours per Week					
X	Y	$(X-\bar{X})$	$(X-\bar{X})^2$	$(Y-\bar{Y})$	$(Y-\bar{Y})^2$	$(X-\bar{X})(Y-\bar{Y})$
55	1	17.2	295.84	–3.5	12.25	–60.2
60	0	22.2	492.84	–4.5	20.25	–99.9
45	3	7.2	51.84	–1.5	2.25	–10.8
35	4	–2.8	7.84	–0.5	.25	1.4
23	7	–14.8	219.04	2.5	6.25	–37
40	2	2.20	4.84	–2.5	6.25	–5.5
22	6	–15.8	249.64	1.5	2.25	–23.7
27	9	–10.8	116.64	4.5	20.25	–48.6
41	5	3.2	10.24	0.5	.25	1.6
30	8	–7.8	60.84	3.5	12.25	–27.3
$\Sigma X = 378$	$\Sigma Y = 45$	0[a]	1,509.60	0[a]	82.50	–310

$$\bar{X} = \frac{\Sigma X}{N} = \frac{378}{10} = 37.8$$

$$\bar{Y} = \frac{\Sigma Y}{N} = \frac{45}{10} = 4.5$$

$$s_X^2 = \frac{\Sigma(X-\bar{X})^2}{N-1} = \frac{1,509.60}{9} = 167.73$$

$$s_X = \sqrt{167.73} = 12.95$$

$$s_Y^2 = \frac{\Sigma(Y-\bar{Y})^2}{N-1} = \frac{82.50}{9} = 9.17$$

$$s_Y = \sqrt{9.17} = 3.03$$

$$s_{XY} = \frac{\Sigma(X-\bar{X})(Y-\bar{Y})}{N-1} = \frac{-310}{9} = -34.44$$

[a]Answers may differ due to rounding; however, the exact value of these column totals, properly calculated, will always be equal to zero.

Figure 10.5 The Best-Fitting Line for Age and Internet Hours per Week, N = 10

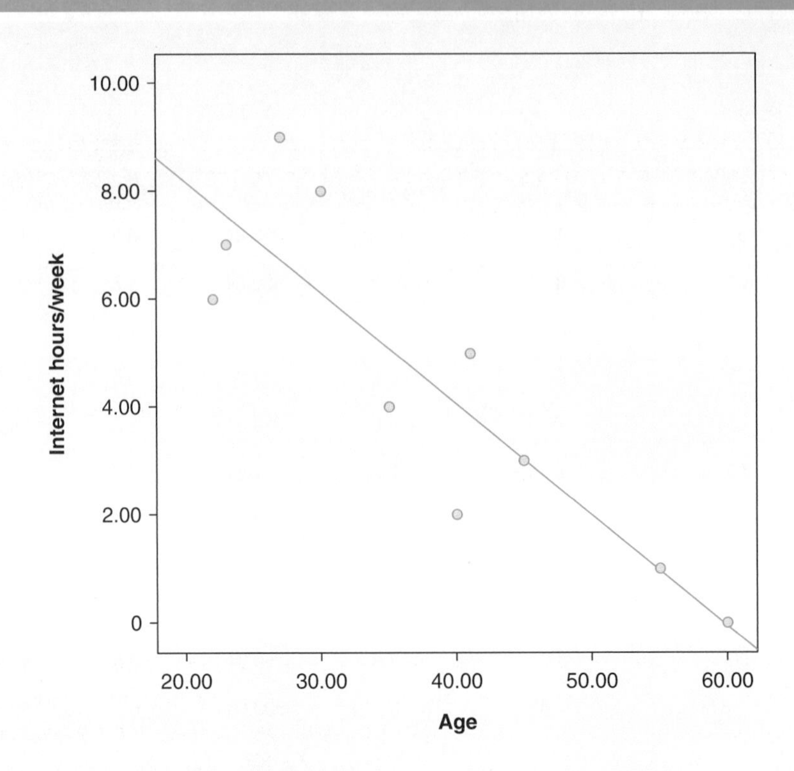

linearly related. It also illustrates that these variables are negatively associated; that is, as age increases, the number of hours of Internet access decreases. (Compare Figure 10.5 with Figure 10.4. Notice how the regression lines are in opposite direction—one positive, the other negative.)

For a more systematic analysis of the association, we will estimate the least squares regression equation for these data. Table 10.3 shows the calculations necessary to find *a* and *b* for our data on age and hours spent weekly on the Internet.

Now, let's substitute the values for the covariance and the variance from Table 10.3 to calculate *b:*

$$b = \frac{s_{xy}}{s_x^2} = \frac{-34.44}{167.73} = -.205 = -.21$$

Interpreting the slope, we can say that with each 1-year increase in age, Internet hours per week will decline by .21. This indicates a negative relationship between age and Internet hours. Once *b* has been calculated, we can solve for *a*, the intercept:

$$a = \bar{Y} - b(\bar{X}) = 4.5 - (-.21)(37.8) = 12.44$$

The prediction equation is therefore

$$\hat{Y} = 12.44 - .21(X)$$

This equation can be used to obtain a predicted value for Internet hours per week given respondent's age. In Figure 10.5, the regression line is plotted over our original scatter diagram.

METHODS FOR ASSESSING THE ACCURACY OF PREDICTIONS

So far, we calculated two regression equations that help us predict Internet usage per week based on educational attainment or age. In both cases, our predictions are far from perfect. If we examine Figures 10.4 and 10.5, we can see that we fail to make accurate predictions in every case. Although some of the individual points lie fairly close to the regression line, not all lie directly on the line—an indication that some prediction error was made. We have a model that helps us make predictions, but how can we assess the accuracy of these predictions?

We saw earlier that one way to judge our accuracy is to review the scatterplot. The closer the observations are to the regression line, the better the fit between the predictions and the actual observations. Still, we need a more systematic method for making such a judgment. We need a measure that tells us how accurate a prediction the regression model provides. The coefficient of determination, or r^2, is such a measure. The coefficient of determination measures the improvement in the prediction error based on our use of the linear prediction equation. The coefficient of determination is a PRE measure of association. Recall from Chapter 8 that PRE measures adhere to the following formula:

$$\text{PRE} = \frac{E_1 - E_2}{E_1}$$

where

E_1 = prediction errors made when the independent variable is ignored

E_2 = prediction errors made when the prediction is based on the independent variable

Applying this to the regression model, we have two prediction rules and two measures of error. The first prediction rule is in the absence of information on X, predict $\bar{Y}$. The error of prediction is defined as $Y - \bar{Y}$. The second rule of prediction uses X and the regression equation to predict $\hat{Y}$. The error of prediction is defined as $Y - \hat{Y}$.

To calculate these two measures of error for all the cases in our sample, we square the deviations and sum them. Thus, for the deviations from the mean of Y, we have

$$\sum\left(Y-\bar{Y}\right)^2$$

The sum of the squared deviations from the mean is called the *total sum of squares*, or *SST*:

$$SST=\sum\left(Y-\bar{Y}\right)^2$$

To measure deviation from the regression line, or $\hat{Y}$, we have

$$\sum\left(Y-\hat{Y}\right)^2$$

The sum of squared deviations from the regression line is denoted as the *residual sum of squares*, or *SSE*:

$$SSE=\sum\left(Y-\hat{Y}\right)^2$$

(We discussed this error term, the residual sum of squares, earlier in the chapter.)

The predictive value of the linear regression equations can be assessed by the extent to which the **residual sum of squares (*SSE*)** is smaller than the total sum of squares, *SST*. By subtracting *SSE* from *SST*, we obtain the **regression sum of squares (*SSR*)**, which reflects improvement in the prediction error resulting from our use of the linear prediction equation. *SSR* is defined as

Residual sum of squares (*SSE*): Sum of squared differences between observed and predicted Y.

Regression sum of squares (*SSR*): Reflects the improvement in the prediction error resulting from using the linear prediction equation, *SST* (sum of squared total) – *SSE* (residual sum of squares).

$$SSR=SST-SSE$$

Let's calculate r^2 for our regression model.

Calculating Prediction Errors

Figure 10.6 displays the regression line we calculated for educational attainment (X) and the Internet hours per week (Y) for 10 individuals, highlighting the prediction of Y for the person with 16 years of education, Subject A. Suppose we didn't know the actual Y, the number of Internet hours per week. Suppose further that we did not have knowledge of X, Subject A's years of education. Because the mean minimizes the sum of the squared errors for a set of scores, our best guess for Y would be $\bar{Y}$, or 4.5 hours. The horizontal line in Figure 10.6 represents this mean. Now, let's compare actual Y, 6 hours, with this prediction:

$$Y-\bar{Y}=6-4.5=1.5$$

Figure 10.6 Error Terms for Subject A

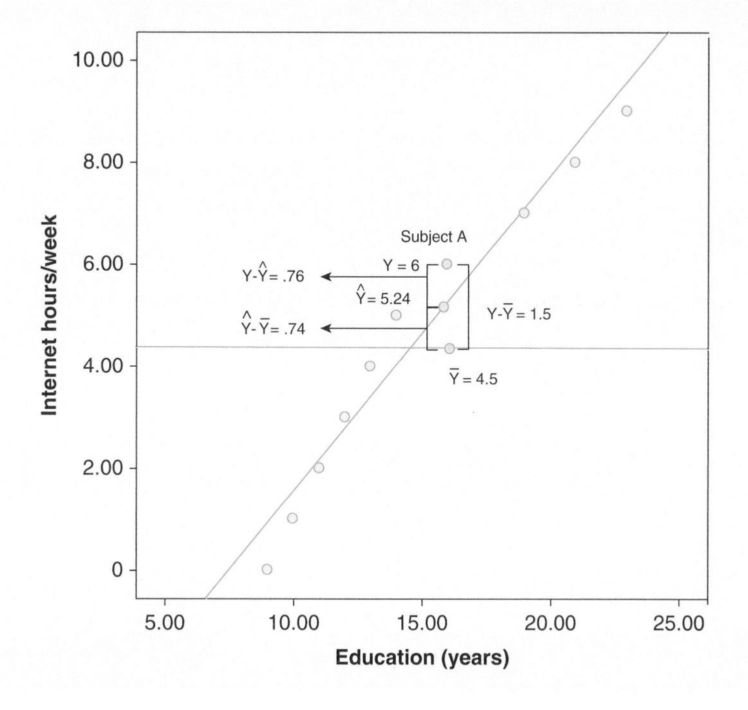

With an error of 1.5, our prediction of the average score for Subject A is not accurate.

Let's see if our predictive power can be improved by using our knowledge of *X*—the years of education—and its linear relationship with *Y*—Internet hours per week. If we insert Subject A's 16 years of education into our prediction equation, as follows:

$$\hat{Y} = -4.68 + .62(X)$$

$$\hat{Y} = -4.68 + .62(16) = 5.24$$

We can now recalculate our new error of prediction by comparing the predicted $\hat{Y}$ with the actual *Y*:

$$Y - \hat{Y} = 6 - 5.24 = .76$$

Although this prediction is by no means perfect, it is a slight improvement of .74 (1.5 – 0.76 = 0.74) over our earlier prediction. This improvement is illustrated in Figure 10.6.

Note that this improvement is the same as $\hat{Y} - \bar{Y} = 5.24 - 4.5 = .74$. This quantity represents the improvement in the prediction error resulting from our use of the linear prediction equation.

Let's calculate these terms for our data on educational attainment (X) and Internet use (Y).

We already have from Table 10.3 the total sum of squares:

$$SST = \sum\left(Y - \bar{Y}\right)^2 = 82.50$$

To calculate the errors sum of squares, we will calculate the predicted $\hat{Y}$ for each individual, subtract it from the observed Y, square the differences, and sum these for all 10 individuals. These calculations are presented in Table 10.4.

The residual sum of squares is thus

$$SSE = \sum\left(Y - \hat{Y}\right)^2 = 3.59$$

SSR is then given as

$$SSR = SST - SSE = 82.50 - 3.59 = 78.91$$

We have all the elements we need to construct a PRE measure. Because *SST* measures the prediction errors when the independent variable is ignored, we can define

Table 10.4 Worksheet for Calculating Errors Sum of Squares (SSE)

(1)	(2)	(3)	(4)	(5)
Educational Attainment	Internet Hours per Week	Predicted Y		
X	Y	$\hat{Y}$	$Y - \hat{Y}$	$(Y - \hat{Y})^2$
10	1	1.52	−0.52	0.27
9	0	0.90	−0.90	0.81
12	3	2.76	0.24	0.06
13	4	3.38	0.62	0.38
19	7	7.10	−0.10	0.01
11	2	2.14	−0.14	0.02
16	6	5.24	−0.76	0.58
23	9	9.58	−0.58	0.34
14	5	4.00	1	1.0
21	8	8.34	−0.34	0.12
$\Sigma X = 148$	$\Sigma Y = 45$			$\Sigma(Y - \hat{Y})^2 = 3.59$

$$E_1 = SST$$

Similarly, because *SSE* measures the prediction errors resulting from using the independent variable, we can define

$$E_2 = SSE$$

We are now ready to define the coefficient of determination r^2. It measures the PRE associated with using the linear regression equation as a rule for predicting *Y*:

$$\text{PRE} = \frac{E_1 - E_2}{E_1} = \frac{\sum(Y - \bar{Y})^2 - \sum(Y - \hat{Y})^2}{\sum(Y - \bar{Y})^2} \tag{10.5}$$

For our example,

$$r^2 = \frac{82.50 - 3.59}{82.50} = \frac{78.91}{82.50} = 0.96$$

The **coefficient of determination (r^2)** reflects the proportion of the total variation in the dependent variable, *Y*, explained by the independent variable, *X*. An r^2 of 0.96 means that by using educational attainment and the linear prediction rule to predict Internet hours per week, we have reduced the error of prediction by 96%. We can also say that the independent variable (*educational attainment*) explains about 96% of the variation in the dependent variable (*Internet hours per week*), as illustrated in Figure 10.7.

Coefficient of determination (r^2): A PRE measure reflecting the proportional reduction of error that results from using the linear regression model. It reflects the proportion of the total variation in the dependent variable, *Y*, explained by the independent variable, *X*.

Figure 10.7 A Pie Graph Approach to r^2

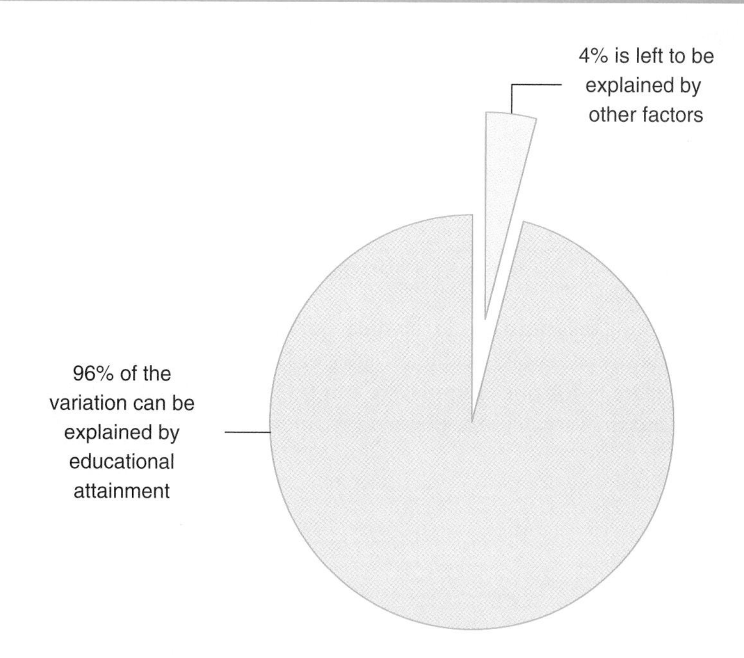

Figure 10.8 Examples Showing r^2 (a) Near 1.0 and (b) Near 0

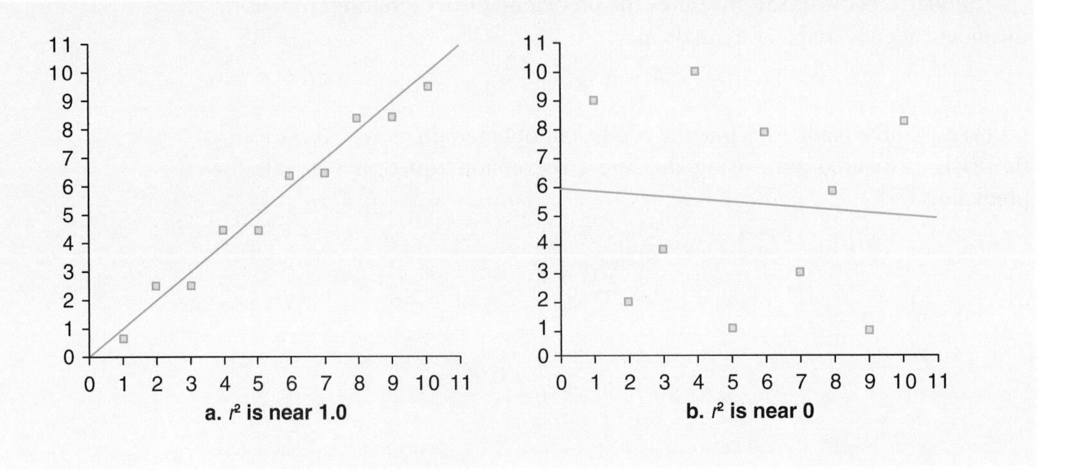

The coefficient of determination ranges from 0.0 to 1.0. An r^2 of 1.0 means that by using the linear regression model, we have reduced uncertainty by 100%. It also means that the independent variable accounts for 100% of the variation in the dependent variable. With an r^2 of 1.0, all the observations fall along the regression line, and the prediction error is equal to 0.0. An r^2 of 0.0 means that using the regression equation to predict Y does not improve the prediction of Y. Figure 10.8 shows r^2 values near 0.0 and near 1.0. In Figure 10.8a, where r^2 is approximately 1.0, the regression model provides a good fit. In contrast, a very poor fit is evident in Figure 10.8b, where r^2 is near zero. An r^2 near zero indicates either poor fit or a well-fitting line with a b of zero.

Calculating r^2

Another method for calculating r^2 uses the following equation:

$$r^2 = \frac{[\text{Covariance}(X,Y)]^2}{[\text{Variance}(X)][\text{Variance}(Y)]} = \frac{s_{XY}^2}{s_X^2 s_Y^2} \tag{10.6}$$

This formula tells us to divide the square of the covariance of X and Y by the product of the variance of X and the variance of Y.

To calculate r^2 for our example, we can go back to Table 10.2, where the covariance and the variances for the two variables have already been calculated:

$$s_{XY} = 14.22$$

$$s_X^2 = 23.07$$

$$s_Y^2 = 9.17$$

Therefore,

$$r^2 = \frac{14.22^2}{23.07(9.17)} = \frac{202.21}{211.55} = .96$$

Since we are working with actual values for educational attainment, its metric, or measurement, values are different from the metric values for the dependent variable, Internet hours PER week. While this hasn't been an issue until now, we must account for this measurement difference if we elect to use the variances and covariance to calculate r^2 (Formula 10.6). The remedy is quite simple: All we have to do is multiply our obtained r^2, 0.96, by 100 to obtain 96. Why multiply the obtained r^2 by 100?

We can multiply r^2 by 100 to obtain the percentage of variation in the dependent variable explained by the independent variable. An r^2 of 0.96 means that by using educational attainment and the linear prediction rule to predict *Y*, Internet hours per week, we have reduced uncertainty of prediction by 96%. We can also say that the independent variable explains 96% of the variation in the dependent variable, as illustrated in Figure 10.7.

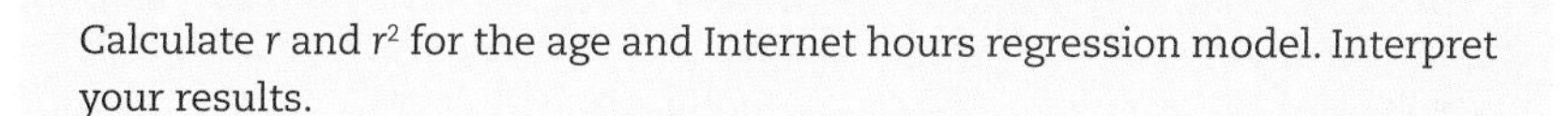

Calculate r and r^2 for the age and Internet hours regression model. Interpret your results.

TESTING THE SIGNIFICANCE OF r^2 USING ANOVA

Like other descriptive statistics, r^2 is an estimate based on sample data. Once r^2 is obtained, we should assess the probability that the linear relationship between median household income and the percentage of state residents with a bachelor's degree, as expressed in r^2, is really zero in the population (given the observed sample coefficient). In other words, we must test r^2 for statistical significance. ANOVA (analysis of variance), presented earlier in Chapter 9, can easily be applied to determine the statistical significance of the regression model as expressed in r^2. In fact, when you look closely, ANOVA and regression analysis can look very much the same. In both methods, we attempt to account for variation in the dependent variable in terms of the independent variable, except that in ANOVA, the independent variable is a categorical variable (nominal or ordinal, e.g., *gender* or *social class*), and with regression, it is an interval-ratio variable (e.g., *income measured in dollars*).

With ANOVA, we decomposed the total variation in the dependent variable into portions explained (*SSB*) and unexplained (*SSW*) by the independent variable. Next, we calculated the mean squares between (SSB / df_b) and mean squares within (SSW / df_w). The statistical test, *F*, is the ratio of the mean squares between to the mean squares within as shown in Formula 10.7.

$$F = \frac{\textit{Mean squares between}}{\textit{Mean squares within}} = \frac{SSB \,/\, df_b}{SSW \,/\, df_w} \tag{10.7}$$

With regression analysis, we decompose the total variation in the dependent variable into portions explained, *SSR*, and unexplained, *SSE*. Similar to ANOVA, the mean squares regression and the mean squares residual are calculated by dividing each sum of squares by its corresponding degrees of freedom (*df*). The degrees of freedom associated with *SSR* (df_r) are equal to *K*, which refers to the number of independent variables in the regression equation.

$$\text{Mean squares regression} = \frac{SSR}{df_r} = \frac{SSR}{K} \tag{10.8}$$

For *SSE*, degrees of freedom (df_e) is equal to $[N - (K + 1)]$, with *N* equal to the sample size.

$$\text{Mean squares residual} = \frac{SSE}{df_e} = \frac{SSE}{\left[N - (K+1)\right]} \tag{10.9}$$

In Table 10.5, for example, we present the ANOVA summary table for educational attainment and Internet hours.

Mean squares regression: An average computed by dividing the regression sum of squares (*SSR*) by its corresponding degrees of freedom.

Mean squares residual: An average computed by dividing the residual sum of squares (*SSE*) by its corresponding degrees of freedom.

In the table, under the heading Source of Variation are displayed the regression, residual, and total sums of squares. The column marked *df* shows the degrees of freedom associated with both the regression and residual sum of squares. In the bivariate case, *SSR* has 1 degree of freedom associated with it. The degrees of freedom associated with *SSE* is $[N - (K + 1)]$, where *K* refers to the number of independent variables in the regression equation. In the bivariate case, with one independent variable—*median household income*—*SSE* has $N - 2$ degrees of freedom associated with it $[N - (1 + 1)]$. Finally, the **mean squares regression** (*MSR*) and the **mean squares residual** (*MSE*) are calculated by dividing each sum of squares by its corresponding degrees of freedom. For our example,

$$MSR = \frac{SSR}{1} = \frac{78.91}{1} = 78.91$$

Table 10.5 ANOVA Summary Table for Educational Attainment and Internet Hours per Week

Source of Variation	Sum of Squares	*df*	Mean Squares	*F*
Regression	78.91	1	78.91	175.36
Residual	3.59	8	0.45	
Total	82.5	9		

$$MSE = \frac{SSE}{N-2} = \frac{3.57}{8} = 0.45$$

The F statistic together with the mean squares regression and the mean squares residual compose the obtained F ratio or F statistic. The F statistic is the ratio of the mean squares regression to the mean squares residual:

$$F = \frac{\text{Mean squares regression}}{\text{Mean squares residual}} = \frac{SSR / df_r}{SSE / df_e} \tag{10.10}$$

The F ratio, thus, represents the size of the mean squares regression relative to the size of the mean squares residual. The larger the mean squares regression relative to the mean squares residual, the larger the F ratio and the more likely that r^2 is significantly larger than zero in the population. We are testing the null hypothesis that r^2 is zero in the population.

The F ratio of our example is

$$F = \frac{Mean\ squares\ regression}{Mean\ squares\ residual} = \frac{78.91}{0.45} = 175.36$$

Making a Decision

To determine the probability of obtaining an F statistic of 175.36, we rely on Appendix E, Distribution of F. Appendix E lists the corresponding values of the F distribution for various degrees of freedom and two levels of significance, .05 and .01. We will set alpha at .05, and thus, we will refer to the table marked "$p < .05$." Note that Appendix E includes two *df*s. For the numerator, df_1 refers to the df_r associated with the mean squares regression; for the denominator, df_2 refers to the df_e associated with the mean squares residual. For our example, we compare our obtained F (175.36) to the F critical. When the *df*s are 1 (numerator) and 8 (denominator), and $\alpha < 0.5$, the F critical is 5.32. Since our obtained F is larger than the F critical ($175.36 > 5.32$), we can reject the null hypothesis that r^2 is zero in the population. We conclude that the linear relationship between educational attainment and Internet hours per week as expressed in r^2 is probably greater than zero in the population (given our observed sample coefficient).

LEARNING CHECK 10.4

Test the null hypothesis that there is a linear relationship between Internet hours and age. The mean squares regression is 63.66 with 1 degree of freedom. The mean squares residual is 2.355 with 8 degrees of freedom. Calculate the F statistic and assess its significance.

Pearson's Correlation Coefficient (r)

Pearson's correlation coefficient (r): The square root of r^2; it is a measure of association for interval-ratio variables, reflecting the strength and direction of the linear association between two interval-ratio variables. It can be positive or negative in sign.

The square root of r^2, or r—known as **Pearson's correlation coefficient**—is most often used as a measure of association between two interval-ratio variables:

$$r = \sqrt{r^2} \qquad (10.11)$$

Pearson's r is usually computed directly by using the following definitional formula:

$$r = \frac{[\text{Covariance}(X,Y)]}{[\text{Standard deviation}(X)]\,[\text{Standard deviation}(Y)]} = \frac{s_{XY}}{s_X s_Y}$$

Thus, r is defined as the ratio of the covariance of X and Y to the product of the standard deviations of X and Y.

Characteristics of Pearson's r

Pearson's r is a measure of relationship or association for interval-ratio variables. Like gamma (introduced in Chapter 8), it ranges from 0.01 to ±1.0, with 0.0 indicating no association between the two variables. An r of +1.0 means that the two variables have a perfect positive association; −1.0 indicates that it is a perfect negative association. The absolute value of r indicates the strength of the linear association between two variables. (Refer to A Closer Look 8.2 for an interpretational guide.) Thus, a correlation of −0.75 demonstrates a stronger association than a correlation of 0.50. Figure 10.9 illustrates a strong positive relationship, a strong negative relationship, a moderate positive relationship, and a weak negative relationship.

Unlike the b coefficient, r is a symmetrical measure. That is, the correlation between X and Y is identical to the correlation between Y and X.

To calculate r for our example of the relationship between *educational attainment* and *Internet hours per week*, let's return to Table 10.2, where the covariance and the standard deviations for X and Y have already been calculated:

$$r = \frac{s_{XY}}{s_X s_Y} = \frac{14.22}{(4.80)(3.03)} = 0.98$$

A correlation coefficient of 0.98 indicates that there is a strong positive linear relationship between educational attainment and Internet hours per week.

Note that we could have taken the square root of r^2 to calculate r, because $r = \sqrt{r^2}$ or $\sqrt{0.96} = 0.98$. Similarly, if we first calculate r, we can obtain r^2 simply by squaring r (be careful not to lose the sign of r^2).

Figure 10.9 Scatter Diagrams Illustrating Weak, Moderate, and Strong Relationships as Indicated by the Absolute Value of *r*

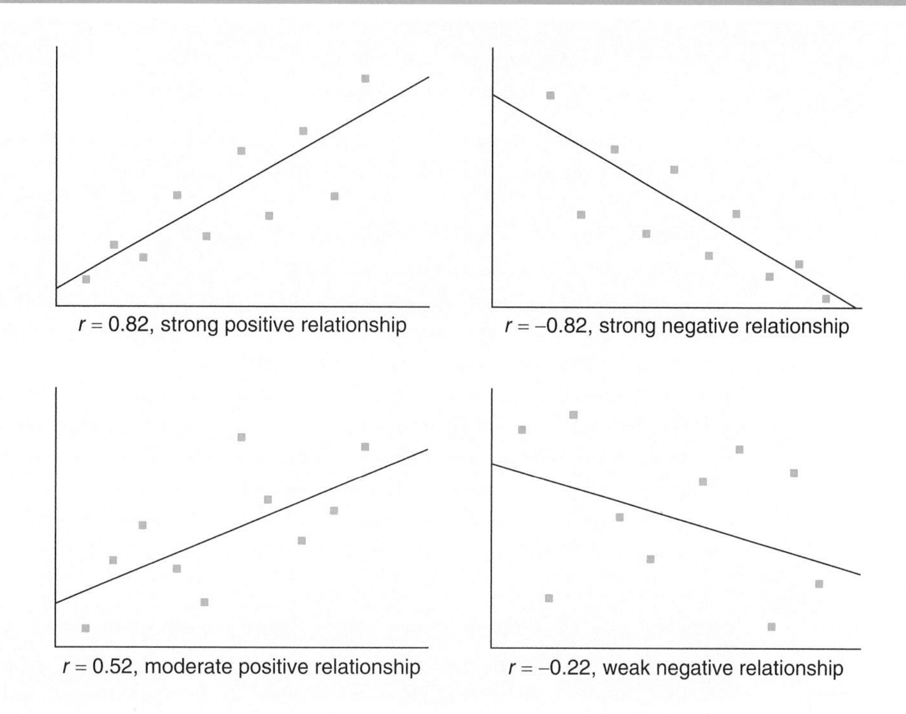

STATISTICS IN PRACTICE: MULTIPLE REGRESSION AND ANOVA

Thus far, we have used examples that involve only two interval-ratio variables: (1) a dependent variable and (2) an independent variable. Multiple regression is an extension of bivariate regression, allowing us to examine the effect of two or more independent variables on the dependent variable.[3]

The general form of the multiple regression equation involving two independent variables is

$$\hat{Y} = a + b_1^* X_1 + b_2^* X_2 \tag{10.12}$$

where

$\hat{Y}$ = the predicted score on the dependent variable

X_1 = the score on independent variable X_1

X_2 = the score on independent variable X_2

a = the Y-intercept, or the value of Y when both X_1 and X_2 are equal to zero

b_1^* = the partial slope of Y and X_1, the change in Y with a unit change in X_1, when the other independent variable X_2 is controlled

b_2^* = the partial slope of Y and X_2, the change in Y with a unit change in X_2, when the other independent variable X_1 is controlled

Partial slopes: The amount of change in Y for a unit change in a specific independent variable while controlling for the other independent variable(s).

Notice how the slopes are referred to as **partial slopes**. Partial slopes reflect the amount of change in Y for a unit change in a specific independent variable while controlling or holding constant the value of the other independent variables.

To illustrate, let's combine our investigation of Internet hours per week, educational attainment, and age. We hypothesize that individuals with higher levels of education will have higher levels of Internet use per week and that older individuals have lower hours of Internet use. We will estimate the multiple regression model data using SPSS. SPSS output is presented in Figure 10.10.

The partial slopes are reported in the Coefficients table, under the column labeled B. The intercept is also in column B, on the (Constant) row. Putting it all

Figure 10.10 SPSS Regression Output for Internet Hours per Week, Educational Attainment, and Age

Model Summary

Model	R	R Square	Adjusted R Square	Std. Error of the Estimate
1	.988[a]	.977	.970	.52203

a. Predictors: (Constant), educ, age

ANOVA[a]

Model		Sum of Squares	df	Mean Square	F	Sig.
1	Regression	80.592	2	40.296	147.870	.000[b]
	Residual	1.908	7	.273		
	Total	82.500	9			

a. Dependent Variable: internet

b. Predictors: (Constant), educ, age

Coefficients[a]

Model		Unstandardized Coefficients		Standardized Coefficients	t	Sig.
		B	Std. Error	Beta		
1	(Constant)	-.605	1.717		-.352	.735
	age	-.057	.023	-.245	-2.477	.042
	educ	.491	.062	.779	7.883	.000

a. Dependent Variable: internet

together, the multiple regression equation that incorporates both educational attainment and respondent age as predictors of Internet hours per week is

$$\hat{Y} = -.605 + .491(X_1) + -.057(X_2)$$

where

$\hat{Y}$ = number of Internet hours per week

X_1 = educational attainment

X_2 = age

This equation tells us that Internet hours increase by 0.49 per each year of education (X_1), holding age (X_2) constant. On the other hand, Internet hours decrease by 0.06 with each year increase in age (X_2) when we hold educational attainment (X_1) constant. Controlling for the effect of one variable, while examining the effect of the other, allows us to separate out the effects of each predictor independently of the other. For example, given two individuals with the same years of education, the person who might be a year older than the other is expected to use the Internet 0.06 hours less. Or given two individuals of the same age, the person who has 1 more year of education will have 0.49 hours more of Internet use than the other.

Finally, the value of *a* (–0.60) reflects Internet hours per week when both education and age are equal to zero. Although this *Y*-intercept doesn't lend itself to a meaningful interpretation, the value of *a* is a baseline that must be added to the equation for Internet hours to be properly estimated.

When a regression model includes more than one independent variable, it is likely that the units of measurement will vary. A multiple regression model could include income (dollars), highest degree (years), and number of children (individuals), making it difficult to compare their effects on the dependent variable. The **standardized slope coefficient** or **beta** (represented by the Greek letter, β) converts the values of each score into a *Z* score, standardizing the units of measurement so we can interpret their relative effects. Beta, also referred to as beta weights, ranges from 0 to ±1.0. The largest β value (whether negative or positive) identifies the independent variable with the strongest effect. Beta is reported in the SPSS Coefficient table, under the column labeled "Standardized Coefficients/Beta."

Standardized slope coefficient (or beta): The slope between the dependent variable and a specific independent variable when all scores are standardized or expressed as *Z* scores. Beta scores range from 0 to ±1.0.

A standardized multiple regression equation can be written as

$$\hat{Y} = a + \beta_1 X_1 + \beta_2 X_2 \qquad (10.13)$$

Based on this data example, the equation is

$$\hat{Y} = -.605 + .779X_1 - .245X_2$$

We can conclude that education has the strongest effect on Internet hours as indicated by the β value of 0.779 (compared with the beta of −.245 for age).

Multiple coefficient of determination (R^2): Measure that reflects the proportion of the total variation in the dependent variable that is explained jointly by two or more independent variables.

Like bivariate regression, multiple regression analysis yields a **multiple coefficient of determination**, symbolized as R^2 (corresponding to r^2 in the bivariate case). R^2 measures the PRE that results from using the linear regression model. It reflects the proportion of the total variation in the dependent variable that is explained jointly by two or more independent variables. We obtained an R^2 of 0.977 (in the Model Summary table, in the column labeled R Square). This means that by using educational attainment and age, we reduced the error of predicting Internet hours by 97.7 or 98%. We can also say that the independent variables, *educational attainment* and *age*, explain 98% of the variation in Internet hours per week.

Including respondent age in our regression model did not improve the prediction of Internet hours per week. As we saw earlier, educational attainment accounted for 96% of the variation in Internet hours per week. The addition of age to the prediction equation resulted in a 2% increase in the percentage of explained variation.

Pearson's multiple correlation coefficient (R): Measure of the linear relationship between the independent variable and the combined effect of two or more independent variables.

As in the bivariate case, the square root of R^2, or R, is **Pearson's multiple correlation coefficient**. It measures the linear relationship between the dependent variable and the combined effect of two or more independent variables. For our model, $R = 0.988$ or 0.99. This indicates that there is a strong relationship between the dependent variable and both independent variables.

The ANOVA summary table for multiple regression is nearly identical to the one for bivariate linear regression, except that the degrees of freedom are adjusted to reflect the number of independent variables in the model.

We conducted an ANOVA test to assess the probability that the linear relationship between Internet hours per week, educational attainment, and age as expressed by R^2 is really zero. The results of this test are reported in Figure 10.10. The obtained F statistic of 147.87 is shown in this table. With 2 and 7 degrees of freedom, we would need an F of 9.55 to reject the null hypothesis that $R^2 = 0$ at the .01 level. Since our obtained F exceeds that value (147.87 > 9.55), we can reject the null hypothesis with $p < .01$.

LEARNING CHECK 10.5

Use the prediction equation describing the relationship between Internet hours per week and both educational attainment and age to calculate Internet hours per week for someone with 20 years of education who is 35 years old.

SPSS can also produce a correlation matrix, a table that presents the Pearson's correlation efficient for all pairs of variables in the multiple regression model. A correlation matrix provides a baseline summary of the relationships between variables, identifying relationships or hypotheses that are usually the main research objective. Extensive correlation matrices are often presented in social science literature, but in this example, we have three pairs: (1) Internet hours with educational attainment, (2) Internet hours with age, and (3) educational attainment with age. Refer to Figure 10.11.

The matrix reports variable names in columns and rows. Note the diagonal from the upper left corner to the lower right corner reporting a correlation value of 1 (there are three 1s). This is the correlation of each variable with itself. This diagonal splits the matrix in half, creating mirrored correlations. We're interested in the intersection of the row and column variables, the cells that report their correlation coefficient for each pair. For example, the correlation coefficient for Internet and age, –0.878, is reported twice at the upper right-hand corner and at the lower left-hand corner. The other two correlations are also reported twice.

We calculated the correlation coefficients for Internet hours with educational attainment and Internet hours with age earlier in this chapter. The negative correlation between Internet hours and age is confirmed in Figure 10.10. We conclude that there is a strong negative relationship (–0.813) between these two variables. We also know that there is a strong positive correlation of 0.978 between Internet hours and educational attainment. The matrix also reports the significance of each correlation.

Figure 10.11 Correlation Matrix for Internet Hours per Week, Educational Attainment, and Age

Correlations

		internet	educ	age
internet	Pearson Correlation	1	.978**	-.878**
	Sig. (2-tailed)		.000	.001
	N	10	10	10
educ	Pearson Correlation	.978**	1	-.813**
	Sig. (2-tailed)	.000		.004
	N	10	10	10
age	Pearson Correlation	-.878**	-.813**	1
	Sig. (2-tailed)	.001	.004	
	N	10	10	10

**. Correlation is significant at the 0.01 level (2-tailed).

A CLOSER LOOK 10.3

Spurious Correlations and Confounding Effects

It is important to note that the existence of a correlation only denotes that the two variables are associated (they occur together or covary) and not that they are causally related. The well-known phrase "correlation is not causation" points to the fallacy of inferring that one variable causes the other based on the correlation between the variables. Such relationship is sometimes said to be spurious because both variables are influenced by a causally prior control variable, and there is no causal link between them. We can also say that a relationship between the independent and dependent variables is confounded by a third variable.

There are numerous examples in the research literature of spurious or confounded relationships. For instance, in a 2004 article, Michael Benson and his colleagues[4,5] discuss the issue of domestic violence as a correlate of race. Studies and reports have consistently found that rates of domestic abuse are higher in communities with a higher percentage of African American residents. Would this correlation indicate that race and domestic violence are causally related? To suggest that African Americans are more prone to engage in domestic violence would be erroneous if not outright racist. Benson and colleagues argue that the correlation between race and domestic violence is confounded by the level of economic distress in the community. Economically distressed communities are typically occupied by a higher percentage of African Americans. Also, rates of domestic violence tend to be higher in such communities. We can say that the relationship between race and domestic violence is confounded by a third variable—level of economic distress in the community.

Similarly, to test for the confounding effect of community economic distress on the relationship between race and domestic violence, Benson and Fox[6] calculated rates of domestic violence for African Americans and whites in communities with high and low levels of economic distress. They found that the relationship between race and domestic violence is not significant when the level of economic distress is constant. That is, the difference in the base rate of domestic violence for African Americans and whites is reduced by almost 50% in communities with high distress levels. In communities with low distress levels (and high income), the rate of domestic violence of African Americans is virtually identical to that of whites. The results showed that the correlation between race and domestic violence is accounted for in part by the level of economic distress of the community.

Uncovering spurious or confounded relations between an independent and a dependent variable can also be accomplished by using multiple regression. Multiple regression, an extension of bivariate regression, helps us examine the effect of an independent variable on a dependent variable while holding constant one or more additional variables.

READING THE RESEARCH LITERATURE: ACADEMIC INTENTIONS AND SUPPORT

Katherine Purswell, Ani Yazedjian, and Michelle Toews (2008)[7] used regression analysis to examine academic intentions (intention to perform specific behaviors related to learning engagement and positive academic behaviors), parental support, and peer support as predictors of self-reported academic behaviors (e.g., speaking in class, completed assignments on time during their freshman year) of first- and continuing-generation college students. The researchers apply social capital theory, arguing that relationships with others (parents and peers) would predict positive academic behaviors.

They estimated three separate multiple regression models for first-generation students (Group 1), students with at least one parent with college experience but with no degree (Group 2), and students with at least one parent with a bachelor's degree or higher (Group 3). The regression models are presented in Table 10.6. All of the variables included in the analysis are ordinal measures, with responses coded on a *strongly disagree* to *strongly agree* scale.

Each model is presented with partial and standardized slopes. No intercepts are reported. The multiple correlation coefficient and F statistic are also reported for each model. The asterisk indicates significance at the .05 level.

The researchers summarize the results of each model:

> The regression model was significant for all three groups ($p < .05$). For FGCS (first generation college students), the model predicted 24% of the variance in behavior. However, intention was the only significant predictor for this group. For the second group, the model predicted 18% of the variance, with peer support significantly predicting academic behavior. Finally, the model predicted 23% of the variance in behavior for those in the third group with all three independent variables—intention, parental support, and peer support—predicting academic behavior.[8]

Table 10.6 Regression Analyses Predicting Behavior by Intention, Parental Support, and Peer Support

	First-Generation Students (N = 44)		Group 2 (N = 82)		Group 3 (N = 203)	
	b	β	b	β	b	β
Intention	.75*	.49	.14	.15	.53*	.48
Parental support	.00	.00	.07	.04	.06*	.13
Peer support	−.02	−.02	.26*	.26	−.20*	−.16
R^2	.24		.18		.23	
F	3.82*		5.77*		18.74*	

Source: Adapted from Katherine Purswell, Ani Yazedjian, and Michelle Toews, "Students' Intentions and Social Support as Predictors of Self-Reported Academic Behaviors: A Comparison of First- and Continuing-Generation College Students," *Journal of College Student Retention* 10, no. 2 (2008): 200.

*$p < .05$.

DATA AT WORK

Shinichi Mizokami: Professor

Photo courtesy of Shinichi Mizokami

Dr. Mizokami is a professor of psychology and pedagogy at Kyoto University, Japan. Pedagogy is a discipline that examines educational theories and teaching methods. His current research involves two areas of study: (1) student learning and development and (2) identity formation in adolescence and young adulthood.

In 2013, his research team launched a 10-year transition survey with 45,000 second-year high school students. He uses multiple regression techniques to examine students' transition from school to work. "My team administers the surveys with the questions regarding what attitudes and distinctions competent students have or what activities they are engaged in. We analyze the data controlling the variables of gender, social class, major, kinds of university (doctoral, master's, or baccalaureate university), and find the results. In the multiple regression analysis, we carefully look at the bivariate tables and correlations between the used variables, and go back and forth between those descriptive statistics and the results [of the] multiple regression analysis."

He would be pleased to learn that you are enrolled in an undergraduate statistics course. According to Mizokami, "Many people will not have enough time to learn statistics after they start to work, so it may be worthwhile to study it in undergraduate education. Learning statistics can expand the possibilities of your job and provide many future advantages. . . . This can happen not only in academic fields but also in business. Good luck!"

MAIN POINTS

- A scatter diagram (also called scatterplot) is a quick visual method used to display relationships between two interval-ratio variables.
- Equations for all straight lines have the same general form:

$$\hat{Y} = a + b(X)$$

- The best-fitting regression line is that line where the residual sum of squares, or Σe^2, is at a minimum. Such a line is called the least squares line, and the technique that produces this line is called the least squares method.
- The coefficient of determination (r^2) and Pearson's correlation coefficient (r) measure how well the regression model fits the data. Pearson's r indicates the strength of the association between the two variables. The coefficient of determination is a PRE measure, identifying the reduction of error based on the regression model.

- The general form of the multiple regression equation involving two independent variables is $\hat{Y} = a + b_1^* X_1 + b_2^* X_2$. The multiple coefficient of determination (R^2) measures the proportional reduction of error based on the multiple regression model.
- The standardized multiple regression equation is $\hat{Y} = a + \beta_1 X_1 + \beta_2 X_2$. The beta coefficients allow us to assess the relative strength of all the independent variables.

KEY TERMS

bivariate regression 304
coefficient of determination (r^2) 321
correlation 304
deterministic (perfect) linear relationship 305
least squares line (best-fitting line) 309
least squares method 309
linear relationship 305
mean squares regression 324
mean squares residual 324
multiple coefficient of determination (R^2) 330
multiple regression 304
partial slopes 328
Pearson's correlation coefficient (r) 326
Pearson's multiple correlation coefficient (R) 330
regression 304
regression sum of squares (SSR) 318
residual sum of squares (SSE) 318
scatter diagram (scatterplot) 304
slope (b) 307
standardized slope coefficient (or beta) 329
Y-intercept (a) 307

DIGITAL RESOURCES

Access key study tools at https://edge.sagepub.com/ssdsess4e

- eFlashcards of the glossary terms
- Datasets and codebooks
- SPSS and Excel walk-through videos
- SPSS and Excel demonstrations and problems to accompany each chapter
- Appendix F: Basic Math Review

CHAPTER EXERCISES

1. Concerns over climate change, pollution, and a growing population have led to the formation of social action groups focused on environmental policies nationally and around the globe. A large number of these groups are funded through donor support. Based on the following eight countries, examine

the data to determine the extent of the relationship between simply being concerned about the environment and giving money to environmental groups.

Country	Percentage Concerned	Percentage Donating Money
United States	33.8	22.8
Austria	35.5	27.8
The Netherlands	30.1	44.8
Slovenia	50.3	10.7
Russia	29.0	1.6
Philippines	50.1	6.8
Spain	35.9	7.4
Denmark	27.2	22.3

Source: International Social Survey Programme, 2000.

a. Construct a scatterplot of the two variables, placing percentage concerned about the environment on the horizontal or *X*-axis and the percentage donating money to environmental groups on the vertical or *Y*-axis.

b. Does the relationship between the two variables seem linear? Describe the relationship.

c. Find the value of the Pearson correlation coefficient that measures the association between the two variables and offer an interpretation.

2. In this exercise, we will investigate the relationships between adolescent fertility rate and female labor force participation in South America. Data are presented for 2014.

Country	Adolescent Fertility Rate	Female Labor Force Participation Rate
Argentina	40.4	63.9
Bolivia	44.6	71.1
Brazil	43.8	67.3
Chile	40.8	48.1
Colombia	42.6	51.7

Country	Adolescent Fertility Rate	Female Labor Force Participation Rate
Ecuador	40.4	76.2
Paraguay	39.2	58.0
Peru	45.3	49.7
Uruguay	44.5	56.5
Venezuela	39.9	79.7

Source: World Bank, *Health Nutrition and Population Statistics,* 2014.

a. Construct a scatterplot for adolescent fertility rate and labor force participation rate. Do you think the scatterplot can be characterized by a linear relationship?

b. Calculate the coefficient of determination and correlation coefficient.

c. Describe the relationship between the variables based on your calculations.

3. Let's examine the relationship between a country's gross national product (GNP) and the percentage of respondents willing to pay higher prices for goods to protect the environment. The following table displays information for five countries selected at random.

   a. Calculate the correlation coefficient between a country's GNP and the percentage of its residents willing to pay higher prices to protect the environment. What is its value?

   b. Provide an interpretation for the coefficient.

Country	GNP per Capita	Percentage Willing to Pay
United States	29.24	44.9
Ireland	18.71	53.3
The Netherlands	24.78	61.2
Norway	34.31	40.7
Sweden	25.58	32.6

Source: International Social Survey Programme, 2000.

4. In 2010, a U.S. Census Bureau report revealed that approximately 14.3% of all Americans were living below the poverty line in 2009. This figure is higher than in 2000, when the poverty rate was 12.2%. Individuals and

families living below the poverty line face many obstacles, the least of which is access to health care. In many cases, those living below the poverty line are without any form of health insurance. Using data from the U.S. Census Bureau, analyze the relationship between living below the poverty line and access to health care for a random sample of 12 states. (The health insurance data are pre–Affordable Care Act implementation.)

State	Percentage Below Poverty Line (2009)	Percentage Without Health Insurance (2009)
Alabama	17.9	16.9
California	14.2	20
Idaho	14.3	15.2
Louisiana	17.3	16
New Jersey	9.4	15.8
New York	14.2	14.8
Pennsylvania	12.5	11.4
Rhode Island	11.5	12.3
South Carolina	17.1	17.0
Texas	17.2	26.1
Washington	12.3	12.9
Wisconsin	12.4	9.5

Source: U.S. Census Bureau, *The 2012 Statistical Abstract*, 2011, Tables 709 and 156.

a. Construct a scatterplot, predicting the percentage without health insurance with the percentage living below the poverty level. Does it appear that a straight-line relationship will fit the data?

b. Calculate the regression equation with percentage of the population without health insurance as the dependent variable, and draw the regression line on the scatterplot. What is its slope? What is the intercept? Has your opinion changed about whether a straight line seems to fit the data? Are there any states that fall far from the regression line? Which one(s)?

5. We test the hypothesis that as an individual's years of education increases, the individual will have fewer children. Based on a subsample from the GSS 2018, we present a scatterplot and regression output for the variables EDUC and CHILDS. Interpret the results.

Scatterplot of Number of Children by Education

Model Summary

Model	R	R Square	Adjusted R Square	Std. Error of the Estimate
1	.225[a]	.051	.050	1.661

a. Predictors: (Constant), Highest year of school completed

ANOVA[a]

Model		Sum of Squares	df	Mean Square	F	Sig.
1	Regression	163.764	1	163.764	59.366	.000[b]
	Residual	3067.529	1112	2.759		
	Total	3231.293	1113			

a. Dependent Variable: Number of children

b. Predictors: (Constant), Highest year of school completed

Coefficients[a]

Model		Unstandardized Coefficients		Standardized Coefficients	t	Sig.
		B	Std. Error	Beta		
1	(Constant)	3.605	.229		15.763	.000
	Highest year of school completed	-.126	.016	-.225	-7.705	.000

a. Dependent Variable: Number of children

Linear Regression Output Specifying the Relationship Between Education and Number of Children

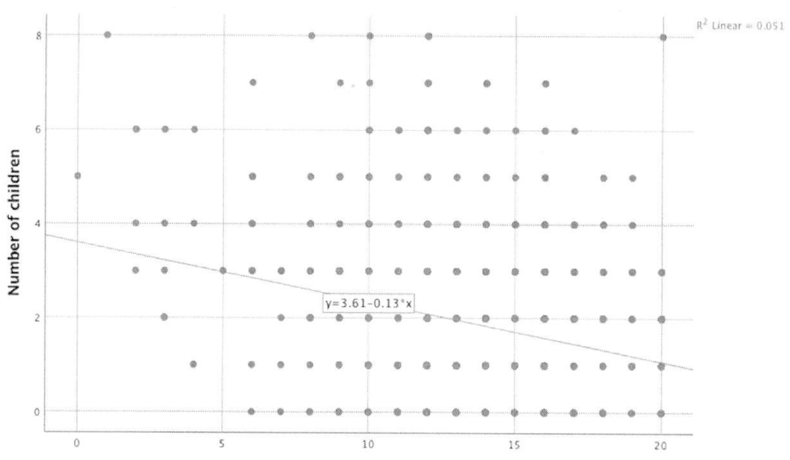

6. We present SPSS output examining the relationship between education (measured in years) and television viewing per day (measured in hours) based on a GSS 2018 subsample. As we predicted in the SPSS Demonstration, we hypothesize that as educational attainment increases, hours of television viewing will decrease, indicating a negative relationship between the two variables. Discuss the significance of the overall model based on F and its p values. Is the relationship between education and television viewing significant?

Linear Regression Output for Hours Spent per Day Watching Television and Education

Model Summary

Model	R	R Square	Adjusted R Square	Std. Error of the Estimate
1	.204[a]	.042	.040	3.176

a. Predictors: (Constant), Highest year of school completed

ANOVA[a]

Model		Sum of Squares	df	Mean Square	F	Sig.
1	Regression	322.181	1	322.181	31.945	.000[b]
	Residual	7402.683	734	10.085		
	Total	7724.864	735			

a. Dependent Variable: Hours per day watching TV

b. Predictors: (Constant), Highest year of school completed

Coefficients[a]

Model		Unstandardized Coefficients		Standardized Coefficients	t	Sig.
		B	Std. Error	Beta		
1	(Constant)	6.017	.527		11.412	.000
	Highest year of school completed	-.215	.038	-.204	-5.652	.000

a. Dependent Variable: Hours per day watching TV

7. Based on the following SPSS output, describe the regression model for educational attainment and amount of money given to charity based on GSS 2014.

   a. Assess the significance of the overall model based on its F and p values. What is the relationship between the two variables?

   b. Calculate the predicted charitable amount for a respondent with 14 years of education and for a respondent with 20 years of education.

Linear Regression Output for Amount of Money Given to Charity and Education

Model Summary

Model	R	R Square	Adjusted R Square	Std. Error of the Estimate
1	.169[a]	.028	.026	6108.241

a. Predictors: (Constant), educ HIGHEST YEAR OF SCHOOL COMPLETED

ANOVA[a]

Model		Sum of Squares	df	Mean Square	F	Sig.
1	Regression	394689548	1	394689548	10.578	.001[b]
	Residual	1.351E+10	362	37310613.5		
	Total	1.390E+10	363			

a. Dependent Variable: VALGIVEN TOTAL DONATIONS PAST YEAR R AND IMMEDIATE FAMILY

b. Predictors: (Constant), educ HIGHEST YEAR OF SCHOOL COMPLETED

Coefficients[a]

Model		Unstandardized Coefficients		Standardized Coefficients	t	Sig.
		B	Std. Error	Beta		
1	(Constant)	-2213.620	1293.760		-1.711	.088
	educ HIGHEST YEAR OF SCHOOL COMPLETED	304.106	93.500	.169	3.252	.001

a. Dependent Variable: VALGIVEN TOTAL DONATIONS PAST YEAR R AND IMMEDIATE FAMILY

8. Research on social mobility, status, and educational attainment has provided convincing evidence on the relationship between parents' and children's socioeconomic achievement. For this exercise, refer to the scatterplot and regression output from a GSS 2018 subsample to describe the relationship between mother's education and respondent's education.

Scatterplot of Respondent Level of Education by Mother's Level of Education

Linear Regression Output Specifying the Relationship Between Respondent's Education by Mother's Education

Model Summary

Model	R	R Square	Adjusted R Square	Std. Error of the Estimate
1	.411[a]	.169	.168	2.746

a. Predictors: (Constant), Mothers highest degree

ANOVA[a]

Model		Sum of Squares	df	Mean Square	F	Sig.
1	Regression	1563.928	1	1563.928	207.348	.000[b]
	Residual	7693.360	1020	7.543		
	Total	9257.288	1021			

a. Dependent Variable: Highest year of school completed

b. Predictors: (Constant), Mothers highest degree

Coefficients[a]

Model		Unstandardized Coefficients		Standardized Coefficients		
		B	Std. Error	Beta	t	Sig.
1	(Constant)	12.433	.124		100.037	.000
	Mothers highest degree	1.090	.076	.411	14.400	.000

a. Dependent Variable: Highest year of school completed

9. We further explore the relationship between respondent's education and mother's education, computing regression models separately for males and females.

a. Calculate the regression equation for each.

b. What is the predicted value of respondent's education when mother's education is
15 years?

c. For which gender group is the relationship between respondent's education and mother's education strongest? Explain.

Linear Regression Output Specifying the Relationship Between Respondent Level of Education by Mother's Level of Education: Males Only

Model Summary[a]

Model	R	R Square	Adjusted R Square	Std. Error of the Estimate
1	.404[b]	.163	.162	2.694

a. Respondents sex = MALE

b. Predictors: (Constant), Mothers highest degree

ANOVA[a,b]

Model		Sum of Squares	df	Mean Square	F	Sig.
1	Regression	619.947	1	619.947	85.409	.000[c]
	Residual	3171.989	437	7.259		
	Total	3791.936	438			

a. Respondents sex = MALE

b. Dependent Variable: Highest year of school completed

c. Predictors: (Constant), Mothers highest degree

Coefficients[a,b]

		Unstandardized Coefficients		Standardized Coefficients		
Model		B	Std. Error	Beta	t	Sig.
1	(Constant)	12.447	.191		65.189	.000
	Mothers highest degree	1.055	.114	.404	9.242	.000

a. Respondents sex = MALE

b. Dependent Variable: Highest year of school completed

Linear Regression Output Specifying the Relationship Between Respondent Level of Education by Mother's Level of Education: Females Only

Model Summary[a]

Model	R	R Square	Adjusted R Square	Std. Error of the Estimate
1	.416[b]	.173	.172	2.789

a. Respondents sex = FEMALE

b. Predictors: (Constant), Mothers highest degree

ANOVA[a,b]

Model		Sum of Squares	df	Mean Square	F	Sig.
1	Regression	945.352	1	945.352	121.529	.000[c]
	Residual	4519.492	581	7.779		
	Total	5464.844	582			

a. Respondents sex = FEMALE

b. Dependent Variable: Highest year of school completed

c. Predictors: (Constant), Mothers highest degree

Coefficients[a,b]

Model		Unstandardized Coefficients		Standardized Coefficients		
		B	Std. Error	Beta	t	Sig.
1	(Constant)	12.423	.164		75.752	.000
	Mothers highest degree	1.117	.101	.416	11.024	.000

a. Respondents sex = FEMALE

b. Dependent Variable: Highest year of school completed

10. In Exercise 6, we examined the relationship between years of education and hours of television watched per day. We saw that as education increases, hours of television viewing decreases. The number of children a family has could also affect how much television is viewed per day. Having children may lead to more shared and supervised viewing and thus increases the number of viewing hours. The following SPSS output displays the relationship between television viewing (measured in hours per day) and both education (measured in years) and number of children. We hypothesize that whereas more education may lead to less viewing, the number of children has the opposite effect: Having more children will result in more hours of viewing per day.

    a. What is the *b* coefficient for education? For number of children? Interpret each coefficient. Is the relationship between each independent variable and hours of viewing as hypothesized?

    b. Using the multiple regression equation with both education and number of children as independent variables, calculate the number of hours of television viewing for a person with 16 years of education and two children. Using the equation from Exercise 6, how do the results compare between a person with 16 years of education (number of children not included in the equation) and a person with 16 years of education with two children?

c. Compare the r^2 value from Exercise 6 with the R^2 value from this regression. Does using education and number of children jointly reduce the amount of error involved in predicting hours of television viewed per day?

Multiple Regression Output Specifying the Relationship Between Education, Number of Children, and Hours Spent per Day Watching Television

Model Summary

Model	R	R Square	Adjusted R Square	Std. Error of the Estimate
1	.213[a]	.046	.043	3.173

a. Predictors: (Constant), Number of children, Highest year of school completed

ANOVA[a]

Model		Sum of Squares	df	Mean Square	F	Sig.
1	Regression	351.146	2	175.573	17.439	.000[b]
	Residual	7359.542	731	10.068		
	Total	7710.688	733			

a. Dependent Variable: Hours per day watching TV

b. Predictors: (Constant), Number of children, Highest year of school completed

Coefficients[a]

Model		Unstandardized Coefficients		Standardized Coefficients	t	Sig.
		B	Std. Error	Beta		
1	(Constant)	5.596	.593		9.438	.000
	Highest year of school completed	-.201	.039	-.190	-5.105	.000
	Number of children	.118	.071	.062	1.657	.098

a. Dependent Variable: Hours per day watching TV

11. We return to our chapter analysis of Internet hours per week (WWWHR), educational attainment (EDUC), and respondent age (AGE), presenting the multiple regression model and correlation matrix based on GSS 2018 data.

a. What is the b coefficient for education? For age? Interpret each coefficient. Is the relationship between education and Internet hours as hypothesized in our chapter example? For age and Internet hours?

b. Using the multiple regression equation with both education and age as independent variables, calculate the number of Internet hours per week for a person with 16 years of education and 55 years of age.

c. Using the standardized multiple regression equation, identify which independent variable has the strongest effect on Internet hours per week.

d. Interpret the multiple coefficient of determination.

e. Interpret each correlation coefficient (based on the correlation matrix).

Multiple Regression Output for Internet Hours per Week, Educational Attainment, and Age

Model Summary

Model	R	R Square	Adjusted R Square	Std. Error of the Estimate
1	.319[a]	.101	.099	16.402

a. Predictors: (Constant), Age of respondent, Highest year of school completed

ANOVA[a]

Model		Sum of Squares	df	Mean Square	F	Sig.
1	Regression	19173.948	2	9586.974	35.637	.000[b]
	Residual	169748.495	631	269.015		
	Total	188922.443	633			

a. Dependent Variable: Www hours per week

b. Predictors: (Constant), Age of respondent, Highest year of school completed

Coefficients[a]

Model		Unstandardized Coefficients		Standardized Coefficients		
		B	Std. Error	Beta	t	Sig.
1	(Constant)	21.827	3.746		5.827	.000
	Highest year of school completed	.515	.237	.082	2.171	.030
	Age of respondent	-.309	.037	-.311	-8.235	.000

a. Dependent Variable: Www hours per week

Correlation Matrix for Internet Hours per Week, Educational Attainment, and Age

Correlations

		Www hours per week	Highest year of school completed	Age of respondent
Www hours per week	Pearson Correlation	1	.068	-.309**
	Sig. (2-tailed)		.089	.000
	N	638	637	635
Highest year of school completed	Pearson Correlation	.068	1	-.088**
	Sig. (2-tailed)	.089		.003
	N	637	1116	1113
Age of respondent	Pearson Correlation	-.309**	-.088**	1
	Sig. (2-tailed)	.000	.003	
	N	635	1113	1114

**. Correlation is significant at the 0.01 level (2-tailed).

12. We revisit Katherine Purswell, Ani Yazedjian, and Michelle Toews's (2008)[9] research regarding the relationship between academic intentions (intention to perform specific behaviors related to learning engagement and positive academic behaviors), parental support, and peer support and self-reported academic behaviors (e.g., speaking in class, completed assignments on time during their freshman year) of first- and continuing-generation college students.

They estimated three separate models for first-generation students (Group 1), students with at least one parent with college experience but with no degree (Group 2), and students with at least one parent with a bachelor's degree or higher (Group 3). The correlation matrix is presented below.

All of the variables included in the analysis are ordinal measures, with responses coded on a *strongly disagree* to *strongly agree* scale.

Intercorrelations Between Variables Based on Parental Education Groups

	1	2	2	3
First Generation Students ($n = 44$)				
1. Intention	—	.34*	.11	.48**
2. Parental support		—	.24	.15
3. Peer support			—	.06
4. Behavior				—
Group 2 ($n = 82$)				
1. Intention	—	.30**	.49**	.32**
2. Parental support		—	.31	.27*
3. Peer support			—	.38**
4. Behavior				—
Group 3 ($n = 203$)				
1. Intention	—	.18*	.37**	.44*
2. Parental support		—	.24**	.17*
3. Peer support			—	.04
4. Behavior				—

Source: Adapted from Katherine Purswell, Ani Yazedjian, and Michelle Toews, "Students' Intentions and Social Support as Predictors of Self-Reported Academic Behaviors: A Comparison of First- and Continuing-Generation College Students," *Journal of College Student Retention* 10, no. 2 (2008): 199.

$*p < .05$. $**p < .01$.

a. Which group has the most significant correlations? Which group has the least?

b. Interpret the correlation for intention and behavior for the three groups. For which group is the relationship the strongest?

c. The correlation for peer support and intention is highest for which group? Explain.

13. We expand on the model for TVHOURS presented in Exercise 10, adding work hours (HRS1) and respondent age (AGE) as independent variables. Data are based on the GSS 2018.

Model Summary

Model	R	R Square	Adjusted R Square	Std. Error of the Estimate
1	.155[a]	.024	.015	2.298

a. Predictors: (Constant), Number of hours worked last week, Number of children, Highest year of school completed, Age of respondent

ANOVA[a]

Model		Sum of Squares	df	Mean Square	F	Sig.
1	Regression	54.032	4	13.508	2.558	.038[b]
	Residual	2191.251	415	5.280		
	Total	2245.283	419			

a. Dependent Variable: Hours per day watching TV

b. Predictors: (Constant), Number of hours worked last week, Number of children, Highest year of school completed, Age of respondent

Coefficients[a]

Model		Unstandardized Coefficients		Standardized Coefficients		
		B	Std. Error	Beta	t	Sig.
1	(Constant)	3.888	.715		5.438	.000
	Highest year of school completed	-.083	.039	-.107	-2.122	.034
	Number of children	-.143	.086	-.091	-1.657	.098
	Age of respondent	.009	.009	.058	1.072	.284
	Number of hours worked last week	-.014	.007	-.096	-1.957	.051

a. Dependent Variable: Hours per day watching TV

a. Write the multiple regression equation for the model. Interpret the slope for each independent variable.

b. Based on their beta scores, rank the independent variables according to the strength of their effect on TVHOURS (from the highest to the lowest).

c. Interpret the multiple coefficient of determination.

APPENDIX A

Table of Random Numbers

A Table of 14,000 Random Units

Line/ Col.	(1)	(2)	(3)	(4)	(5)	(6)	(7)	(8)	(9)	(10)	(11)	(12)	(13)	(14)
1	10480	15011	01536	02011	81647	91646	69179	14194	62590	36207	20969	99570	91291	90700
2	22368	46573	25595	85393	30995	89198	27982	53402	93965	34095	52666	19174	39615	99505
3	24130	48360	22527	97265	76393	64809	15179	24830	49340	32081	30680	19655	63348	58629
4	42167	93093	06243	61680	07856	16376	39440	53537	71341	57004	00849	74917	97758	16379
5	37570	39975	81837	16656	06121	91782	60468	81305	49684	60672	14110	06927	01263	54613
6	77921	06907	11008	42751	27756	53498	18602	70659	90655	15053	21916	81825	44394	42880
7	99562	72905	56420	69994	98872	31016	71194	18738	44013	48840	63213	21069	10634	12952
8	96301	91977	05463	07972	18876	20922	94595	56869	69014	60045	18425	84903	42508	32307
9	89579	14342	63661	10281	17453	18103	57740	84378	25331	12566	58678	44947	05585	56941
10	85475	36857	43342	53988	53060	59533	38867	62300	08158	17983	16439	11458	18593	64952
11	28918	69578	88231	33276	70997	79936	56865	05859	90106	31595	01547	85590	91610	78188
12	63553	40961	48235	03427	49626	69445	18663	72695	52180	20847	12234	90511	33703	90322
13	09429	93969	52636	92737	88974	33488	36320	17617	30015	08272	84115	27156	30613	74952
14	10365	61129	87529	85689	48237	52267	67689	93394	01511	26358	85104	20285	29975	89868
15	07119	97336	71048	08178	77233	13916	47564	81056	97735	85977	29372	74461	28551	90707
16	51085	12765	51821	51259	77452	16308	60756	92144	49442	53900	70960	63990	75601	40719
17	02368	21382	52404	60268	89368	19885	55322	44819	01188	65255	64835	44919	05944	55157
18	01011	54092	33362	94904	31273	4146	18594	29852	71585	85030	51132	01915	92747	64951
19	52162	53916	46369	58586	23216	14513	83149	98736	23495	64350	94738	17752	35156	35749
20	07056	97628	33787	09998	42698	06691	76988	13602	51851	46104	88916	19509	25625	58104
21	48663	91245	85828	14346	09172	30168	90229	04734	59193	22178	30421	61666	99904	32812
22	54164	58492	22421	74103	47070	25306	76468	26384	58151	06646	21524	15227	96909	44592
23	32639	32363	05597	24200	13363	38005	94342	28728	35806	06912	17012	64161	18296	22851
24	29334	27001	87637	87308	58731	00256	45834	15398	46557	41135	10367	07684	36188	18510
25	02488	33062	28834	07351	19731	92420	60952	61280	50001	67658	32586	86679	50720	94953
26	81525	72295	04839	96423	24878	82651	66566	14778	76797	14780	13300	87074	79666	95725

(*Continued*)

(Continued)

Line/ Col.	(1)	(2)	(3)	(4)	(5)	(6)	(7)	(8)	(9)	(10)	(11)	(12)	(13)	(14)
27	29676	20591	68086	26432	46901	20849	89768	81536	86645	12659	92259	57102	80428	25280
28	00742	57392	39064	66432	84673	40027	32832	61362	98947	96067	64760	64584	96096	98253
29	05366	04213	25669	26422	44407	44048	37937	63904	45766	66134	75470	66520	34693	90449
30	91921	26418	64117	94305	26766	25940	39972	22209	71500	64568	91402	42416	07844	69618
31	00582	04711	87917	77341	42206	35126	74087	99547	81817	42607	43808	76655	62028	76630
32	00725	69884	62797	56170	86324	88072	76222	36086	84637	93161	76038	65855	77919	88006
33	69011	65797	95876	55293	18988	27354	26575	08625	40801	59920	29841	80150	12777	48501
34	25976	57948	29888	88604	67917	48708	18912	82271	65424	69774	33611	54262	85963	03547
35	09763	83473	73577	12908	30883	18317	28290	35797	05998	41688	34952	37888	38917	88050
36	91567	42595	27958	30134	04024	86385	29880	99730	55536	84855	29080	09250	79656	73211
37	17955	56349	90999	49127	20044	59931	06115	20542	18059	02008	73708	83317	36103	42791
38	46503	18584	18845	49618	02304	51038	20655	58727	28168	15475	56942	53389	20562	87338
39	92157	89634	94824	78171	84610	82834	09922	25417	44137	48413	25555	21246	35509	20468
40	14577	62765	35605	81263	39667	47358	56873	56307	61607	49518	89656	20103	77490	18062
41	98427	07523	33362	64270	01638	92477	66969	98420	04880	45585	46565	04102	46880	45709
42	34914	63976	88720	82765	34476	17032	87589	40836	32427	70002	70663	88863	77775	69348
43	70060	28277	39475	46473	23219	53416	94970	25832	69975	94884	19661	72828	00102	66794
44	53976	54914	06990	67245	68350	82948	11398	42878	80287	88267	47363	46634	06541	97809
45	76072	29515	40980	07391	58745	25774	22987	80059	39911	96189	41151	14222	60697	59583
46	90725	52210	83974	29992	65831	38857	50490	83765	55657	14361	31720	57375	56228	41546
47	64364	67412	33339	31926	14883	24413	59744	92351	97473	89286	35931	04110	23726	51900
48	08962	00358	31662	25388	61642	34072	81249	35648	56891	69352	48373	45578	78547	81788
49	95012	68379	93526	70765	10593	04542	76463	54328	02349	17247	28865	14777	62730	92277
50	15664	10493	20492	38391	91132	21999	59516	81652	27195	48223	46751	22923	32261	85653
51	16408	81899	04153	53381	79401	21438	83035	92350	36693	31238	59649	91754	72772	02338
52	18629	81953	05520	91962	04739	13092	97662	24822	94730	6496	35090	04822	86772	98289
53	73115	35101	47498	87637	99016	71060	88824	71013	18735	20286	23153	72924	35165	43040
54	57491	16703	23167	49323	45021	33132	12544	41035	80780	45393	44812	12515	98931	91202
55	30405	83946	23792	14422	15059	45799	22716	19792	09983	74353	68668	30429	70735	25499
56	16631	35006	85900	98275	32388	52390	16815	69298	82732	38480	73817	32523	41961	44437
57	96773	20206	42559	78985	05300	22164	24369	54224	35083	19687	11052	91491	60383	19746
58	38935	64202	14349	82674	66523	44133	00697	35552	35970	19124	63318	29686	03387	59846
59	31624	76384	17403	53363	44167	64486	64758	75366	76554	31601	12614	33072	60332	92325
60	78919	19474	23632	27889	47914	02584	37680	20801	72152	39339	34806	08930	85001	87820

Line/ Col.	(1)	(2)	(3)	(4)	(5)	(6)	(7)	(8)	(9)	(10)	(11)	(12)	(13)	(14)
61	03931	33309	57047	74211	63445	17361	62825	39908	05607	91284	68833	25570	38818	46920
62	74426	33278	43972	10119	89917	15665	52872	73823	73144	88662	88970	74492	51805	99378
63	09066	00903	20795	95452	92648	45454	09552	88815	16553	51125	79375	97596	16296	66092
64	42238	12426	87025	14267	20979	04508	64535	31355	86064	29472	47689	05974	52468	16834
65	16153	08002	26504	41744	81959	65642	74240	56302	00033	67107	77510	70625	28725	34191
66	21457	40742	29820	96783	29400	21840	15035	34537	33310	06116	95240	15957	16572	06004
67	21581	57802	02050	89728	17937	37621	47075	42080	97403	48626	68995	43805	33386	21597
68	55612	78095	83197	33732	05810	24813	86902	60397	16489	03264	88525	42786	05269	92532
69	44657	66999	99324	51281	84463	60563	79312	93454	68876	25471	93911	25650	12682	73572
70	91340	84979	46949	81973	37949	61023	43997	15263	80644	43942	89203	71795	99533	50501
71	91227	21199	31935	27022	84067	05462	35216	14486	29891	68607	41867	14951	91696	85065
72	50001	38140	66321	19924	72163	9538	12151	06878	91903	18749	34405	56087	82790	70925
73	65390	05224	72958	28609	81406	39147	25549	48542	42627	45233	57202	94617	23772	07896
74	27504	96131	83944	41575	10573	08619	64482	73923	36152	05184	94142	25299	84387	34925
75	37169	94851	39117	89632	00959	16487	65536	49071	39782	17095	02330	74301	00275	48280
76	11508	70225	51111	38351	19444	66499	71945	05422	13442	78675	84081	66938	93654	59894
77	37449	30362	06694	54690	04052	53115	62757	95348	78662	11163	81651	50245	34971	52924
78	46515	70331	85922	38329	57015	15765	97161	17869	45349	61796	66345	81073	49106	79860
79	30986	81223	42416	58353	21532	30502	32305	86482	05174	07901	54339	58861	74818	46942
80	63798	64995	46583	09765	44160	78128	83991	42865	92520	83531	80377	35909	81250	54238
81	82486	84846	99254	67632	43218	50076	21361	64816	51202	88124	41870	52689	51275	83556
82	21885	32906	92431	09060	64297	51674	64126	62570	26123	05155	59194	52799	28225	85762
83	60336	98782	07408	53458	13564	59089	26445	29789	85205	41001	12535	12133	14645	23541
84	43937	46891	24010	25560	86355	33941	25786	54990	71899	15475	95434	98227	21824	19585
85	97656	63175	89303	16275	07100	92063	21942	18611	47348	20203	18534	03862	78095	50136
86	03299	01221	05418	38982	55758	92237	26759	86367	21216	98442	08303	56613	91511	75928
87	79626	06486	03574	17668	07785	76020	79924	25651	83325	88428	85076	72811	22717	50585
88	85636	68335	47539	03129	65651	11977	02510	26113	99447	68645	34327	15152	55230	93448
89	18039	14367	61337	06177	12143	46609	32989	74014	64708	00533	35398	58408	13261	47908
90	08362	15656	60627	36478	65648	16764	53412	09013	07832	41574	17639	82163	60859	75567
91	79556	29068	04142	16268	15387	12856	66227	38358	22478	73373	88732	09443	82558	05250
92	92608	82674	27072	32534	17075	27698	98204	63863	11951	34648	88022	56148	34925	57031
93	23982	25835	40055	67006	12293	02753	14827	22235	35071	99704	37543	11601	35503	85171
94	09915	96306	05908	97901	28395	14186	00821	80703	70426	75647	76310	88717	37890	40129
95	50937	33300	26695	62247	69927	76123	50842	43834	86654	70959	79725	93872	28117	19233
96	42488	78077	69882	61657	34136	79180	97526	43092	04098	73571	80799	76536	71255	64239
97	46764	86273	63003	93017	31204	36692	40202	35275	57306	55543	53203	18098	47625	88684
98	03237	45430	55417	63282	90816	17349	88298	90183	36600	78406	06216	95787	42579	90730
99	86591	81482	52667	61583	14972	90053	89534	76036	49199	43716	97548	04379	46370	28672
100	38534	01715	94964	87288	65680	43772	39560	12918	86537	62738	19636	51132	25739	56947

Source: William H. Beyer, ed., *Handbook for Probability and Statistics,* 2nd ed.

APPENDIX B

The Standard Normal Table

The values in column A are Z scores. Column B lists the proportion of area between the mean and a given Z. Column C lists the proportion of area beyond a given Z. Only positive Z scores are listed. Because the normal curve is symmetrical, the areas for negative Z scores will be exactly the same as the areas for positive Z scores.

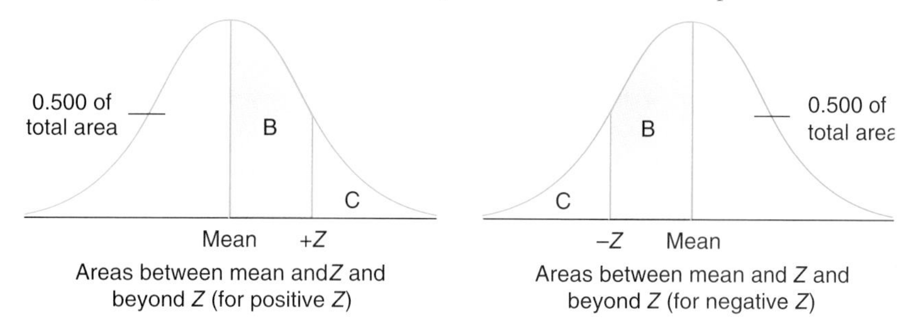

The Standard Normal Table

A Z	B Area Between Mean and Z	C Area Beyond Z	A Z	B Area Between Mean and Z	C Area Beyond Z	A Z	B Area Between Mean and Z	C Area Beyond Z
0.00	0.0000	0.5000	0.11	0.0438	0.4562	0.22	0.0871	0.4129
0.01	0.0040	0.4960	0.12	0.0478	0.4522	0.23	0.0910	0.4090
0.02	0.0080	0.4920	0.13	0.0517	0.4483	0.24	0.0948	0.4052
0.03	0.0120	0.4880	0.14	0.0557	0.4443	0.25	0.0987	0.4013
0.04	0.0160	0.4840	0.15	0.0596	0.4404	0.26	0.1026	0.3974
0.05	0.0199	0.4801	0.16	0.0636	0.4364	0.27	0.1064	0.3936
0.06	0.0239	0.4761	0.17	0.0675	0.4325	0.28	0.1103	0.3897
0.07	0.0279	0.4721	0.18	0.0714	0.4286	0.29	0.1141	0.3859
0.08	0.0319	0.4681	0.19	0.0753	0.4247	0.30	0.1179	0.3821
0.09	0.0359	0.4641	0.20	0.0793	0.4207	0.31	0.1217	0.3783
0.10	0.0398	0.4602	0.21	0.0832	0.4168	0.32	0.1255	0.3745

(Continued)

(Continued)

A Z	B Area Between Mean and Z	C Area Beyond Z	A Z	B Area Between Mean and Z	C Area Beyond Z	A Z	B Area Between Mean and Z	C Area Beyond Z
0.33	0.1293	0.3707	0.64	0.2389	0.2611	0.95	0.3289	0.1711
0.34	0.1331	0.3669	0.65	0.2422	0.2578	0.96	0.3315	0.1685
0.35	0.1368	0.3632	0.66	0.2454	0.2546	0.97	0.3340	0.1660
0.36	0.1406	0.3594	0.67	0.2486	0.2514	0.98	0.3365	0.1635
0.37	0.1443	0.3557	0.68	0.2517	0.2483	0.99	0.3389	0.1611
0.38	0.1480	0.3520	0.69	0.2549	0.2451	1.00	0.3413	0.1587
0.39	0.1517	0.3483	0.70	0.2580	0.2420	1.01	0.3438	0.1562
0.40	0.1554	0.3446	0.71	0.2611	0.2389	1.02	0.3461	0.1539
0.41	0.1591	0.3409	0.72	0.2642	0.2358	1.03	0.3485	0.1515
0.42	0.1628	0.3372	0.73	0.2673	0.2327	1.04	0.3508	0.1492
0.43	0.1664	0.3336	0.74	0.2703	0.2297	1.05	0.3531	0.1469
0.44	0.1700	0.3300	0.75	0.2734	0.2266	1.06	0.3554	0.1446
0.45	0.1736	0.3264	0.76	0.2764	0.2236	1.07	0.3577	0.1423
0.46	0.1772	0.3228	0.77	0.2794	0.2206	1.08	0.3599	0.1401
0.47	0.1808	0.3192	0.78	0.2823	0.2177	1.09	0.3621	0.1379
0.48	0.1844	0.3156	0.79	0.2852	0.2148	1.10	0.3643	0.1357
0.49	0.1879	0.3121	0.80	0.2881	0.2119	1.11	0.3665	0.1335
0.50	0.1915	0.3085	0.81	0.2910	0.2090	1.12	0.3686	0.1314
0.51	0.1950	0.3050	0.82	0.2939	0.2061	1.13	0.3708	0.1292
0.52	0.1985	0.3015	0.83	0.2967	0.2033	1.14	0.3729	0.1271
0.53	0.2019	0.2981	0.84	0.2995	0.2005	1.15	0.3749	0.1251
0.54	0.2054	0.2946	0.85	0.3023	0.1977	1.16	0.3770	0.1230
0.55	0.2088	0.2912	0.86	0.3051	0.1949	1.17	0.3790	0.1210
0.56	0.2123	0.2877	0.87	0.3078	0.1992	1.18	0.3810	0.1190
0.57	0.2157	0.2843	0.88	0.3106	0.1894	1.19	0.3830	0.1170
0.58	0.2190	0.2810	0.89	0.3133	0.1867	1.20	0.3849	0.1151
0.59	0.2224	0.2776	0.90	0.3159	0.1841	1.21	0.3869	0.1131
0.60	0.2257	0.2743	0.91	0.3186	0.1814	1.22	0.3888	0.1112
0.61	0.2291	0.2709	0.92	0.3212	0.1788	1.23	0.3907	0.1093
0.62	0.2324	0.2676	0.93	0.3238	0.1762	1.24	0.3925	0.1075
0.63	0.2357	0.2643	0.94	0.3264	0.1736	1.25	0.3944	0.1056

A Z	B Area Between Mean and Z	C Area Beyond Z	A Z	B Area Between Mean and Z	C Area Beyond Z	A Z	B Area Between Mean and Z	C Area Beyond Z
1.26	0.3962	0.1038	1.58	0.4429	0.0571	1.90	0.4713	0.0287
1.27	0.3980	0.1020	1.59	0.4441	0.0559	1.91	0.4719	0.0281
1.28	0.3997	0.1003	1.60	0.4452	0.0548	1.92	0.4726	0.0274
1.29	0.4015	0.0985	1.61	0.4463	0.0537	1.93	0.4732	0.0268
1.30	0.4032	0.0968	1.62	0.4474	0.0526	1.94	0.4738	0.0262
1.31	0.4049	0.0951	1.63	0.4484	0.0516	1.95	0.4744	0.0256
1.32	0.4066	0.0934	1.64	0.4495	0.0505	1.96	0.4750	0.0250
1.33	0.4082	0.0918	1.65	0.4505	0.0495	1.97	0.4756	0.0244
1.34	0.4099	0.0901	1.66	0.4515	0.0485	1.98	0.4761	0.0239
1.35	0.4115	0.0885	1.67	0.4525	0.0475	1.99	0.4767	0.0233
1.36	0.4131	0.0869	1.68	0.4535	0.0465	2.00	0.4772	0.0228
1.37	0.4147	0.0853	1.69	0.4545	0.0455	2.01	0.4778	0.0222
1.38	0.4612	0.0838	1.70	0.4554	0.0466	2.02	0.4783	0.0217
1.39	0.4177	0.0823	1.71	0.4564	0.0436	2.03	0.4788	0.0212
1.40	0.4192	0.0808	1.72	0.4573	0.0427	2.04	0.4793	0.0207
1.41	0.4207	0.0793	1.73	0.4582	0.0418	2.05	0.4798	0.0202
1.42	0.4222	0.0778	1.74	0.4591	0.0409	2.06	0.4803	0.0197
1.43	0.4236	0.0764	1.75	0.4599	0.0401	2.07	0.4808	0.0192
1.44	0.4251	0.0749	1.76	0.4608	0.0392	2.08	0.4812	0.0188
1.45	0.4265	0.0735	1.77	0.4616	0.0384	2.09	0.4817	0.0183
1.46	0.4279	0.0721	1.78	0.4625	0.0375	2.10	0.4821	0.0179
1.47	0.4292	0.0708	1.79	0.4633	0.0367	2.11	0.4826	0.0174
1.48	0.4306	0.0694	1.80	0.4641	0.0359	2.12	0.4830	0.0170
1.49	0.4319	0.0681	1.81	0.4649	0.0351	2.13	0.4834	0.0166
1.50	0.4332	0.0668	1.82	0.4656	0.0344	2.14	0.4838	0.0162
1.51	0.4345	0.0655	1.83	0.4664	0.0336	2.15	0.4842	0.0158
1.52	0.4357	0.0643	1.84	0.4671	0.0329	2.16	0.4846	0.0154
1.53	0.4370	0.0630	1.85	0.4678	0.0322	2.17	0.4850	0.0150
1.54	0.4382	0.0618	1.86	0.4686	0.0314	2.18	0.4854	0.0146
1.55	0.4394	0.0606	1.87	0.4693	0.0307	2.19	0.4857	0.0143
1.56	0.4406	0.0594	1.88	0.4699	0.0301	2.20	0.4861	0.0139
1.57	0.4418	0.0582	1.89	0.4706	0.0294	2.21	0.4864	0.0136

(*Continued*)

(Continued)

A Z	B Area Between Mean and Z	C Area Beyond Z	A Z	B Area Between Mean and Z	C Area Beyond Z	A Z	B Area Between Mean and Z	C Area Beyond Z
2.22	0.4868	0.0132	2.53	0.4943	0.0057	2.84	0.4977	0.0023
2.23	0.4871	0.0129	2.54	0.4945	0.0055	2.85	0.4978	0.0022
2.24	0.4875	0.0125	2.55	0.4946	0.0054	2.86	0.4979	0.0021
2.25	0.4878	0.0122	2.56	0.4948	0.0052	2.87	0.4979	0.0021
2.26	0.4881	0.0119	2.57	0.4949	0.0051	2.88	0.4980	0.0020
2.27	0.4884	0.0116	2.58	0.4951	0.0049	2.89	0.4981	0.0019
2.28	0.4887	0.0113	2.59	0.4952	0.0048	2.90	0.4981	0.0019
2.29	0.4890	0.0110	2.60	0.4953	0.0047	2.91	0.4982	0.0018
2.30	0.4893	0.0107	2.61	0.4955	0.0045	2.92	0.4982	0.0018
2.31	0.4896	0.0104	2.62	0.4956	0.0044	2.93	0.4983	0.0017
2.32	0.4898	0.0102	2.63	0.4957	0.0043	2.94	0.4984	0.0016
2.33	0.4901	0.0099	2.64	0.4959	0.0041	2.95	0.4984	0.0016
2.34	0.4904	0.0096	2.65	0.4960	0.0040	2.96	0.4985	0.0015
2.35	0.4906	0.0094	2.66	0.4961	0.0039	2.97	0.4985	0.0015
2.36	0.4909	0.0091	2.67	0.4962	0.0038	2.98	0.4986	0.0014
2.37	0.4911	0.0089	2.68	0.4963	0.0037	2.99	0.4986	0.0014
2.38	0.4913	0.0087	2.69	0.4964	0.0036	3.00	0.4986	0.0014
2.39	0.4916	0.0084	2.70	0.4965	0.0035	3.01	0.4987	0.0013
2.40	0.4918	0.0082	2.71	0.4966	0.0034	3.02	0.4987	0.0013
2.41	0.4920	0.0080	2.72	0.4967	0.0033	3.03	0.4988	0.0012
2.42	0.4922	0.0078	2.73	0.4968	0.0032	3.04	0.4988	0.0012
2.43	0.4925	0.0075	2.74	0.4969	0.0031	3.05	0.4989	0.0011
2.44	0.4927	0.0073	2.75	0.4970	0.0030	3.06	0.4989	0.0011
2.45	0.4929	0.0071	2.76	0.4971	0.0029	3.07	0.4989	0.0011
2.46	0.4931	0.0069	2.77	0.4972	0.0028	3.08	0.4990	0.0010
2.47	0.4932	0.0068	2.78	0.4973	0.0027	3.09	0.4990	0.0010
2.48	0.4934	0.0066	2.79	0.4974	0.0026	3.10	0.4990	0.0010
2.49	0.4936	0.0064	2.80	0.4974	0.0026	3.11	0.4991	0.0009
2.50	0.4938	0.0062	2.81	0.4975	0.0025	3.12	0.4991	0.0009
2.51	0.4940	0.0060	2.82	0.4976	0.0024	3.13	0.4991	0.0009
2.52	0.4941	0.0059	2.83	0.4977	0.0023	3.14	0.4992	0.0008

A	B	C	A	B	C	A	B	C
Z	Area Between Mean and Z	Area Beyond Z	Z	Area Between Mean and Z	Area Beyond Z	Z	Area Between Mean and Z	Area Beyond Z
3.15	0.4992	0.0008	3.29	0.4995	0.0005	3.43	0.4997	0.0003
3.16	0.4992	0.0008	3.30	0.4995	0.0005	3.44	0.4997	0.0003
3.17	0.4992	0.0008	3.31	0.4995	0.0005	3.45	0.4997	0.0003
3.18	0.4993	0.0007	3.32	0.4995	0.0005	3.46	0.4997	0.0003
3.19	0.4993	0.0007	3.33	0.4996	0.0004	3.47	0.4997	0.0003
3.20	0.4993	0.0007	3.34	0.4996	0.0004	3.48	0.4997	0.0003
3.21	0.4993	0.0007	3.35	0.4996	0.0004	3.49	0.4998	0.0002
3.22	0.4994	0.0006	3.36	0.4996	0.0004	3.50	0.4998	0.0002
3.23	0.4994	0.0006	3.37	0.4996	0.0004	3.60	0.4998	0.0002
3.24	0.4994	0.0006	3.38	0.4996	0.0004	3.70	0.4999	0.0001
3.25	0.4994	0.0006	3.39	0.4997	0.0003	3.80	0.4999	0.0001
3.26	0.4994	0.0006	3.40	0.4997	0.0003	3.90	0.4999	<0.0001
3.27	0.4995	0.0005	3.41	0.4997	0.0003	4.00	0.4999	<0.0001
3.28	0.4995	0.0005	3.42	0.4997	0.0003			

APPENDIX C

Distribution of *t*

	Level of Significance for One-Tailed Test					
	.10	.05	.025	.01	.005	.0005
	Level of Significance for Two-Tailed Test					
df	.20	.10	.05	.02	.01	.001
1	3.078	6.314	12.706	31.821	63.657	636.619
2	1.886	2.920	4.303	6.965	9.925	31.598
3	1.638	2.353	3.182	4.541	5.841	12.941
4	1.533	2.132	2.776	3.747	4.604	8.610
5	1.476	2.015	2.571	3.365	4.032	6.859
6	1.440	1.943	2.447	3.143	3.707	5.959
7	1.415	1.895	2.365	2.998	3.499	5.405
8	1.397	1.860	2.306	2.896	3.355	5.041
9	1.383	1.833	2.262	2.821	3.250	4.781
10	1.372	1.812	2.228	2.764	3.169	4.587
11	1.363	1.796	2.201	2.718	3.106	4.437
12	1.356	1.782	2.179	2.681	3.055	4.318
13	1.350	1.771	2.160	2.650	3.012	4.221
14	1.345	1.761	2.145	2.624	2.977	4.140
15	1.341	1.753	2.131	2.602	2.947	4.073
16	1.337	1.746	2.120	2.583	2.921	4.015
17	1.333	1.740	2.110	2.567	2.898	3.965
18	1.330	1.734	2.101	2.552	2.878	3.922

(*Continued*)

(Continued)

	Level of Significance for One-Tailed Test					
	.10	.05	.025	.01	.005	.0005
	Level of Significance for Two-Tailed Test					
df	.20	.10	.05	.02	.01	.001
19	1.328	1.729	2.093	2.539	2.861	3.883
20	1.325	1.725	2.086	2.528	2.845	3.850
21	1.323	1.721	2.080	2.518	2.831	3.819
22	1.321	1.717	2.074	2.508	2.819	3.792
23	1.319	1.714	2.069	2.500	2.807	3.767
24	1.318	1.711	2.064	2.492	2.797	3.745
25	1.316	1.708	2.060	2.485	2.787	3.725
26	1.315	1.706	2.056	2.479	2.779	3.707
27	1.314	1.703	2.052	2.473	2.771	3.690
28	1.313	1.701	2.048	2.467	2.763	3.674
29	1.311	1.699	2.045	2.462	2.756	3.659
30	1.310	1.697	2.042	2.457	2.750	3.646
40	1.303	1.684	2.021	2.423	2.704	3.551
60	1.296	1.671	2.000	2.390	2.660	3.460
120	1.289	1.658	1.980	2.358	2.617	3.373
∞	1.282	1.645	1.960	2.326	2.576	3.291

Source: Abridged from R. A. Fisher and F. Yates, *Statistical Tables for Biological, Agricultural and Medical Research*, 6th ed. Copyright ©R. A. Fisher and F. Yates 1963. Reprinted by permission of Pearson Education Limited.

APPENDIX D

Distribution of Chi-Square

df	.99	.98	.95	.90	.80	.70	.50	.30	.20	.10	.05	.02	.01	.001
1	.03157	.03628	.00393	.0158	.0642	.148	.455	1.074	1.642	2.706	3.841	5.412	6.635	10.827
2	.0201	.0404	.103	.211	.446	.713	1.386	2.408	3.219	4.605	5.991	7.824	9.210	13.815
3	.115	.185	.352	.584	1.005	1.424	2.366	3.665	4.642	6.251	7.815	9.837	11.341	16.268
4	.297	.429	.711	1.064	1.649	2.195	3.357	4.878	5.989	7.779	9.488	11.668	13.277	18.465
5	.554	.752	1.145	1.610	2.343	3.000	4.351	6.064	7.289	9.236	11.070	13.388	15.086	20.517
6	.872	1.134	1.635	2.204	3.070	3.828	5.348	7.231	8.558	10.645	12.592	15.033	16.812	22.457
7	1.239	1.564	2.167	2.833	3.822	4.671	6.346	8.383	9.803	12.017	14.067	16.622	18.475	24.322
8	1.646	2.032	2.733	3.490	4.594	5.527	7.344	9.524	11.030	13.362	15.507	18.168	20.090	26.125
9	2.088	2.532	3.325	4.168	5.380	6.393	8.343	10.656	12.242	14.684	16.919	19.679	21.666	27.877
10	2.558	3.059	3.940	4.865	6.179	7.267	9.342	11.781	13.442	15.987	18.307	21.161	23.209	29.588
11	3.053	3.609	4.575	5.578	6.989	8.148	10.341	12.899	14.631	17.275	19.675	22.618	24.725	31.264
12	3.571	4.178	5.226	6.304	7.807	9.034	11.340	14.011	15.812	18.549	21.026	24.054	26.217	32.909
13	4.107	4.765	5.892	7.042	8.634	9.926	12.340	15.119	16.985	19.812	22.362	25.472	27.688	34.528
14	4.660	5.368	6.571	7.790	9.467	10.821	13.339	16.222	18.151	21.064	23.685	26.873	29.141	36.123
15	5.229	5.985	7.261	8.547	10.307	11.721	14.339	17.322	19.311	22.307	24.996	28.259	30.578	37.697
16	5.812	6.614	7.962	9.312	11.152	12.624	15.338	18.418	20.465	23.542	26.296	29.633	32.000	39.252
17	6.408	7.255	8.672	10.085	12.002	13.531	16.338	19.511	21.615	24.769	27.587	30.995	33.409	40.790
18	7.015	7.906	9.390	10.865	12.857	14.440	17.338	20.601	22.760	25.989	28.869	32.346	34.805	42.312
19	7.633	8.567	10.117	11.651	13.716	15.352	18.338	21.689	23.900	27.204	30.144	33.687	36.191	43.820
20	8.260	9.237	10.851	12.443	14.578	16.266	19.337	22.775	25.038	28.412	31.410	35.020	37.566	45.315
21	8.897	9.915	11.591	13.240	15.445	17.182	20.337	23.858	26.171	29.615	32.671	36.343	38.932	46.797
22	9.542	10.600	12.338	14.041	16.314	18.101	21.337	24.939	27.301	30.813	33.924	37.659	40.289	48.268
23	10.196	11.293	13.091	14.848	17.187	19.021	22.337	26.018	28.429	32.007	35.172	38.968	41.638	49.728
24	10.856	11.992	13.848	15.659	18.062	19.943	23.337	27.096	29.553	33.196	36.415	40.270	42.980	51.179
25	11.524	12.697	14.611	16.473	18.940	20.867	24.337	28.172	30.675	34.382	37.652	41.566	44.314	52.620
26	12.198	13.409	15.379	17.292	19.820	21.792	25.336	29.246	31.795	35.563	38.885	42.856	45.642	54.052
27	12.879	14.125	16.151	18.114	20.703	22.719	26.336	30.319	32.912	36.741	40.113	44.140	46.963	55.476
28	13.565	14.847	16.928	18.939	21.588	23.647	27.336	31.391	34.027	37.916	41.337	45.419	48.278	56.893
29	14.256	15.574	17.708	19.768	22.475	24.577	28.336	32.461	35.139	39.087	42.557	46.693	49.588	58.302
30	14.953	16.306	18.493	20.599	23.364	25.508	29.336	33.530	36.250	40.256	43.773	47.962	50.892	59.703

Source: R. A. Fisher & F. Yates, *Statistical Tables for Biological, Agricultural and Medical Research*, 6th ed.

APPENDIX E

Distribution of F

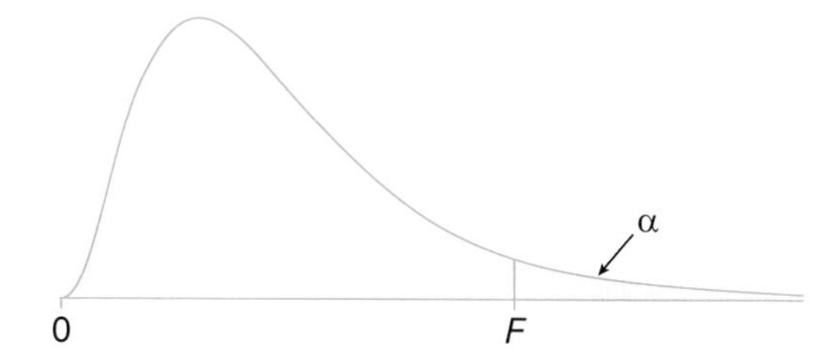

	$\alpha = .05$									
	df_1									
df_2	1	2	3	4	5	6	8	12	24	∞
1	161.4	199.5	215.7	224.6	230.2	234.0	238.9	243.9	249.0	254.3
2	18.51	19.00	19.16	19.25	19.30	19.33	19.37	19.41	19.45	19.50
3	10.13	9.55	9.28	9.12	9.01	8.94	8.84	8.74	8.64	8.53
4	7.71	6.94	6.59	6.39	6.26	6.16	6.04	5.91	5.77	5.63
5	6.61	5.79	5.41	5.19	5.05	4.95	4.82	4.68	4.53	4.36
6	5.99	5.14	4.76	4.53	4.39	4.28	4.15	4.00	3.84	3.67
7	5.59	4.74	4.35	4.12	3.97	3.87	3.73	3.57	3.41	3.23
8	5.32	4.46	4.07	3.84	3.69	3.58	3.44	3.28	3.12	2.93
9	5.12	4.26	3.86	3.63	3.48	3.37	3.23	3.07	2.90	2.71
10	4.96	4.10	3.71	3.48	3.33	3.22	3.07	2.91	2.74	2.54
11	4.84	3.98	3.59	3.36	3.20	3.09	2.95	2.79	2.61	2.40
12	4.75	3.88	3.49	3.26	3.11	3.00	2.85	2.69	2.50	2.30
13	4.67	3.80	3.41	3.18	3.02	2.92	2.77	2.60	2.42	2.21
14	4.60	3.74	3.34	3.11	2.96	2.85	2.70	2.53	2.35	2.13
15	4.54	3.68	3.29	3.06	2.90	2.79	2.64	2.48	2.29	2.07
16	4.49	3.63	3.24	3.01	2.85	2.74	2.59	2.42	2.24	2.01
17	4.45	3.59	3.20	2.96	2.81	2.70	2.55	2.38	2.19	1.96

(*Continued*)

(Continued)

$\alpha = .05$										
	df_1									
df_2	1	2	3	4	5	6	8	12	24	∞
18	4.41	3.55	3.16	2.93	2.77	2.66	2.51	2.34	2.15	1.92
19	4.38	3.52	3.13	2.90	2.74	2.63	2.48	2.31	2.11	1.88
20	4.35	3.49	3.10	2.87	2.71	2.60	2.45	2.28	2.08	1.84
21	4.32	3.47	3.07	2.84	2.68	2.57	2.42	2.25	2.05	1.81
22	4.30	3.44	3.05	2.82	2.66	2.55	2.40	2.23	2.03	1.78
23	4.28	3.42	3.03	2.80	2.64	2.53	2.38	2.20	2.00	1.76
24	4.26	3.40	3.01	2.78	2.62	2.51	2.36	2.18	1.98	1.73
25	4.24	3.38	2.99	2.76	2.60	2.49	2.34	2.16	1.96	1.71
26	4.22	3.37	2.98	2.74	2.59	2.47	2.32	2.15	1.95	1.69
27	4.21	3.35	2.96	2.73	2.57	2.46	2.30	2.13	1.93	1.67
28	4.20	3.34	2.95	2.71	2.56	2.44	2.29	2.12	1.91	1.65
29	4.18	3.33	2.93	2.70	2.54	2.43	2.28	2.10	1.90	1.64
30	4.17	3.32	2.92	2.69	2.53	2.42	2.27	2.09	1.89	1.62
40	4.08	3.23	2.84	2.61	2.45	2.34	2.18	2.00	1.79	1.51
60	4.00	3.15	2.76	2.52	2.37	2.25	2.10	1.92	1.70	1.39
120	3.92	3.07	2.68	2.45	2.29	2.17	2.02	1.83	1.61	1.25
∞	3.84	2.99	2.60	2.37	2.21	2.09	1.94	1.75	1.52	1.00

$\alpha = .01$										
	df_1									
df_2	1	2	3	4	5	6	8	12	24	∞
1	4052	4999	5403	5625	5764	5859	5981	6106	6234	6366
2	98.49	99.01	99.17	99.25	99.30	99.33	99.36	99.42	99.46	99.50
3	34.12	30.81	29.46	28.71	28.24	27.91	27.49	27.05	26.60	26.12
4	21.20	18.00	16.69	15.98	15.52	15.21	14.80	14.37	13.93	13.46
5	16.26	13.27	12.06	11.39	10.97	10.67	10.27	9.89	9.47	9.02
6	13.74	10.92	9.78	9.15	8.75	8.47	8.10	7.72	7.31	6.88
7	12.25	9.55	8.45	7.85	7.46	7.19	6.84	6.47	6.07	5.65
8	11.26	8.65	7.59	7.01	6.63	6.37	6.03	5.67	5.28	4.86
9	10.56	8.02	6.99	6.42	6.06	5.80	5.47	5.11	4.73	4.31
10	10.04	7.56	6.55	5.99	5.64	5.39	5.06	4.71	4.33	3.91

	$\alpha = .01$									
	df_1									
df_2	1	2	3	4	5	6	8	12	24	∞
11	9.65	7.20	6.22	5.67	5.32	5.07	4.74	4.40	4.02	3.60
12	9.33	6.93	5.95	5.41	5.06	4.82	4.50	4.16	3.78	3.36
13	9.07	6.70	5.74	5.20	4.86	4.62	4.30	3.96	3.59	3.16
14	8.86	6.51	5.56	5.03	4.69	4.46	4.14	3.80	3.43	3.00
15	8.68	6.36	5.42	4.89	4.56	4.32	4.00	3.67	3.29	2.87
16	8.53	6.23	5.29	4.77	4.44	4.20	3.89	3.55	3.18	2.75
17	8.40	6.11	5.18	4.67	4.34	4.10	3.79	3.45	3.08	2.65
18	8.28	6.01	5.09	4.58	4.25	4.01	3.71	3.37	3.00	2.57
19	8.18	5.93	5.01	4.50	4.17	3.94	3.63	3.30	2.92	2.49
20	8.10	5.85	4.94	4.43	4.10	3.87	3.56	3.23	2.86	2.42
21	8.02	5.78	4.87	4.37	4.04	3.81	3.51	3.17	2.80	2.36
22	7.94	5.72	4.82	4.31	3.99	3.76	3.45	3.12	2.75	2.31
23	7.88	5.66	4.76	4.23	3.94	3.71	3.41	3.07	2.70	2.26
24	7.82	5.61	4.72	4.22	3.90	3.67	3.36	3.03	2.66	2.21
25	7.77	5.57	4.68	4.18	3.86	3.63	3.32	2.99	2.62	2.17
26	7.72	5.53	4.64	4.14	3.82	3.59	3.29	2.96	2.58	2.13
27	7.68	5.49	4.60	4.11	3.78	3.56	3.26	2.93	2.55	2.10
28	7.64	5.45	4.57	4.07	3.75	3.53	3.23	2.90	2.52	2.06
29	7.60	5.42	4.54	4.04	3.73	3.50	3.20	2.87	2.49	2.03
30	7.56	5.39	4.51	4.02	3.70	3.47	3.17	2.84	2.47	2.01
40	7.31	5.18	4.31	3.83	3.51	3.29	2.99	2.66	2.29	1.80
60	7.08	4.98	4.13	3.65	3.34	3.12	2.82	2.50	2.12	1.60
120	6.85	4.79	3.95	3.48	3.17	2.96	2.66	2.34	1.95	1.38
∞	6.64	4.60	3.78	3.32	3.02	2.80	2.51	2.18	1.79	1.00

LEARNING CHECK SOLUTIONS

CHAPTER 1

(p. 9)

Learning Check 1.3. Identify the independent and dependent variables in the following hypotheses:

- *Older Americans are more likely to support stricter immigration laws than younger Americans.*
- *People who attend church regularly are more likely to oppose abortion than people who do not attend church regularly.*
- *Elderly women are more likely to live alone than elderly men.*
- *Individuals with postgraduate education are likely to have fewer children than those with less education.*

Independent	Dependent
Age	Support for stricter immigration laws
Church attendance	Opposition to abortion
Gender	Living arrangement
Educational attainment	Number of children

(p. 12)

Learning Check 1.4. Review the definitions of exhaustive and mutually exclusive. Now look at Table 1.2. What other categories could be added to each variable to be exhaustive and mutually exclusive?

Answer:

To the variable gender, we can include a transgender or gender neutral category. For the variable religion, we can include a category for those without any religion. For the marital status category, we could include a divorced category, though this would be covered under the Other category.

CHAPTER 2

(p. 44)

Learning Check 2.6. Examine Table 2.12. Make sure you can identify all the parts just described and that you understand how the numbers were obtained. Can you identify the independent and dependent variables in the table? You will need to know this to convert the frequencies to percentages.

Answer:

The dependent variable is homeownership and the independent variable is race.

CHAPTER 3

(p. 71)

Learning Check 3.1. Listed below are the political party affiliations of 15 individuals. Find the mode.

Democrat	Republican	Democrat	Republican	Republican
Independent	Democrat	Democrat	Democrat	Republican
Independent	Democrat	Independent	Republican	Democrat

Answer:

The mode is "Democrat," because this category has the highest frequency, which is 7.

(p. 76)

Learning Check 3.2. Find the median of the following distribution of an interval-ratio variable: 22, 15, 18, 33, 17, 5, 11, 28, 40, 19, 8, 20.

Answer:

First, we need to arrange the numbers: 5, 8, 11, 15, 17, 18, 19, 20, 22, 28, 33, 40.

$(N + 1)/2 = (12 + 1)/2 = 6.5$. So the median is the average of the sixth and the seventh numbers, which are 18 and 19. The median is 18.5.

$$\text{Median} = \frac{18+19}{2} = 18.5$$

(p. 82)

Learning Check 3.5. The following distribution is the same as the one you used to calculate the median in an earlier Learning Check: 22, 15, 18, 33, 17, 5, 11, 28, 40, 19, 8, 20. Can you calculate the mean? Is it the same as the median, or is it different?

Answer:

$$\text{Mean} = \frac{22+15+18+33+17+5+11+28+40+19+8+20}{12} = 19.67$$

So the mean, 19.67, is larger than the median, 18.5.

(p. 95)

Learning Check 3.9. Why can't we use the range to describe diversity in nominal variables? The range can be used to describe diversity in ordinal variables (e.g., we can say that responses to a question ranged from "somewhat satisfied" to "very dissatisfied"), but it has no quantitative meaning. Why not?

Answer:

In nominal variables, the numbers are used only to represent the different categories of a variable without implying anything about the magnitude or quantitative difference between these categories. Therefore, the range, being a measure of variability that gives the quantitative difference between two values that a variable takes, is not an appropriate measure for nominal variables. Similarly, in ordinal variables, numbers corresponding with the categories of a variable are only used to rank order these categories without having any meaning in terms of the quantitative difference between these categories. Therefore, the range does not convey any quantitative meaning when used to describe the diversity in ordinal variables.

(p. 98)

Learning Check 3.10. Why is the IQR better than the range as a measure of variability, especially when there are extreme scores in the distribution? To answer this question, you may want to examine Figure 3.12.

Answer:

Extreme scores directly impact the range – the difference between the highest and the lowest scores. Therefore, if a distribution has extreme (very high and/or very low) scores, the range does not provide an accurate description of the distribution. IQR, on the other hand, is not affected by extreme scores. Thus, it is a better measure of variability than the range when there are extreme scores in the distribution.

CHAPTER 4

(p. 131)

Learning Check 4.2. How many students obtained a score between 305 and 536?

Answer:

$.4881 \times 2{,}136{,}539 = 1{,}042{,}845$

(p. 133)

Learning Check 4.3. Calculate the proportion of test-takers who earned a SAT ERW score of 400 or less. What is the proportion of students who earned a score of 600 or higher?

Answer:

The proportion who earned a score of 400 or less is .0918 (Z_{400}= −1.33)

The proportion who earned a score of 800 or higher is .0048 (Z_{600} = 2.59)

(p. 135)

Learning Check 4.4. Which score corresponds to the top 5% of SAT ERW test takers?

Answer:

We select a Z score of 1.65, corresponding to .45 (B) and .05 (C) areas. The formula is Y = 536 + (1.65) (102) = 536 + 168.3 = 704.3

(p. 138)

Learning Check 4.5. In a normal distribution, how many standard deviations from the mean is the 95th percentile?

Answer:

The number of standard deviations from the mean is what we call a *Z* score. The *Z* score associated with the 95th percentile is 1.65. So a score at the 95th percentile is 1.65 standard deviations above the mean.

(p. 139)

Learning Check 4.6. What is the raw SAT ERW score associated with the 50th percentile?

Answer:

The raw score associated with the 50th percentile is the median.

(p. 142)

Learning Check 4.7. Review the mean math Z-scores for the variable "conditions at age 7, 11 and 16" (the last column of the table). From ages 7, 11, and 16, was there an improvement in their math scores? Explain.

Answer:

Each group's mean math score is consistently lower than the overall mean, but over time, the distance from the mean is reduced (from −1.026 to −.706). Recall that larger Z scores indicate a greater distance from the overall mean.

(p. 166)

Learning Check 5.4. Suppose a population distribution has a mean $\mu = 150$ and a standard deviation $\sigma = 30$, and you draw a simple random sample of $N = 100$ cases. What is the probability that the mean is between 147 and 153? What is the probability that the sample mean exceeds 153? Would you be surprised to find a mean score of 159? Why? (Hint: To answer these questions, you need to apply what you learned in Chapter 4 about Z scores and areas under the normal curve [Appendix B].) Remember, to translate a raw score into a Z score we used this formula:

$$Z = \frac{Y - \bar{Y}}{s}$$

However, because here we are dealing with a sampling distribution, replace Y with the sample mean $\bar{Y}$, $\bar{Y}$ with the sampling distribution's mean $_{\bar{Y}}$, and σ with the standard error of the mean

$$Z = \frac{\bar{Y} - \mu_{\bar{Y}}}{\sigma / \sqrt{N}}$$

Answer:

Z score equivalent of 147 is

$$\frac{147 - 150}{30 / \sqrt{100}} = \frac{-3}{3} = -1$$

Z score equivalent of 153 is

$$\frac{153 - 150}{30 / \sqrt{100}} = \frac{3}{3} = 1$$

Using the standard normal table (Appendix B), we can see that the probability of the area between the mean and a score 1 standard deviation above or below the mean is 0.3413. So the probability that the mean is between 147 and 153, both of which deviate from the mean by 1 standard deviation, is 0.6826 (0.3413 + 0.3413), or 68.26%.

The probability of the area beyond 1 standard deviation from the mean is 0.1587. So the probability that the mean exceeds 153 is 0.1587, or 15.87%.

Z score equivalent of 159 is

$$\frac{159-150}{30/\sqrt{100}}=\frac{9}{3}=3$$

The probability of the area beyond 3 standard deviations from the mean, according to the standard normal table, is 0.0014. Therefore, it would be surprising to find a mean score of 159, as the probability is very low (0.14%).

CHAPTER 6

(p. 177)

Learning Check 6.1. What is the difference between a point estimate and a confidence interval?

Answer:

When the estimate of a population parameter is a single number, it is called a point estimate. When the estimate is a range of scores, it is called an interval estimate. Confidence intervals are used for interval estimates.

(p. 180)

Learning Check 6.2. To understand the relationship between the confidence level and Z, review the material in Chapter 6. What would be the corresponding Z value for a 98% confidence interval?

Answer:

The appropriate Z value for a 98% confidence interval is 2.33.

(p. 181)

Learning Check 6.3. What is the 90% confidence interval for the mean commuting time? First, find the Z value associated with a 90% confidence level.

Answer:

$$\begin{aligned}90\%\ \text{CI} &= 7.5 \pm 1.65(0.07)\\ &= 7.5 \pm 0.12\\ &= 7.38 \text{ to } 7.62\end{aligned}$$

(p. 186)

Learning Check 6.4. Why do smaller sample sizes produce wider confidence intervals? (See Figure 6.5.) Compare the standard errors of the mean for the three sample sizes.

Answer:

As the sample size gets smaller, the standard error of the mean gets larger, which, in turn, results in a wider confidence interval.

(p. 193)

Learning Check 6.5. Calculate the 95% confidence interval for the CBS News poll results for those who support stricter laws for gun sales.

Answer:

The 95% confidence interval is .63 to .67 or 63% to 67% in favor of stricter gun sale laws.

CHAPTER 7

(p. 217)

Learning Check 7.2. For the following research situations, state your research and null hypotheses:

- *There is a difference between the mean statistics grades of social science majors and the mean statistics grades of business majors.*
- *The average number of children in two-parent black families is lower than the average number of children in two-parent nonblack families.*
- *Grade point averages are higher among girls who participate in organized sports than among girls who do not.*

Answer:

Null Hypothesis	Research Hypothesis
Means are presumed equal for all statements.	Two-tailed test. No direction is stated.
	One-tailed test, left.
	One-tailed test, right.

(p. 221)

Learning Check 7.3. Would you change your decision in the previous example if alpha were.01? Why or why not?

Answer:

For alpha = .01, the t-critical is 2.576. T-obtained (1.95) is less than t-critical. We would fail to reject the null hypothesis.

(p. 223)

Learning Check 7.4. State the null and research hypothesis for this SPSS example. Would you change your decision in the previous example if alpha was .01? Why or why not

Answer:

The null hypothesis is: $\mu_1 = \mu_2$

The research hypothesis is: $\mu_1 \neq \mu_2$

If alpha was set at .01, we would reject the null hypothesis. The t-obtained of –9.478 (equal variances assumed) is significant at the .001 < .01.

(p. 226)

Learning Check 7.5. If alpha was changed to .01, two-tailed test, would your final decision change? Explain.

Answer:

The probability of the obtained Z is .0002 < .01. We would still reject the null hypothesis of no difference.

(p. 227)

Learning Check 7.6. If alpha was changed to .01, one-tailed test, would your final decision change? Explain.

Answer:

The probability of the obtained Z is .0062 < .01. We would still reject the null hypothesis of no difference.

(p. 229)

Learning Check 7.7. Based on Table 8.5, what would be the t *critical at the .05 level for the first indicator, EC Index? Assume a two-tailed test.*

Answer:

The N's are reported as a Note in the bottom of the table. The *df* calculation would be (78 + 113) – 2 = 189. Based on Appendix C, $df = \infty$, *t* critical is 1.960.

CHAPTER 8

(p. 240)

Learning Check 8.1. Construct a bivariate table (in percentages) showing no association between age and first-generation college status.

Answer:

Age and First-Generation College Status

	19 Years or Younger	20 Years or Older	
Firsts	41.9%	41.9%	41.9% (1,934)
Nonfirsts	58.1%	58.1%	58.1% (2,683)
	100.0%	100.0%	4,617

(p. 241)

Learning Check 8.2. Refer to the data in the previous Learning Check. Are the variables age and first-generation college status statistically independent? Write out the research and the null hypotheses for your practice data.

Answer:

Null hypothesis: There is no association between age and first-generation college status.

Research hypothesis: Age and first-generation college status are statistically dependent.

(p. 243)

Learning Check 8.3. Refer to the data in Learning Check 8.1. Calculate the expected frequencies for age and first-generation college status and construct a bivariate table. Are your column and row marginals the same as in the original table?

Answer:

First Generation	19 Years or Younger	20 Years or Older	Total
Firsts	1,138.84 (41.9%)	795.68 (41.9%)	1,934 (41.9%)
Nonfirsts	1,579.16 (58.1%)	1,103.32 (58.1%)	2,683 (58.1%)
Total (*N*)	2,718 (100.0%)	1,899 (100.0%)	4,617 (100.0%)

(p. 245)

Learning Check 8.4. Using the format of Table 8.5, construct a table to calculate chi-square for age and educational attainment.

Answer:

	f_o	f_e	$f_o - f_e$	$(f_o - f_e)^2$	$(f_o - f_e)^2/f_e$
19/Firsts	916	1138.53	–222.53	49519.60	43.49
19/Nonfirsts	1802	1579.47	222.53	49519.60	31.35
20/Firsts	1018	795.47	222.53	49519.60	62.25
20/Nonfirsts	–881	1103.53	–222.53	49519.60	44.87

Chi-square = 181.96, with Yates correction = 181.15.

(p. 247)

Learning Check 8.5. Based on Appendix D, identify the probability for each chi-square value (df *in parentheses)*

Answer:

- 12.307 (15) Between .70 and .50
- 20.337 (21) Exactly .50
- 54.052 (24) Less than .001

(p. 249)

Learning Check 8.6. What decision can you make about the association between age and first-generation college status? Should you reject the null hypothesis at the .05 alpha level or at the .01 level?

Answer:

We would reject the null hypothesis of no difference. Our calculated chi-square is significant at the .05 and the .01 levels. We have evidence that age is related to first-generation college status—older students are more likely to be first-generation students than younger students. Fifty-four percent of students 20 years or older are first-generation students versus 33.7% of students 19 years or younger.

(p. 253)

Learning Check 8.7. For the bivariate table with age and first-generation college status, the value of the obtained chi-square is 181.15 with 1 degree of freedom. Based on Appendix D, we determine that its probability is less than.001. This probability is less than our alpha level of.05. We reject the null hypothesis of no relationship between age and first-generation college status. If we reduce our sample size by half, the obtained chi-square is 90.58. Determine the P *value for 90.58. What decision can you make about the null hypothesis?*

Answer:

Even if we reduce the chi-square by half, we would still reject the null hypothesis.

CHAPTER 9

(p. 280)

Learning Check 9.1.

Identify the independent and dependent variables in Table 9.1.

Answer:

Race/ethnicity is the independent variable, Educational attainment is the dependent variable.

(p. 289)

Learning Check 9.3.

If alpha were changed to .01, would our final decision change?

Answer:

If alpha was changed to .01, the F critical would be 5.01. Our F obtained is greater than the F critical. We would still reject the null hypothesis of no difference.

(p. 290)

Learning Check 9.4. Calculate eta^2 for this model.

Answer:

Based on the output eta^2 = 10.408/1300.902=.008 or .01. If we know political party identification, we can predict attitudes about ethical consumerism with 1% accuracy. This is a very weak prediction model.

(p. 293)

Learning Check 9.5. For the "Sense of Belonging and Satisfaction" ANOVA model, what additional information would you need to determine the F critical?

Answer:

You would need to k and N. In this case, $k = 4$ (number of categories) and $N = 33{,}415$ (total sample size). There are two degrees of freedom to calculate: df(between) $= k - 1 = 4 - 1 = 3$ and df(within) $= N - k = 33{,}415 - 4 = 33{,}411$. For an alpha of. 05, the F critical is 2.60 (based on Appendix E).

CHAPTER 10

(p. 307)

Learning Check 10.1. For each of these four lines, as X goes up by 1 unit, what does Y do? Be sure you can answer this question using both the equation and the line.

Answer:

For the line $Y = 1X$, as X goes up by 1 unit, Y also goes up by 1 unit. In the second line, $Y = 2 + 0.5X$, Y increases by 0.5 units as a result of 1-unit increase in X. The line $Y = 6 - 2X$ tells that every 1-unit increase in X results in 2-unit decrease in Y. Finally, in the fourth line, Y decreases by 0.33 units as a result of 1-unit increase in X.

(p. 314)

Learning Check 10.2. Use the prediction equation to calculate the predicted values of Y if X equals 9, 11, and 14. Verify that the regression line in Figure 10.4 passes through these points.

Answer:

$$\hat{Y} = -4.68 + .62\ (9) = .90$$

$$\hat{Y} = -4.68 + .62\ (11) = 2.14$$

$$\hat{Y} = -4.68 + .62(14) = 4.00$$

(p. 323)

Learning Check 10.3. Calculate r and r^2 for age and internet hours regression model. Interpret your results.

Answer: $r = -.77$ and $r^2 = .88$. The correlation coefficient indicates a strong negative relationship between age and internet hours per week. Using information about respondent's age helps reduce the error in predicting internet hours by 88%.

(p. 325)

Learning Check 10.4. Test the null hypothesis that there is a linear relationship between age and internet hours. The mean squares regression is 63.66 with 1 degree of freedom. The mean squares residual is 2.355 with 8 degrees of freedom. Calculate the F statistic and assess its significance.

Answer: $F = 63.66/2.35 = 27.09$. This exceeds the F-critical (1,8) of 5.32 (alpha = .05). We can reject the null hypothesis and conclude that the linear relationship between age and internet hours per week as expressed in r^2 is greater than zero in the population.

(p. 330)

Learning Check 10.5. Use the prediction equation describing the relationship between internet hours per week and both educational attainment and age to calculate internet hours per week for someone with 20 years of education who is 35 years old.

Answer:

$\hat{Y} = -.605 + .491(20) + -.057(35) = -.605 + 9.82 + -1.995 = 7.22$ internet hours per week

ANSWERS TO ODD-NUMBERED EXERCISES

CHAPTER 1

1. Once our research question, the hypothesis, and the study variables have been selected, we move on to the next stage of the research process—measuring and collecting the data. The choice of a particular data collection method or instrument depends on our study objective. After our data have been collected, we have to find a systematic way to organize and analyze our data and set up some set of procedures to decide what we mean.

3. a. Interval-ratio
   b. Interval-ratio
   c. Nominal
   d. Ordinal
   e. Nominal
   f. Interval-ratio
   g. Ordinal

5. There are many possible variables from which to choose. Some of the most common selections by students will probably be type of occupation or industry, work experience, and educational training or expertise. Students should first address the relationship between these variables and gender. Students may also consider measuring structural bias or discrimination.

7.
   a. Annual income
   b. Gender—nominal; Number of hours worked per week—interval-ratio; Years of education—interval-ratio; Job title—nominal.
   c. This is an application of inferential statistics. The researcher is using information based on her sample to predict the annual income of a larger population of young graduates.

9.

a. Individual age: This variable could be measured as an interval-ratio variable, with actual age in years reported. As discussed in the chapter, interval-ratio variables are the highest level of measurement and can also be measured at the ordinal level.

b. Annual income: This variable could be measured as an interval-ratio variable, with actual dollar earnings reported. Income can also be measured as an ordinal variable with income brackets.

c. Religiosity: This variable could be measured in several ways. For example, as church attendance, the variable could be ordinal (number of times attended church in a month: every week, at least twice a month, less than two times a month, none at all).

d. Student performance: This could be measured as an interval-ratio variable as GPA or test score. It can also be measured as an ordinal variable with performance categories ranging from excellent to poor.

e. Social class: This variable is an ordinal variable, with categories low, working, middle, and upper.

f. Number of children: This variable could be measured in several ways. As an interval-ratio measure, the actual number of children could be reported. As an ordinal measure, the number of children could be measured in categories: 0, 1–2, 3–4, 5 or more. This could also be a nominal measurement—do you have children? Yes or no.

CHAPTER 2

1.

a. Race is a nominal variable. Class is an ordinal variable, since the categories can be ordered. Trauma is an interval variable.

b. Frequency Table for Race

Race	Frequency (*f*)
White	17
Nonwhite	13
Total (*N*)	30

c. White: 17/30 = .57; Nonwhite: 13/30 = .43

3.

Number of Traumas	Frequency (*f*)
0	15
1	11
2	4
Total (*N*)	30

a. Trauma is an interval- or ratio-level variable, since it has a real zero point and a meaningful numeric scale.

b. People in this survey are more likely to have experienced no traumas last year (50% of the group).

c. The proportion who experienced one or more traumas is calculated by first adding 36.7% and 13.3% = 50%. Then divide that number by 100 to obtain the proportion, 0.50, or half the group.

5. Yes, the data suggest those who voted for Donald Trump in the 2016 U.S. presidential election have less confidence in the press than those who didn't vote for Trump. Of those who voted for Trump, 79.2% had "hardly any" confidence in the press, compared to 19.3% of those who voted for Clinton.

7. According to this time-series chart, rates of voting have consistently varied by race and Hispanic origin. The group with the largest increase in voting rates is blacks, from 53% in 1996 to 66.6% in 2012. However, in 2016, the voting rates for blacks declined to 59.6%—the lowest it has been since the 2000 election, when it was 56.9%. Hispanic voting rates have, for the most part, remained stable, although with some fluctuation most notably from the 1992 election (51.6%) to the 1996 election (44%). As noted in the exercise, in the 2012 presidential election, blacks had the highest voting rates for all groups, followed by non-Hispanic whites, non-Hispanic other races, and Hispanics. White voting rates increased by 1.2% from 2012 to 2016. The highest voting rate for whites was in 1992 (70.2%), 1992 for Hispanics (51.6%), 2012 for blacks (66.6%), and 1992 for non-Hispanic other races (54%).

9. If we identify younger Americans as those in the 18 to 29 and 30 to 44 age groups and older Americans in the 45 to 64 and 65 and over categories—the data indicate that as age increases, so does the percent voting in a presidential election—a pattern that has remained fairly stable since the 1980s (the first time point available in this time-series chart). The group with the highest percent of voting in the 2016 election is those 65 and

older, with 70.9% voting. The percentage slightly drops for those in the 45 to 64 age group but is still higher than the reported percentages for the 18 to 29 and 30 to 44 age groups.

11.

a. Based on the student's argument, the independent variable is race and the dependent variable is home ownership.

b. 539/857 = 63%

c. There appears to be a relationship between race and home ownership. While 63% of those surveyed report owning a home, far less black respondents report such (41.3%) than white respondents (67.7%). Furthermore, 58.7% of black respondents rent as compared to 32.3% of white respondents.

13.

a. The dependent variable is housework disagreement.

b. 7,144 is the total sample size.

c. 3,953/7,144 = 55.33%

d. There is a relationship between perceived unfairness of housework divisions and housework disagreement for women. In total, 33.68% of women reported never having housework disagreement. However, women who felt housework divisions were fair to women (43.65%) were more likely to report never having disagreement over housework than women who felt housework divisions were unfair to women (25.62%).

CHAPTER 3

1.

a. Mode = Routine (f = 356)

b. Median = Routine

c. Based on the mode and median for this variable, most respondents indicate that their lives are "routine."

d. A mean score could not be interpreted for this variable. A mean is only appropriate for an interval-ratio variable.

3.

a. Interval-ratio. The mode can be found two ways: by looking either for the highest frequency (14) or the highest percentage (43.8%). The mode is the category that corresponds to the value "40 hours worked last week." The median can be found two ways: by using either the frequencies column or the cumulative percentages.

Using Frequencies	Using Cumulative Percentages
$\frac{N+1}{2} = \frac{32+1}{2} = 16.5\text{th case}$	Notice that 34.4% of the observations fall in or below the "32 hours worked last week" category; 78.1% fall in or below the "40 hours worked last week" category.
Starting with the frequency in the first category (1), add up the frequencies until you find where the 16th and 17th cases fall. Both of these cases correspond to the category "40 hours worked last week," which is the median.	The 50% mark, or the median, is located somewhere within the "40 hours worked last week" category. So the median is "40 hours worked last week."

b. Since the median is merely a synonym for the 50th percentile, we already know that its value is 40 hours worked last week.

25th percentile = (32 × 0.25) = 8th case = 30 hours worked last week

75th percentile = (32 × 0.75) = 24th case = 40 hours worked last week

5.

a.

	Mode	Median
Black	Seldom ($f = 11$)	Sometimes
White	Every day ($f = 84$)	Nearly every day
Hispanic	Every day ($f = 25$)	Most days

b. Teens' breakfast habits vary by race/ethnicity. Out of the three racial/ethnic groups, black students were more likely to report seldom or sometimes eating breakfast. On the other hand, white and Hispanic students eat breakfast more frequently. The mode for white and Hispanic students is every day.

7.

a. The range is 2.30 (4.50 – 2.20). The 25th percentile, 2.45, means that 25% of cases fall below the 2.45 divorce rate per 1,000 population. Likewise, the 75th percentile means that 75% of all cases fall below the 3.60 divorce rate per 1,000 population.

25th percentile	10(0.25) = 2.5th case	So (2.40 + 2.50)/2 = 2.45
75th percentile	10(0.75) = 7.5th case	So (3.60 + 3.60)/2 = 3.60

The IQR is thus 3.60 – 2.45 = 1.15. Both measures of variability are appropriate, but the range is somewhat better, as the value for the IQR is fairly small. In other words, the range gives us a better picture of the variability of divorce rates for all states in our sample.

b.

State	Divorce Rate per 1,000 Population	$Y - \bar{Y}$	$(Y - \bar{Y})^2$
Alaska	3.6	3.6–3.14	.21
Florida	3.6	3.5–3.14	.21
Idaho	3.9	3.9–3.14	.46
Maine	3.2	3.2–3.14	.004
Maryland	2.5	2.5–3.14	.41
Nevada	4.5	4.5–3.14	1.85
New Jersey	2.6	2.6–3.14	.29
Texas	2.2	2.2–3.14	.88
Vermont	2.9	2.9–3.14	.058
Wisconsin	2.4	2.4–3.14	.55
Total	31.40	0.00	4.922

$$\bar{Y} = \frac{\Sigma Y}{N} = \frac{31.40}{10} = 3.14$$

$$s_y = \sqrt{s^2} = \sqrt{\frac{\Sigma(Y - \bar{Y})^2}{N - 1}} = \sqrt{\frac{4.922}{9}} = .74$$

c. Divorce rates may vary by state due to factors such as variation in religiosity, state policy (i.e., no-fault divorce laws), or employment opportunities.

9.

a. The mean numbers of crimes is 302.26 (per 100,000 population) and the standard deviation is 97.125. The mean amount of dollars (in millions) spent on police protection is $290.81 and the standard deviation is $70.32.

b. Because the number of crimes and police protection expenditures is measured according to different scales, it isn't appropriate to directly compare the mean and standard deviation for one variable with the other. But we can talk about each distribution separately. We know from examining the mean and standard deviation for the number of crimes that the standard deviation (97.125) is large, indicating a wide dispersion of scores from the mean. For the number of crimes, states such as Missouri and Vermont contribute more to its variability because they have values far from the mean (both above and below). With respect to police protection expenditures, we can see that there is a large dispersion from the mean, as the standard deviation is $70.32. States such as New York and Indiana contribute more to its variability because they have values far from the mean (both above and below).

c. Among other considerations, we need to consider the economic conditions in each state. A downturn in the local and state economy may play a part in the number of crimes and police expenditures per capita.

11.

a. The range of population increase for the western states is 9.30%. The range of percent increase for the midwestern states is 12.90%. The midwestern states have a much larger range.

b. The IQR for the western states is 3.90%. The IQR for the midwestern states is 3.40%.

c. Recall that the range is based only on two values of the distribution, while the IQR relies on the middle 50% of the distribution. The range is larger for the midwestern states due to the variability in percentage growth from 12.7 (North Dakota) to –0.2 (Illinois). The IQR indicates greater variability in the increase in the population in western states, caused by the intermediate scores of the distribution.

13. Overall, Clinton voters were younger, were more educated, and attended religious services less than Trump voters. The youngest voters were male Clinton voters at 50.49 years ($s = 18.84$), followed by female Clinton voters, 51.93 years ($s = 18.21$). For education, males who voted for Clinton had the highest mean of 14.69 ($s = 2.73$). Males who voted for Trump had 13.80 years of education ($s = 2.75$). Trump voters, both males and females, attended religious services more often than Clinton voters. Mean scores were 3.72 for males ($s = 2.93$) and 3.82 for females ($s = 2.75$), indicating church attendance about several times a year to once a month. The standard deviations indicate a consistency in the distributions of education, age, and religious service attendance across all four groups. The largest standard deviations are for age, ranging from 15.94 to 18.84 years. These wide standard deviations indicate more dispersion around the mean age scores.

CHAPTER 4

1.

a. The Z score for a person who watches more than 8 hr/day:

$$Z = \frac{8 - 2.97}{3.00} = 1.68$$

b. We first need to calculate the Z score for a person who watches 5 hr/day:

$$Z = \frac{5 - 2.97}{3.00} = 0.68$$

The area between Z and the mean is 0.2517. We then need to add 0.50 to 0.2517 to find the proportion of people who watch television less than 5 hrs/day. Thus, we conclude that the proportion of people who watch television less than 5 hrs/day is 0.7517. This corresponds to 762 people (762.22 = 0.7517 × 1,014).

c. 5.97 television hours per day corresponds to a Z score of +1.

$$Y = \overline{Y} + Z(S_Y) = 2.97 + 1(3) = 5.97$$

d. The Z score for a person who watches 1 hr of television per day is –.66. The area between the mean and Z is .2454.

$$\frac{1 - 2.97}{3.00} = -.66$$

The Z score for a person who watches 6 hrs of television per day is 1.01. The area between the mean and Z is .3438.

$$\frac{6 - 2.97}{3.00} = 1.01$$

Therefore, the percentage of people who watch between 1 and 6 hrs of television
per day is 58.92% (0.2454 + 0.3438 = 0.5892 × 100).

3.

a. For an individual with 13.71 years of education, his or her Z score would be

$$Z = \frac{13.71 - 13.71}{3.04} = 0.0$$

b. Since our friend's number of years of education completed is associated with the 60th percentile, we need to solve for Y. However, we must first use the logic of the normal distribution to find Z. For any normal distribution,

50% of all cases will fall above the mean. Since our friend is in the 60th percentile, we know that the area between the mean and our friend's score is 0.10. Similarly, the area beyond our friend's score is 0.40. We can now look in Appendix B column "B" for 0.10 or in column "C" for 0.40. We find that the Z associated with these values is 0.25. Now, we can solve for Y:

$$Y = \bar{Y} + Z(s) = 13.71 + 0.25(3.04) = 14.47$$

c. Since we already know that the proportion between our number of years of education (13.71) and our friend's number of years of education (14.47) is 0.10, we can multiply N (1,498) by this proportion. Thus, 149.8, or rounded, 150 people have between 13.71 and 14.47 years of education.

5.

a. Among working-class respondents:

The Z score for a value of 12 is

$$Z = \frac{12 - 13.05}{2.77} = -0.38$$

The Z score for a value of 16 is

$$Z = \frac{16 - 13.05}{2.77} = 1.06$$

You'll find the area between the Z scores and the mean under column B. The total area between the scores is .1480 + .3554 = .5034. The proportion of working-class respondents with 12 to 16 years of education is 0.5034.

Among upper-class respondents:

The Z score for a value of 12 is

$$Z = \frac{12 - 15.48}{2.76} = -1.26$$

The Z score for a value of 16 is

$$Z = \frac{16 - 15.48}{2.76} = 0.19$$

The area between a Z of −1.26 and the mean is 0.3962. The area between a Z of 0.19 is 0.0753, so the total area between the scores is .3962 + .0753 = .4715. The proportion of upper-class respondents with 12 to 16 years of education is 0.4715.

A slightly higher proportion of working-class respondents than upper-class respondents has 12 to 16 years of education.

b. For working-class respondents:

As previously calculated, the Z score for a value of 16 is 1.06. The area between a Z of 1.06 and the tail of the distribution (column C) is .1446. So the probability of a working-class respondent having more than 16 years of education is 14.46%.

For middle-class respondents:

The Z score for a value of 16 is

$$Z = \frac{16 - 14.56}{2.95} = .49$$

The area between a Z of 0.49 and the tail of the distribution (column C) is 0.3121. So the probability of a middle-class respondent having more than 16 years of education is 31.21%.

c. For lower-class respondents:

The Z score for a value of 10 is

$$Z = \frac{10 - 12.03}{3.24} = -0.63$$

The area beyond Z of 0.63 is .2643. So the probability of a lower-class respondent having less than 10 years of education is .2643 (or 26.43%).

7.

a. The Z score of 150 is 3.33.
b. The area beyond 3.33 is .0004. The percentage of scores above 150 is .04%, a very small percentage.
c. The Z score for 85 is −1.0. The percentage of scores between 85 and 150 is 84.09% (.3413 + .4996 = .8409).
d. Scoring in the 95th percentile means that 95% of the sample scored below this level. Identifying the 95th percentile can be calculated by this formula: 100 + 1.65(15) = 124.75. The IQ score that is associated with the 95th percentile is 124.75.

9.

a. About 0.1894 of the distribution falls above the Z score, so that is the proportion of crime incidents with more than two victims.

$$Z = \frac{2 - 1.28}{0.82} = 0.88$$

b. The area between the mean and the Z score is about 0.1331, so the total area above one victim is 0.50 + 0.1331 = 0.6331, or 63.31%.

$$Z = \frac{1 - 1.28}{0.82} = -0.34$$

c. The area between the mean and the Z score is about 0.4995, so the total area below four victims is 0.50 + 0.4995 = 0.9995.

$$Z = \frac{4 - 1.28}{0.82} = 3.32$$

11.

a. For a team with an APR score of 990

$$Z = \frac{990 - 981}{27.3} = 0.33$$

From Appendix B, the area beyond 0.33 is 0.3707 or about the 67th percentile. The team is not in the upper quartile.

b. The Z value, which corresponds to a cutoff score with an area of about 0.25 toward the tail of the distribution, is 0.67. This is translated into a cutoff score:

$$981 + .67(27.3) = 999.29$$

c. The Z value is 0.67.

13. The 95th percentile corresponds to a Z score of 1.65.

Hungary

11.76 + 1.65 (2.91) = 16.56 years

Czech Republic

12.82 + 1.65 (2.29) = 16.60 years

Denmark

13.93 + 1.65 (5.83) = 23.55 years

France

14.12 + 1.65 (5.73) = 23.57 years

Ireland

15.15 + 1.65 (3.90) = 21.59 years

CHAPTER 5

1.

   a. Although there are problems with the collection of data from all Americans, the census is assumed to be complete, so the mean age would be a parameter.
   b. A statistic because it is estimated from a sample.
   c. A parameter because the school would have information on all its students.
   d. A statistic because it is estimated from a sample.
   e. A parameter because the school has information on all employees.

3. This is not a probability sample. We do not know how the researcher chose the twenty study participants to represent a cross section of the population in New York City.

5. The relationship between the standard error and the standard deviation is $\sigma_{\bar{Y}} = \sigma / \sqrt{N}$. Since σ is divided by $\sqrt{N}$, $\sigma_{\bar{Y}} = \sigma / \sqrt{N}$ must always be smaller than σ, except in the trivial case where $N = 1$. Theoretically, the dispersion of the mean must be less than the dispersion of the raw scores. This implies that the standard error of the mean is less than the standard deviation.

7.

   a. FB polls are definitely not probability samples. Here are just two reasons why: First, the "population" from which data are gathered is limited to the "friends" on one's FB profile—and such is certainly not a representative sample from the larger population. Second, due to FB algorithms, we can't be sure that every FB friend would have the same opportunity to see, and thus participate, in the poll.
   b. The population is our FB friends who come across the poll.

9.

   a. This is not a reliable sample. The students eating lunch on Tuesday are not necessarily representative of all students at the school, and you have no way of calculating the probability of inclusion of any student. Many students might, for example, rarely eat lunch at the student union and, therefore, have no chance of being represented in your sample. The fact that you selected *all* the students eating lunch on Tuesday makes your selection appear to be a census of a population, but that isn't true, either, unless all the students ate at the student union on Tuesday.
   b. This is not a reliable sample. The majority of people in attendance at the rally would probably be in support of open borders. Students against such might not have any chance of selection.

c. Answers will vary, but it seems the best route would be to obtain a list of all registered students from your university's registrar. And then employ a known sampling strategy (simple random sampling, stratified sampling, etc.) to create your sample of 10% of the student body. From there, you would reach out to each person you've selected via e-mail, telephone, mail, and so on and ask them to complete your survey.

11.

a. Mean = 5.3 (53/10); standard deviation = 3.27.

b. Here are 10 means from random samples of size 3: 6.33, 5.67, 3.33, 5.00, 7.33, 2.33, 6.00, 6.33, 7.00, 3.00.

c. The mean of these 10 sample means is 5.23. The standard deviation is 1.76. The mean of the sample means is very close to the mean for the population. The standard deviation of the sample means is much less than the standard deviation for the population. The standard deviation of the means from the samples is an estimate of the standard error of the mean we would find from one random sample of size 3.

CHAPTER 6

1.

a. The estimate at the 90% confidence level is 20.87% to 21.13%. This means that there are 90 chances out of 100 that the confidence interval will contain the true population percentage of victims in the American population.

Due to the large sample size, we converted the proportions to percentages, subtracting from 100, rather than 1.

$$\text{Standard error} = \sqrt{\frac{(21)\,(100 - 21)}{239{,}541}} = 0.08$$

$$\begin{aligned}\text{Confidence interval} &= 21 \pm 1.65(0.08)\\ &= 21 \pm 0.13\\ &= 20.87 \text{ to } 21.13\end{aligned}$$

b. The true percentage of crime victims in the American population is somewhere between 20.79% and 21.21% based on the 99% confidence interval. There are 99 chances out of 100 that the confidence interval will contain the true population percentage of crime victims.

$$\begin{aligned}\text{Confidence interval} &= 21 \pm 2.58(0.08)\\ &= 21 \pm 0.21\\ &= 20.79 \text{ to } 21.21\end{aligned}$$

3.

a. For Canadians:

$$s_p = \sqrt{\frac{(0.51)(1-0.51)}{1{,}004}} = 0.02$$

$$\begin{aligned}\text{Confidence interval} &= 0.51 \pm 1.96(0.02)\\ &= 0.51 \pm 0.04\\ &= 0.47 \text{ to } 0.55\end{aligned}$$

b. For Americans:

$$s_p = \sqrt{\frac{(0.45)(1-0.45)}{1{,}003}} = 0.02$$

$$\begin{aligned}\text{Confidence interval} &= 0.45 \pm 1.96(0.02)\\ &= 0.45 \pm 0.04\\ &= 0.39 \text{ to } 0.49\end{aligned}$$

c. Based upon the calculated 95% confidence interval, the majority of Americans do not believe climate change is a serious problem. The true percentage of Americans who believe climate change is a serious problem is under 50%, somewhere between 39% and 49%, based upon this Pew Research Center sample. On the other hand, it is possible that the majority of Canadians believe climate change is a serious problem. We can be 95% confident that the true percentage of Canadians is somewhere between 47% and 55%.

5.

Due to the large sample size, we converted the proportion to full percentages, subtracting from 100 (rather than 1).

$$\text{Standard error} = \sqrt{\frac{(51)\,(100-51)}{5{,}490}} = 0.67$$

$$\begin{aligned}\text{Confidence interval} &= 51 \pm 1.96(0.67)\\ &= 49.69\% \text{ to } 52.31\%\end{aligned}$$

We set the interval at the 95% confidence level. However, no matter whether the 90%, 95%, or 99% confidence level is chosen, the calculated interval includes values below 50% for the vote for a Republican candidate. Therefore, you should tell your supervisors that it would not be possible to declare a Republican candidate the likely winner of the votes coming from men if there was an election today because it seems quite possible that less than a majority of male voters would support her or him.

7.

a.

$$s_p = \sqrt{\frac{(0.83)(1-0.83)}{1{,}054}} = 0.01$$

$$\begin{aligned}\text{Confidence interval} &= 0.83 \pm 1.96(0.01)\\ &= 0.83 \pm 0.02\\ &= .81 \text{ to } .85\end{aligned}$$

b. Based on our answer in 7a, we know that a 90% confidence interval will be more precise than a 95% confidence interval that has a lower bound of 81% and an upper bound of 85%. Accordingly, a 90% confidence interval will have a lower bound that is greater than 81% and an upper bound that is less than 85%. Additionally, we know that a 99% confidence interval will be less precise than what we calculated in 7a. The lower bound for a 99% confidence interval will be less than 81%, and the upper bound will be greater than 85%.

9.

Country	Mean	Standard Error	Confidence Interval
France	14.12	$5.73/\sqrt{975} = .18$	$14.12 + .18(1.65) = 14.42$ $14.12 - .18(1.65) = 13.82$
Japan	12.48	$2.53/\sqrt{528} = .11$	$12.48 + .11(1.65) = 12.66$ $12.48 - .11(1.65) = 12.30$
Croatia	12.18	$2.71/\sqrt{480} = .12$	$12.18 + .12(1.65) = 12.38$ $12.18 - .12(1.65) = 11.98$
Turkey	9.15	$11.98/\sqrt{783} = .43$	$9.15 + .43(1.65) = 9.86$ $9.15 - .43(1.65) = 8.44$

11.

For Republicans:

$$s_p = \sqrt{\frac{(.18)(1-.18)}{446}} = 0.02$$

$$\begin{aligned}\text{Confidence interval} &= .18 \pm 1.96(0.02)\\ &= .18 \pm .04\\ &= .14 \text{ to } .22\end{aligned}$$

For Democrats:

$$s_p = \sqrt{\frac{(.15)(1-.15)}{522}} = 0.02$$

$$\text{Confidence interval} = .15 \pm 1.96(0.02)$$
$$= .15 \pm .04$$
$$= .11 \text{ to } .19$$

13.

a. For those who identified as liberal:

$$s_p = \sqrt{\frac{(.29)(1-.29)}{1117}} = .01$$

$$\text{Confidence interval} = .29 \pm 1.96(.01)$$
$$= .29 \pm .02$$
$$= .27 \text{ to } .31$$

For those who identified as conservative:

$$s_p = \sqrt{\frac{(.32)(1-.32)}{1117}} = .01$$

$$\text{Confidence interval} = .32 \pm 1.96(.01)$$
$$= .32 \pm .02$$
$$= .30 \text{ to } .34$$

b. For those who identified as moderate:

$$s_p = \sqrt{\frac{(.39)(1-.39)}{1117}} = .01$$

$$\text{Confidence interval} = .39 \pm 1.96(.01)$$
$$= .39 \pm .02$$
$$= .37 \text{ to } .41$$

CHAPTER 7

1.

a. $H0$: $\mu Y = 13.5$ years; $H1$: $\mu Y < 13.5$ years.
b. The Z value obtained is –4.19. The p value for a Z of –4.19 is less than .001 for a one-tailed test. This is less than the alpha of .01, so we reject the null hypothesis and conclude that the doctors at the HMO do have less experience than the population of doctors at all HMOs.

3.

a. Two-tailed test, \$57,652; null hypothesis, \$57,652
b. One-tailed test, $\mu > 3.2$; null hypothesis, $\mu = 3.2$
c. One-tailed test, $\mu_1 < \mu_2$; null hypothesis, $\mu_1 = \mu_2$
d. Two-tailed test, $\mu_1 \neq \mu_2$; null hypothesis, $\mu_1 = \mu_2$
e. One-tailed test, $\mu_1 > \mu_2$; null hypothesis, $\mu_1 = \mu_2$
f. One-tailed test, $\mu_1 < \mu_2$; null hypothesis, $\mu_1 = \mu_2$

5.

a. Research hypothesis, 37.8; null hypothesis, 37.8.
b. The *t* obtained is 23.17 and its *p* level is <.001 (the *t* obtained is greater than the last reported *t* critical of 3.291).

$$t = \frac{48.69 - 37.8}{17.99 / \sqrt{1495}} = \frac{10.89}{.47} = 23.17$$

c. We conclude that we can reject the null hypothesis in favor of the research hypothesis. There is a difference between the mean age of the GSS sample and the mean age of all American adults. Relative to age, the GSS sample is not representative of all American adults (the GSS sample is significantly older).

7.

a. The appropriate test statistic is *t* test for sample means.
b.

$$t = \frac{1.69 - 2.11}{.17} = -2.47$$

$$\text{Standard error} = \sqrt{\frac{86(1.04)^2 + 67(1.05)^2}{(86+67)-2}}\sqrt{\frac{86+67}{(86)(67)}} = (1.05)(.16) = .17$$

The *t* obtained of –2.47 is greater than the *t* critical of –1.645. We reject the null hypothesis of no difference. Strong Democrats have a lower mean score than strong Republicans. Strong Democrats are more likely to strongly agree that the differences in income are too large than strong Republicans.

9.

a. "Less than" indicates a one-tailed test.
b. $Z = -5.00$ with a significance of less than 0.0001. We can reject the null hypothesis and conclude that the proportion of males who feel electing a

woman as president is very important historically is significantly less than the proportion of females who feel electing a woman as president is very important historically (.55 – .65 = –.10).

$$Z = \frac{.55 - .65}{.02} = -5.0$$

c. The significance of –5.00 is less than 0.0001. The decision to reject the null hypothesis does not change.

11. Older individuals, aged 50 to 59 years, gave more money in the past year than younger adults aged 30 to 39 years. However, the difference in giving is not significant. The *t* obtained is –.800 (equal variances assumed) with a probability of .425 (>.05 alpha).

13.

a. There is no significant difference between the average number of relaxation hours for married men and women. We fail to reject the null hypothesis. The *t* obtained of .086 (equal variances assumed) is significant at the .931 level (greater than our alpha of .05).

b. If alpha was changed to .01, we would fail to reject the null hypothesis of no difference. The probability of the *t* obtained is .931 > .01.

CHAPTER 8

1.

a. Degrees of freedom = (3 – 1)(2 – 1) = 2

b. Chi-square = 3.311. The probability of our obtained chi-square (.191) is greater than our .01 alpha level. We fail to reject the null hypothesis and conclude that sex and opinion of whether or not people can be trusted are independent. More specifically, similar percentages of men (34.9%) and women (29.7%) felt that people can be trusted.

c. If α were changed to .05, we would still fail to reject the null hypothesis. The probability of our obtained chi-square is still less than alpha.

d. The lambda is .000. There is no proportional reduction of error using sex to predict whether or not people can be trusted.

3.

a. A higher percentage of white respondents (38.1%), compared to black respondents (23%), reported that people could be trusted.

b. For black respondents, a slightly higher percentage of men (15.1%) compared to women (12.5%) reported that people could be trusted.

c. Whites, $\chi^2 = 2.395$; we fail to reject the null hypothesis.

Blacks, $\chi^2 = .233$; we fail to reject the null hypothesis. *Note:* Two cells have expected counts less than 5. Results should be cautiously interpreted. It would be beneficial to repeat the study with a larger sample size.

5.

a. We will make 2,973 errors, because we predict that all victims fall in the modal category (white). $E_1 = 6{,}084 - 3{,}111 = 2{,}973$.

b. For white offenders, we could make 373 errors; for black offenders, 493 errors; and for other offenders, we would make 42 errors. $E_2 = 908$.

c. The proportional reduction in error is then $(2{,}973 - 908)/2{,}973 = .6946$. This indicates a very strong relationship between the two variables. We can reduce the error in predicting victims' race based upon race of offender by 69.46%.

7.

Race/First-Generation College Status	f_o	f_e	$f_o - f_e$	$(f_o - f_e)^2$	$\frac{(f_o - f_e)^2}{f_e}$
White/first	1,742	1,749.6	–7.6	57.76	0.03
White/nonfirst	2,392	2,384.4	7.6	57.76	0.02
Black/first	102	93.5	8.5	72.25	0.77
Black/nonfirst	119	127.5	–8.5	72.25	0.57
Native American/first	41	36.4	4.6	21.16	0.58
Native American/nonfirst	45	49.6	–4.6	21.16	0.43
Hispanic/first	19	18.6	0.4	0.16	0.01
Hispanic/nonfirst	25	25.4	–0.4	0.16	0.01
Asian American/first	6	11.9	–5.9	34.81	2.93
Asian American/nonfirst	22	16.1	5.9	34.81	2.16
		$\chi^2 = 7.51$			

Chi-square = 7.51, with 4 degrees of freedom $[(2 - 1)(5 - 1) = 4]$.

We would fail to reject the null hypothesis. The probability of our obtained chi-square lies somewhere between 0.20 and 0.10, above our alpha level.

9. We would reject the null hypothesis. The chi-square obtained of 52.047 is significant at the .032 level (< .05 alpha). There is a relationship between degree and church attendance for these French respondents. Overall, as educational attainment increases, church attendance decreases.

11. Because both the independent variable (MARITAL) and dependent variable (PRES16) are nominal variables, we would use lambda to describe the association. Given that Voted for Clinton or Trump is the DV, we can see that the lambda value is .089, indicating a weak relationship between the two variables. We reduce only 8.9% of our prediction error by using MARITAL to predict PRES16.

13. Gender: The model is significant at the .01 level, indicating a significant relationship between the variables. Although males contribute to more violent onset, in proportional terms, females exhibit a higher prevalence rate—18.32% of females exhibit violent onset compared with 11.71% of males.

Age at first offense: The model is significant at the .01 level, indicating a significant relationship between age at first offense and violent onset. Violent onset is more likely among the group 14 years and older (14.74%) than those less than 14 years of age at first onset (9.67%).

CHAPTER 9

1.

$\bar{Y}_1 = 2.875$	$\bar{Y}_2 = 2.250$	$\bar{Y}_3 = 2.00$	$\bar{Y}_4 = 1.375$
$\Sigma Y_1 = 23$	$\Sigma Y_2 = 18$	$\Sigma Y_3 = 16$	$\Sigma Y_4 = 11$
$\Sigma Y_1^2 = 71$	$\Sigma Y_2^2 = 44$	$\Sigma Y_3^2 = 38$	$\Sigma Y_4^2 = 17$
$n_1 = 8$	$n_2 = 8$	$n_3 = 8$	$n_4 = 8$
$\bar{Y} = 2.125$			
$N = 32$			

$$SSB = 8(2.875 - 2.125)^2 + 8(2.250 - 2.125)^2 + 8(2.00 - 2.125)^2 + 8(1.375 - 2.125)^2$$

$$= 8(0.5625) + 8(.015625) + 8(.015625) + 8(.5625)$$

$$= 4.5 + .125 + .125 + 4.5$$

$$\mathbf{SSB = 9.25}$$

$$df_b = 4 - 1$$

$$\mathbf{df_b = 3}$$

Mean square between = 9.25/3 = 3.08

$$SSW = (71 + 44 + 38 + 17) - [(23^2/8) + (18^2/8) + (16^2/8) + (11^2/8)]$$

$$= 170 - (66.125 + 40.5 + 32 + 15.125)$$

$$= 170 - 153.75$$

$$\mathbf{SSW = 16.25}$$

$$df_w = 32 - 4$$

$$\mathbf{df_w = 28}$$

Mean square within = 16.25/28 = 0.58

$$F = 3.08/0.58$$

$$\mathbf{F = 5.31}$$

Decision: If we set alpha at 0.05, F critical would be 2.95 ($df_1 = 3$ and $df_2 = 28$). Based on our F obtained of 5.31, we would reject the null hypothesis and conclude that at least one of the means is significantly different from the others. Upper-class respondents rate their health the highest (1.375), followed by middle- and working-class respondents (2.00 and 2.25, respectively) and lower-class respondents (2.875) on a scale where 1 = *excellent*, 4 = *poor*.

3.

$\bar{Y}_1 = 1.6$	$\bar{Y}_2 = 1.4$	$\bar{Y}_3 = 0.6$
$\Sigma Y_1 = 16$	$\Sigma Y_2 = 14$	$\Sigma Y_3 = 6$
$\Sigma Y_1^2 = 30$	$\Sigma Y_2^2 = 24$	$\Sigma Y_3^2 = 8$
$n_1 = 10$	$n_2 = 10$	$n_3 = 10$

$$\bar{Y} = 1.2$$

$$N = 30$$

$$SSB = 10(1.6 - 1.2)^2 + 10(1.4 - 1.2)^2 + 10(0.6 - 1.2)^2$$

$$= 10(0.16) + 10(0.04) + 10(0.36)$$

$$= 1.6 + 0.4 + 3.6$$

$$\mathbf{SSB = 5.6}$$

$$df_b = 3 - 1$$

$$\mathbf{df_b = 2}$$

Mean square between = 5.6/2 = 2.8

$$SSW = (30 + 24 + 8) - [(16^2/10) + (14^2/10) + (6^2/10)]$$

$$= 62 - (25.6 + 19.6 + 3.6)$$

$$= 62 - 48.8$$

$$\mathbf{SSW = 13.2}$$

$$df_w = 30 - 3$$

$$\mathbf{df_w = 27}$$

Mean square within = 13.2/27 = 0.488889

$$F = 2.8/0.49$$

$$\mathbf{F = 5.71}$$

Decision: If we set alpha at 0.01, F critical would be 5.49 ($df_1 = 2$ and $df_2 = 27$). Based on our F obtained of 5.71, we would reject the null hypothesis and conclude that at least one of the means is significantly different from the others. Respondents with no degree rate their church attendance highest (1.6), followed by respondents with a secondary degree (1.4) and then respondents with a university degree (0.6).

5.

$\bar{Y}_1 = 0.8$	$\bar{Y}_3 = 1.75$	$\bar{Y}_3 = 1.75$
$\Sigma Y_1 = 4$	$\Sigma Y_2 = 7$	$\Sigma Y_3 = 16$
$\Sigma Y_1^2 = 6$	$\Sigma Y_2^2 = 15$	$\Sigma Y_3^2 = 54$
$n_1 = 5$	$n_2 = 4$	$n_3 = 5$
	$\bar{Y} = 1.93$	
	$N = 14$	

$$SSB = 5(.8 - 1.93)^2 + 4(1.75 - 1.93)^2 + 5(3.20 - 1.93)^2$$

$$= 5(1.2769) + 4(.0324) + 5(1.6129)$$

$$= 6.3845 + 0.1296 + 8.0645$$

$$SSB = 14.58$$

$$df_b = 3 - 1$$

$$df_b = 2$$

Mean square between = 14.58/2 = 7.29

$$SSW = (6 + 15 + 54) - [(4^2/5) + (7^2/4) + (16^2/5)]$$

$$= 75 - (3.2 + 12.25 + 51.2)$$

$$= 75 - 66.65$$

$$SSW = 8.35$$

$$df_w = 14 - 3$$

$$df_w = 11$$

$$\text{Mean square within} = 8.35/11 = 0.76$$

$$F = 7.29/0.76$$

$$F = 9.59$$

Decision: If we set alpha at .05, *F* critical would be 3.98 ($df_1 = 2$ and $df_2 = 11$). Based on our *F* obtained of 9.59, we would reject the null hypothesis and conclude that at least one of the means is significantly different from the others. The average number of moving violations is the highest for large-city respondents (3.2); medium-sized city residents are next (1.75), followed last by small-town respondents (0.8).

7. For each sociocultural resource, we would reject the null hypothesis. For social support, the obtained *F* ratio is 12.17, $p < .001$. Whites report the highest level of social support (2.85) while non-Cuban Hispanics have the lowest (2.58). For religious attendance, the obtained *F* ratio is 56.43, $p < .001$. Church attendance is highest for African Americans and non-Cuban Hispanics in the sample (3.94 and 3.37 on the 5-point scale).

9. Based on alpha = .01, we reject the null hypothesis of no difference. The average donation amount does vary by educational degree. The group with the highest average donation amount is graduate degree ($5,590.61) followed by bachelor degree ($3,397.40). The group with the lowest donation amount was less than high school graduates ($593.85).

11.

   a. Yes, agreement to the statement does vary by respondent's educational attainment. The ANOVA model is significant at the .001 level (< .01 alpha). The lowest mean score is for graduate degree respondents, 1.63 (between strongly agree and agree).

   b. Eta-squared is 8.474/252.312 = .03. Only 3% of the variation in ADVFRONT can be explained by educational attainment.

13.

   a. $df_b = k - 1 = 5 - 1 = 4$; $df_w = N - k = 254 - 5 = 249$

   b. We would reject the null hypothesis for the three models. Students' perception of mentoring does vary by racial/ethnic identity. The most significant model is for the statement, “There are peer mentors who can advise me.” Native American students have the highest level of agreement, followed by African American students. The lowest average score is for Asian students. The model for “I mentor other students” is significant at the .006 level. Native American students have the highest level of agreement, followed by African American students. The lowest average score is for Asian students. Finally, the model for “There are persons of color in administrative roles from whom I would seek mentoring at this institution” is significant at the .008 level. Native American students have the highest average level of agreement, followed by multiethnic students. The lowest score was reported by Hispanic students.

CHAPTER 10

1.

a. On the scatterplot below, the regression line has been plotted to make it easier to see the relationship between the two variables.

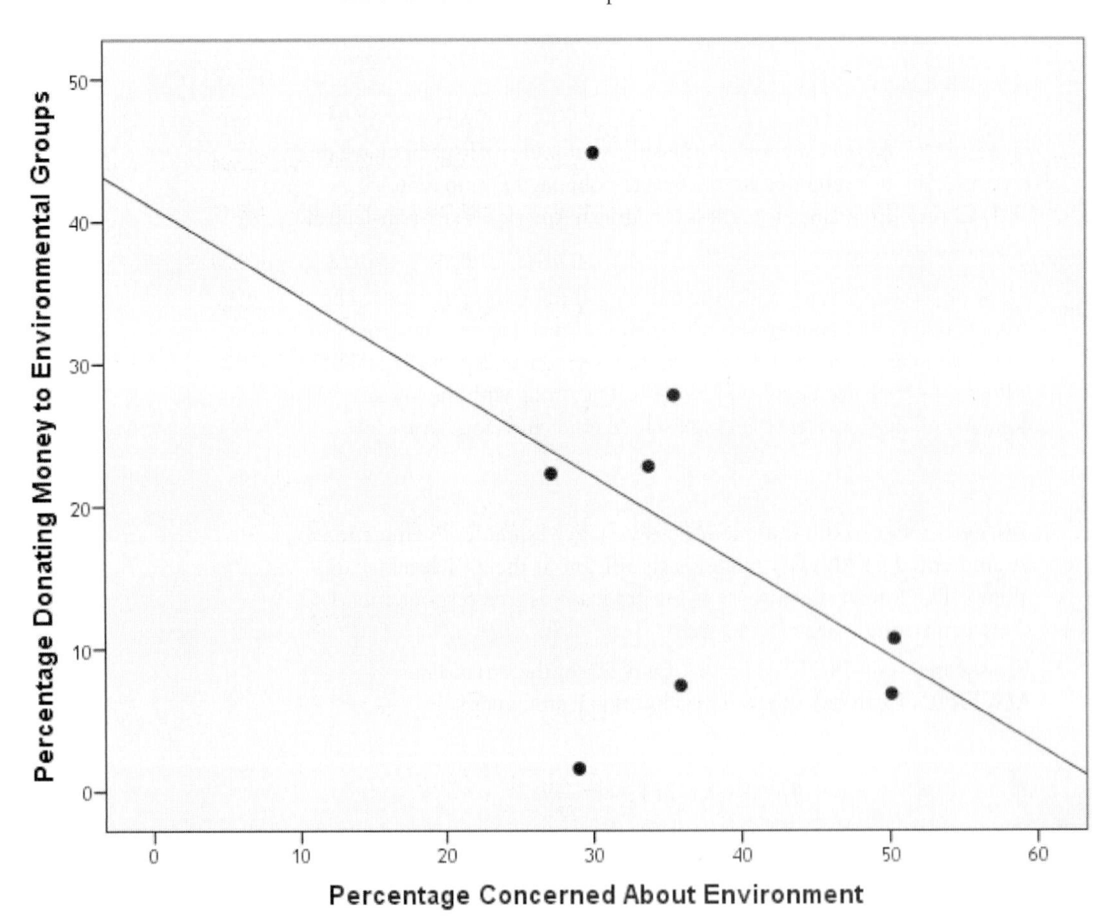

b. The scatterplot shows there is a general linear relationship between the two variables. There is not a lot of scatter about the straight line describing the relationship. As the percentage of respondents concerned about the environment increases, the percentage of respondents donating money to environmental groups decreases.

c. The Pearson correlation coefficient between the two variables is –.40. This is consistent with the scatterplot indicating a negative relationship between being concerned about the environment and actually donating money to environmental groups.

	(1)	(2)	(3)	(4)	(5)	(6)	(7)
	Percentage Concerned	Percentage Donating					
Country	X	Y	$(X-\bar{X})$	$(X-\bar{X})^2$	$(Y-\bar{Y})$	$(Y-\bar{Y})^2$	$(X-\bar{X})(Y-\bar{Y})$
United States	33.8	22.8	–2.69	7.24	4.77	22.75	–12.83
Austria	35.5	27.8	–0.99	0.98	9.77	95.45	–9.67
Netherlands	30.1	44.8	–6.39	40.83	26.77	716.63	–171.06
Slovenia	50.3	10.7	13.81	190.72	–7.33	53.73	–101.23
Russia	29.0	1.6	–7.49	56.10	–16.43	269.94	123.06
Philippines	50.1	6.8	13.61	185.23	–11.23	126.11	–152.84
Spain	35.9	7.4	–0.59	0.35	–10.63	113.00	6.27
Denmark	27.2	22.3	–9.29	86.30	4.27	18.23	–39.67
	$\Sigma X = 291.9$	$\Sigma Y = 144.2$	–0.02[a]	567.75	0.04[a]	1,415.84	–357.97

$$\text{Mean } X = \bar{X} = \frac{\Sigma X}{N} = \frac{291.9}{8} = 36.49$$

$$\text{Mean } Y = \bar{Y} = \frac{\Sigma Y}{N} = \frac{144.2}{8} = 18.03$$

$$\text{Variance}(Y) = s_Y^2 = \frac{\Sigma(Y-\bar{Y})^2}{N-1} = \frac{1{,}415.84}{7} = 202.26$$

$$\text{Standard Deviation}(Y) = s_Y = \sqrt{202.26} = 14.22$$

$$\text{Variance}(X) = s_X^2 = \frac{\Sigma(X-\bar{X})^2}{N-1} = \frac{567.75}{7} = 81.11$$

$$\text{Standard Deviation}(X) = s_X = \sqrt{81.11} = 9.01$$

	(1)	(2)	(3)	(4)	(5)	(6)	(7)
	Percentage Concerned	Percentage Donating					
Country	X	Y	$(X-\bar{X})$	$(X-\bar{X})^2$	$(Y-\bar{Y})$	$(Y-\bar{Y})^2$	$(X-\bar{X})(Y-\bar{Y})$

$$\text{Covariance}(X,Y) = s_{XY} = \frac{\Sigma(X-\bar{X})(Y-\bar{Y})}{N-1} = \frac{-357.97}{7} = -51.14$$

$$r = \frac{s_{XY}}{s_X s_Y} = \frac{-51.14}{(9.01)(14.22)} = -0.40^{a}$$

[a]Answers may differ slightly due to rounding.

3.

a. The correlation coefficient is –0.45.

	(1)	(2)	(3)	(4)	(5)	(6)	(7)
Country	GNP per Capita	Percentage Willing to Pay					
	X	Y	$(X-\bar{X})$	$(X-\bar{X})^2$	$(Y-\bar{Y})$	$(Y-\bar{Y})^2$	$(X-\bar{X})(Y-\bar{Y})$
United States	29.24	44.9	2.72	7.40	–1.64	2.69	–4.46
Ireland	18.71	53.3	–7.81	61.00	6.76	45.70	–52.80
Netherlands	24.78	61.2	–1.74	3.03	14.66	214.92	–25.51
Norway	34.31	40.7	7.79	60.68	–5.84	34.11	–45.49
Sweden	25.58	32.6	–0.94	0.88	–13.94	194.32	13.10
	$\Sigma X =$ 132.62	$\Sigma Y = 232.7$	–0.02[a]	132.99	0.04[a]	491.74	–115.16

$$\text{Mean } X = \bar{X} = \frac{\Sigma X}{N} = \frac{132.62}{5} = 26.52$$

$$\text{Mean } Y = \bar{Y} = \frac{\Sigma Y}{N} = \frac{232.7}{5} = 46.54$$

$$\text{Variance } (X) = s_X^2 = \frac{\Sigma(X-\bar{X})^2}{N-1} = \frac{132.99}{74} = 33.25$$

$$\text{Standard deviation } (X) = s_X = \sqrt{33.25} = 5.77$$

Country	(1) GNP per Capita	(2) Percentage Willing to Pay	(3)	(4)	(5)	(6)	(7)
	X	Y	$(X-\bar{X})$	$(X-\bar{X})^2$	$(Y-\bar{Y})$	$(Y-\bar{Y})^2$	$(X-\bar{X})(Y-\bar{Y})$

$$\text{Variance}(Y) = s_Y^2 = \frac{\Sigma(Y-\bar{Y})^2}{N-1} = \frac{491.74}{4} = 122.94$$

$$\text{Standard deviation}(Y) = s_Y = \sqrt{122.94} = 11.09$$

$$\text{Covariance}(X,Y) = s_{XY} = \frac{\Sigma(X-\bar{X})(Y-\bar{Y})}{N-1} = \frac{-115.16}{4} = -28.79$$

$$r = \frac{s_{XY}}{s_X s_Y} = \frac{-28.79}{(5.77)(11.09)} = -0.45^{a}$$

[a]Answers may differ slightly due to rounding.

b. A correlation coefficient of –0.45 means that relatively high values of GNP are moderately negatively associated with low values of percentage of residents willing to pay higher prices to protect the environment.

5. The analysis reveals a negative relationship between years of education and number of children. The bivariate regression equation is $\hat{Y} = 3.605 + -.126X$. For each year increase in education, the number of children is predicted to decrease by .126. The model explains just 5% of the variance; however, based on the ANOVA F obtained, we can reject the null hypothesis that $r^2 = 0$.

7.

a. The regression analysis confirms a positive relationship between years of education and total donations given in the past year. The F obtained is 10.578 (significant at .001). We can conclude that the relationship between the two variables is significant.

b. For respondent with 14 years of education: $2,043.86

For respondent with 20 years of education: $3,868.50

9.

a. For males: $\hat{Y} = 12.447 + 1.055X$

For females: $\hat{Y} = 12.423 + 1.117X$

b. For males, mother with 15 years of education: $12.447 + 1.055(15) = 28.272$

For females, mother with 15 years of education: $12.423 + 1.117(15) = 29.178$

c. The model for females has a slightly higher r^2. Mother's education explains 17.3% of the variance in female respondent education compared to the 16.3% explained for male respondent education. Based on the F-obtained statistic, both models are significant.

11.

a. Both hypotheses are confirmed.

   The slope for education is .515. Holding age constant, for each year increase in education, Internet hours per week increases by .515.

   The slope for age is –.309. Holding years of education constant, for each year increase in age, Internet hours per week decreases by .309.

b. $\hat{Y} = 21.827 + .515(X_1) + -.309(X_2)$

   $\hat{Y} = 21.827 + .515(16) + -.309(55) = 13.072$ Internet hours per week

c. $\hat{Y} = 21.827 + .082(X_1) + -.311(X_2)$

   Age has the strongest effect on Internet hours per week (beta = .311).

d. The R square is .101. Education and age explain 10.1% of the variance in predicting Internet hours per week. This is a weak prediction model.

e. The correlation between Internet hours per week and age of respondent is –.309, indicating a weak negative relationship. The correlation between Internet hours per week and education is .068, indicating a weak positive relationship. Finally, the correlation between age and education is –.088, a weak negative relationship. All three correlations are significant at the .01 level.

13.

a. $\hat{Y} = 3.888 + -.083(X_1) + -.143(X_2) + .009(X_3) + -.014(X_4)$

   (X_1 = education, X_2 = child, X_3 = age, X_4 = hours worked per week)

   Holding all the other independent variables constant:

   For each year of increase in education, television viewing should decrease by .083 hours.

   For each additional child, television viewing decreases by .143 hours.

   For each additional year of age, television viewing increases by .009 hours.

   For each additional hour of work, television viewing decreases by .014 hours.

b.

   Education, –.107

   Hours worked last week, –.096

   Number of children, –.091

   Age, .058

c. Together, these four independent variables reduce the error in predicting TVHOURS by 2.4%. This is a weak prediction model.

GLOSSARY

Alpha (α). The level of probability at which the null hypothesis is rejected. It is customary to set alpha at the .05, .01, or .001 level.

Analysis of variance (ANOVA). An inferential statistics technique designed to test for the significant relationship between two variables in two or more groups or samples.

Asymmetrical measure of association. A measure of association whose value may vary depending on which variable is considered the independent variable and which the dependent variable.

Bar graph. A graph showing the differences in frequencies or percentages among categories of a nominal or an ordinal variable. The categories are displayed as rectangles of equal width with their height proportional to the frequency or percentage of the category.

Beta (β). See standardized slope coefficient.

Between-group sum of squares (*SSB*). The sum of squared deviations between each sample mean to the overall mean score.

Bivariate analysis. A statistical method designed to detect and describe the relationship between two variables.

Bivariate regression. A regression model that examines the effect of one independent variable on the values of a dependent variable.

Bivariate table. A table that displays the distribution of one variable across the categories of another variable.

Cell. The intersection of a row and a column in a bivariate table.

Central limit theorem. If all possible random samples of size N are drawn from a population with a mean μ_Y and a standard deviation, σ_Y then as N becomes larger, the sampling distribution of sample means becomes approximately normal, with mean $\mu^{\bar{Y}}$ and standard deviation, $\sigma_{\bar{Y}} = \sigma_Y / \sqrt{N}$.

Chi-square (obtained). The test statistic that summarizes the differences between the observed (f_o) and the expected (f_e) frequencies in a bivariate table.

Chi-square test. An inferential technique designed to test for a significant relationship between two variables organized in a bivariate table.

Coefficient of determination (r^2). A PRE measure reflecting the proportional reduction of error that results from using the linear regression model. It reflects the proportion of the total variation in the dependent variable, Y, explained by the independent variable, X.

Column variable. A variable whose categories are the columns of a bivariate table.

Confidence interval (CI). A range of values defined by the confidence level within which the population parameter is estimated to fall.

Confidence level. The likelihood, expressed as a percentage or a probability, that a specified interval will contain the population parameter.

Control variable. An additional variable considered in a bivariate relationship. The variable is controlled for when we take into account its effect on the variables in the bivariate relationship.

Correlation. A measure of association used to determine the existence and strength of the relationship between interval-ratio variables.

Cramer's *V*. A chi-square-related measure of association for nominal variables. Cramer's V is based on the value of chi-square and ranges between 0 and 1.

Cross-tabulation. A technique for analyzing the relationship between two variables that have been organized in a table.

Cumulative frequency distribution. A distribution showing the frequency at or below each category (class interval or score) of the variable.

Cumulative percentage distribution. A distribution showing the percentage at or below each category (class interval or score) of the variable.

Data. Information represented by numbers, which can be the subject of statistical analysis.

Degrees of freedom (*df*). The number of scores that are free to vary in calculating a statistic.

Dependent variable. The variable to be explained (the "effect").

Descriptive statistics. Procedures that help us organize and describe data collected from either a sample or a population.

Deterministic (perfect) linear relationship. A relationship between two interval-ratio variables in which all the observations (the dots) fall along a straight line. The line provides a predicted value of Y (the vertical axis) for any value of X (the horizontal axis).

Dichotomous variable. A variable that has only two values.

Elaboration. A process designed to further explore a bivariate relationship; it involves the introduction of control variables.

Empirical research. A research based on evidence that can be verified by using our direct experience.

Estimation. A process whereby we select a random sample from a population and use a sample statistic to estimate a population parameter.

Expected frequencies (f_e). The cell frequencies that would be expected in a bivariate table if the two variables were statistically independent.

F critical. F-test statistic that corresponds to the alpha level, df_w, and df_b.

F obtained. The F-test statistic that is calculated.

F ratio or F statistic. The test statistic for ANOVA, calculated by the ratio of mean square to mean square within.

Frequency distribution. A table reporting the number of observations falling into each category of the variable.

Gamma. A symmetrical measure of association suitable for use with ordinal variables or with dichotomous nominal variables. It can vary from 0.0 to ±1.0 and provides us with an indication of the strength and direction of the association between the variables. Gamma is also referred to as Goodman and Kruskal's gamma.

Histogram. A graph showing the differences in frequencies or percentages among categories of an interval-ratio variable. The categories are displayed as contiguous bars, with width proportional to the width of the category and height proportional to the frequency or percentage of that category.

Hypothesis. A statement predicting the relationship between two or more observable attributes.

Independent variable. The variable expected to account for (the "cause" of) the dependent variable.

Inferential statistics. The logic and procedures concerned with making predictions or inferences about a population from observations and analyses of a sample.

Interquartile range (IQR). The width of the middle 50% of the distribution. It is defined as the difference between the lower and upper quartiles (Q_1 and Q_3). IQR can be calculated for interval-ratio and ordinal data.

Interval-ratio level of measurement. Measurements for all cases are expressed in the same units and equally spaced. Interval-ratio values can be rank-ordered.

Kendall's tau-*b*. A symmetrical measure of association suitable for use with ordinal variables. Unlike gamma, it accounts for pairs tied on the independent and dependent variable. It can vary from 0.0 to ±1.0. It provides an indication of the strength and direction of the association between the variables.

Lambda. An asymmetrical measure of association, lambda is suitable for use with nominal variables and may range from 0.0 to 1.0. It provides us with an indication of the strength of an association between the independent and dependent variables.

Least squares line (best-fitting line). A line where the residual sum of squares, or Σe^2, is at a minimum.

Least squares method. The technique that produces the least squares line.

Left-tailed test. A one-tailed test in which the sample outcome is hypothesized to be at the left tail of the sampling distribution.

Linear relationship. A relationship between two interval-ratio variables in which the observations displayed in a scatter diagram can be approximated with a straight line.

Line graph. A graph showing the differences in frequencies or percentages among categories of an interval-ratio variable. Points representing the frequencies of each category are placed above the midpoint of the category and are joined by a straight line.

Marginals. The row and column totals in a bivariate table.

Margin of error. The radius of a confidence interval.

Mean. A measure typically used to describe central tendency in interval-ratio variables. The arithmetic average obtained by adding up all the scores and dividing by the total number of scores.

Mean square between. Sum of squares between divided by its corresponding degrees of freedom.

Mean squares regression. An average computed by dividing the regression sum of squares (*SSR*) by its corresponding degrees of freedom.

Mean squares residual. An average computed by dividing the residual sum of squares (*SSE*) by its corresponding degrees of freedom.

Mean square within. Sum of squares within divided by its corresponding degrees of freedom.

Measure of association. A single summarizing number that reflects the strength of a relationship, indicates the usefulness of predicting the dependent variable from the independent variable, and often shows the direction of the relationship.

Measures of central tendency. Numbers that describe what is average or typical of the distribution.

Measures of variability. Numbers that describe diversity or variability in the distribution.

Median. A measure of central tendency. The score that divides the distribution into two equal parts so that half the cases are above and half below.

Mode. A measure of central tendency. The category or score with the highest frequency (or percentage) in the distribution of main points.

Multiple coefficient of determination (R^2). Measure that reflects the proportion of the total variation in the dependent variable that is explained jointly by two or more independent variables.

Multiple regression. A regression model that examines the effects of several independent variables on the values of one dependent variable.

Negatively skewed distribution. A distribution with a few extremely low values.

Negative relationship. A bivariate relationship between two variables measured at the ordinal level or higher in which the variables vary in opposite directions.

Nominal level of measurement. Numbers or other symbols are assigned to a set of categories for the purpose of naming, labeling, or classifying the observations. Nominal categories cannot be rank-ordered.

Normal distribution. A bell-shaped and symmetrical theoretical distribution with the mean, the median, and the mode all coinciding at its peak and with the frequencies gradually decreasing at both ends of the curve.

Null hypothesis (H_0). A statement of "no difference," which contradicts the research hypothesis and is always expressed in terms of population parameters.

Observed frequencies (f_o). The cell frequencies actually observed in a bivariate table.

One-tailed test. A type of hypothesis test that involves a directional hypothesis. It specifies that the values of one group are either larger or smaller than some specified population value.

One-way ANOVA. Analysis of variance application with one dependent variable and one independent variable.

Ordinal level of measurement. Numbers are assigned to rank-ordered categories ranging from low to high or high to low.

Parameter. A measure (e.g., mean or standard deviation) used to describe the population distribution.

Partial slopes. The amount of change in Y for a unit change in a specific independent variable while controlling for the other independent variable(s).

Pearson's correlation coefficient (r). The square root of r^2; it is a measure of association for interval-ratio variables, reflecting the strength and direction of the linear association between two interval-ratio variables. It can be positive or negative in sign.

Pearson's multiple correlation coefficient (R). Measure of the linear relationship between the independent variable and the combined effect of two or more independent variables.

Percentage. A relative frequency obtained by dividing the frequency in each category by the total number of cases and multiplying by 100.

Percentage distribution. A table showing the percentage of observations falling into each category of the variable.

Percentile. A score below which a specific percentage of the distribution falls.

Pie chart. A graph showing the differences in frequencies or percentages among categories of a nominal or an ordinal variable. The categories are displayed as segments of a circle whose pieces add up to 100% of the total frequencies.

Point estimate. A sample statistic used to estimate the exact value of a population parameter.

Population. The total set of individuals, objects, groups, or events in which the researcher is interested.

Positively skewed distribution. A distribution with a few extremely high values.

Positive relationship. A bivariate relationship between two variables measured at the ordinal level or higher in which the variables vary in the same direction.

Probability. A quantitative measure that a particular event will occur.

Probability sampling. A method of sampling that enables the researcher to specify for each case in the population the probability of its inclusion in the sample.

Proportion. A relative frequency obtained by dividing the frequency in each category by the total number of cases.

Proportional reduction of error (PRE). A measure that tells us how much we can improve predicting the value of a dependent variable based on information about an independent variable.

***p* value.** The probability associated with the obtained value of *Z*.

Range. A measure of variation in interval-ratio variables. It is the difference between the highest (maximum) and the lowest (minimum) scores in the distribution.

Rate. A number obtained by dividing the number of actual occurrences in a given time period by the number of possible occurrences.

Regression. A linear prediction model using one or more independent variables to predict the values of a dependent variable.

Regression sum of squares (*SSR*). Reflects the improvement in the prediction error resulting from using the linear prediction equation, *SST* (sum of squared total) – *SSE* (residual sum of squares).

Research hypothesis (H_1). A statement reflecting the substantive hypothesis. It is always expressed in terms of population parameters, but its specific form varies from test to test.

Research process. A set of activities in which social scientists engage to answer questions, examine ideas, or test theories.

Residual sum of squares (*SSE*). Sum of squared differences between observed and predicted *Y*.

Right-tailed test. A one-tailed test in which the sample outcome is hypothesized to be at the right tail of the sampling distribution.

Row variable. A variable whose categories are the rows of a bivariate table.

Sample. A subset of cases selected from a population.

Sampling. The process of identifying and selecting the subset of the population for study.

Sampling distribution. The sampling distribution is a theoretical probability distribution of all possible sample values for the statistics in which we are interested.

Sampling distribution of the difference between means. A theoretical probability distribution that would be obtained by calculating all the possible mean differences $(\bar{Y}_1 - \bar{Y}_2)$ that would be obtained by drawing all the possible independent random samples of size N_1 and N_2 from two populations where N_1 and N_2 are each greater than 50.

Sampling distribution of the mean. A theoretical probability distribution of sample means that would be obtained by drawing from the population all possible samples of the same size.

Sampling error. The discrepancy between a sample estimate of a population parameter and the real population parameter.

Scatter diagram (scatterplot). A visual method used to display a relationship between two interval-ratio variables.

Simple random sample. A sample designed in such a way as to ensure that (a) every member of the population has an equal chance of being chosen and (b) every combination of N members has an equal chance of being chosen.

Skewed distribution. A distribution with a few extreme values on one side of the distribution.

Slope (b). The amount of change in a dependent variable per unit change in an independent variable.

Spurious relationship. A relationship in which both the independent and dependent variables are influenced by a causally prior-control variable, and there is no causal link between them. The relationship between the independent and dependent variables is said to be "explained away" by the control variable.

Standard deviation. A measure of variation for interval-ratio and ordinal variables; it is equal to the square root of the variance.

Standard error of the mean. The standard deviation of the sampling distribution of the mean. It describes how much dispersion there is in the sampling distribution of the mean.

Standardized slope coefficient (or beta). The slope between the dependent variable and a specific independent variable when all scores are standardized or expressed as Z scores. Beta scores range from 0 to ± 1.0.

Standard normal distribution. A normal distribution represented in standard (Z) scores.

Standard normal table. A table showing the area (as a proportion, which can be translated into a percentage) under the standard normal curve corresponding to any Z score or its fraction.

Standard (Z) score. The number of standard deviations that a given raw score is above or below the mean.

Statistic. A specific measure used to describe the sample distribution.

Statistical hypothesis testing. A procedure that allows us to evaluate hypotheses about population parameters based on sample statistics.

Statistical independence. The absence of association between two cross-tabulated variables. The percentage distributions of the dependent variable within each category of the independent variable are identical.

Statistics. A set of procedures used by social scientists to organize, summarize, and communicate numerical information.

Symmetrical distribution. The frequencies at the right and left tails of the distribution are identical; each half of the distribution is the mirror image of the other.

Symmetrical measure of association. A measure of association whose value will be the same when either variable is considered the independent variable or the dependent variable.

t distribution. A family of curves, each determined by its degrees of freedom (df). It is used when the population standard deviation is unknown and the standard error is estimated from the sample standard deviation.

Theory. A set of assumptions and propositions used to explain, predict, and understand social phenomena.

Time-series chart. A graph displaying changes in a variable at different points in time. It shows time (measured in units such as years or months) on the horizontal axis and the frequencies (percentages or rates) of another variable on the vertical axis.

Total sum of squares (SST). The total variation in scores, calculated by adding SSB (between-group sum of squares) and SSW (within-group sum of squares).

t statistic (obtained). The test statistic computed to test the null hypothesis about a population mean when the population standard deviation is unknown and is estimated using the sample standard deviation.

Two-tailed test. A type of hypothesis test that involves a nondirectional research hypothesis. We are equally interested in whether the values are less than or greater than one another. The sample outcome may be located at both the low and high ends of the sampling distribution.

Type I error. The probability associated with rejecting a null hypothesis when it is true.

Type II error. The probability associated with failing to reject a null hypothesis when it is false.

Unit of analysis. The object of research, such as individuals, groups, organizations, or social artifacts.

Univariate frequency table. A table that displays the distribution of one variable.

Variable. A property of people or objects that takes on two or more values.

Variance. A measure of variation for interval-ratio and ordinal variables; it is the average of the squared deviations from the mean.

Within-group sum of squares (SSW). Sum of squared deviations within each group, calculated between each individual score and the sample mean.

Y-intercept (a). The point where the regression line crosses the Y-axis and where $X = 0$.

Z statistic (obtained). The test statistic computed by converting a sample statistic (such as the mean) to a Z score. The formula for obtaining Z varies from test to test.

NOTES

CHAPTER 1

1. U.S. Bureau of Labor Statistics, *Economic News Release: Usual Weekly Earnings Summary*, January 17, 2019.
2. Institute for Women's Policy Research, "The Gender Wage Gap by Occupation 2017," 2018. Retrieved from https://iwpr.org/wp-content/uploads/2018/04/C467_2018-Occupational-Wage-Gap.pdf
3. Anna Leon-Guerrero, *Social Problems: Community, Policy and Social Action* (Thousand Oaks, CA: Sage, 2015).
4. Chava Frankfort-Nachmias and David Nachmias, *Research Methods in the Social Sciences* (New York: Worth, 2000), p. 56.
5. Barbara Reskin and Irene Padavic, *Women and Men at Work* (Thousand Oaks, CA: Pine Forge Press, 2002), pp. 65, 144.
6. Frankfort-Nachmias and Nachmias, *Research Methods in the Social Sciences*, p. 50.
7. Ibid., p. 52.
8. Patricia Hill Collins, "Toward a New Vision: Race, Class and Gender as Categories of Analysis and Connection" (keynote address at Integrating Race and Gender Into the College Curriculum, a workshop sponsored by the Center for Research on Women, Memphis State University, Memphis, TN, 1989).
9. Adela García-Aracil, "Gender Earnings Gap Among Young European Higher Education Graduates," *Higher Education* 53, no. 4 (2007): 431–455.

CHAPTER 2

1. Anna Leon-Guerrero, *Social Problems: Community, Policy and Social Action* (Thousand Oaks, CA: Sage, 2018).
2. Jennifer Medina, "New Suburban Dream Born of Asia and Southern California" (*New York Times*, April 29, 2012), p. A9.
3. Gary Hytrek and Kristine Zentgraf, *America Transformed: Globalization, Inequality and Power* (New York: Oxford University Press, 2007).
4. Jynnah Radford and Abby Budiman, "Statistical Portrait of the Foreign-Born Population in the United States," 2018. Retrieved from https://www.pewhispanic.org/2018/09/14/facts-on-u-s-immigrants
5. Full consideration of the question of detecting the presence of a bivariate relationship requires the use of inferential statistics.
6. The relationship between home ownership and race noted here may not necessarily hold true in other (larger) samples.
7. Harry Moody, *Aging: Concepts and Controversies* (Thousand Oaks, CA: Sage, 2010), p. xxiii.
8. U.S. Census Bureau, *Marital Status and Living Arrangements: March 1996*, Current Population Reports, P20-496, 1998, p. 5.
9. U.S. Census Bureau, *65+ in America*, Current Population Reports, Special Studies, P23-190, 1996, pp. 2–3.
10. Edward R. Tufte, *The Visual Display of Quantitative Information* (Cheshire, CT: Graphics Press, 1983), p. 53.

CHAPTER 3

1. U.S. Census Bureau, *Census Bureau Reports at Least 350 Languages Spoken in U.S. Homes*, 2015. Retrieved from http://www.census.gov/newsroom/pressreleases/2015/cb15-185.html
2. U.S. Bureau of Labor Statistics, *Usual Weekly Earnings of Wage and Salary Workers Fourth Quarter 2018*, 2019. Retrieved from https://www.bls.gov/news.release/pdf/wkyeng.pdf
3. Federal Bureau of Investigation. *Uniform Crime Report—Hate Crime Statistics 2017*. Table 12, Agency Hate Crime Reporting by State, 2017. The states associated with the number of reported hate crimes are Texas, California, Florida, New York, Pennsylvania, Illinois, Michigan, New Jersey, and North Carolina (as listed).
4. This rule was adapted from David Knoke and George W. Bohrnstedt, *Basic Statistics* (New York: Peacock Publishers, 1991), pp. 56–57.
5. The rates presented in Table 3.4 are computed for aggregate units (states) of different sizes. The mean of 526.5 is referred

to as an unweighted mean. It is not the same as the incarceration rate for the population in the combined states.

6. Carmen DeNavas-Walt and Bernadette Proctor, *Income and Poverty in the United States: 2014*, Current Population Reports, P60-252, 2015.
7. Johnneta B. Cole, "Commonalities and Differences," in *Race, Class, and Gender*, eds. Margaret L. Andersen and Patricia Hill Collins (Belmont, CA: Wadsworth, 1998), pp. 128–129.
8. Ibid., pp. 129–130.
9. Peter Dreier, John Mollenkopf, and Todd Swanstrom, *Place Matters: Metropolitics for the Twenty-First Century*, 2nd ed. (Lawrence: University Press of Kansas, 2004).
10. Silvia Domínguez, *Getting Ahead: Social Mobility, Public Housing, and Immigrant Networks* (New York: New York University Press, 2011).
11. Gustavo Lopez, Kristen Bialik, and Jynnah Radford, *Key Findings About U.S. Immigrants*, November 30, 2018, Pew Research Center.
12. Douglas S. Massey, *Categorically Unequal: The American Stratification System* (New York: Russell Sage Foundation, 2007).
13. The extreme values at either end are referred to as outliers. SPSS will include outliers in box plots and in the calculation of the IQR; however, SPSS extends whiskers from the box edges to 1.5 times the box width (the IQR). If there are additional values beyond 1.5 times the IQR, SPSS displays the individual cases. It is important to keep this in mind when examining the shape of a distribution from a box plot.
14. $N - 1$ is used in the formula for computing variance because usually we are computing from a sample with the intention of generalizing to a larger population. $N - 1$ in the formula gives a better estimate and is also the formula used in SPSS.
15. A good discussion of the relationship between the standard deviation and the mean can be found in Stephen Gould, "The Median Isn't the Message," *Discover Magazine*, June 1985.
16. Herman J. Loether and Donald G. McTavish, *Descriptive and Inferential Statistics: An Introduction* (Boston: Allyn and Bacon, 1980), pp. 160–161.
17. Myron Pope, "Community College Mentoring Minority Student Perception," *Community College Review* 30, no. 3 (2002): 31–45.
18. Ibid., pp. 42–43.
19. Ibid., p. 43.

CHAPTER 4

1. Margot Jackson, "Cumulative Inequality in Child Health and Academic Achievement," *Journal of Health and Social Behavior* 56, no. 2 (2015): 262–280.
2. Ibid.
3. Ibid.

CHAPTER 5

1. U.S. Department of Education, National Center for Education Statistics, *Digest of Education Statistics, 2014*, 2016.
2. This discussion has benefited from a more extensive presentation on the aims of sampling in Richard Maisel and Caroline Hodges Persell, *How Sampling Works* (Thousand Oaks, CA: Pine Forge Press, 1996).
3. The discussion in these sections is based on Chava Frankfort-Nachmias and David Nachmias, *Research Methods in the Social Sciences* (New York: Worth, 2007), pp. 167–177.
4. The population of the 20 individuals presented in Table 5.3 is considered a finite population. A finite population consists of a finite (countable) number of elements (observations). Other examples of finite populations include all women in the labor force in 2008 and all public hospitals in New York City. A population is considered infinite when there is no limit to the number of elements it can include. Examples of infinite populations include all women in the labor force, in the past or the future. Most samples studied by social scientists come from finite populations. However, it is also possible to form a sample from an infinite population.
5. Here we are using an idealized example in which the sampling distribution is actually computed. However, please bear in mind that in practice, one never computes a sampling distribution because it is also infinite.
6. Sarah Begley, "Hillary Clinton Leads by 2.8 Million in Final Popular Vote Count," *Time*, December 20, 2016.
7. Andrew Mercer, Claudia Deane, and Kyley McGeeney, "Why 2016 Election Polls Missed Their Mark," *Pew Research Center*, November 9, 2016.
8. Retrieved from https://www.bbc.com/news/election-us-2016-37595321
9. Retrieved from https://abcnews.go.com/US/donald-trump-wrong-groping-comments/story?id=42660651

CHAPTER 6

1. Bradley Jones, "Majority of Americans Continue to Say Immigrants Strengthen the U.S.," Pew Research Center, January 31, 2019. Retrieved from https://www.pewresearch

.org/fact-tank/2019/01/31/majority-of-americans-continue-to-say-immigrants-strengthen-the-u-s/

2. CBS News, "Poll: Support for Stricter Gun Laws Rises, Divisions Arming Teachers," February 23, 2018. Retrieved from https://www.cbsnews.com/news/poll-support-for-stricter-gun-laws-rises-divisions-on-arming-teachers/
3. U.S. Census Bureau, "Hispanic Heritage Month 2018," September 12, 2018. Retrieved from https://www.census.gov/content/dam/Census/library/visualizations/2018/comm/hispanic-fff-2018.pdf
4. Marta Tienda, "The Ghetto Underclass: Social Science Perspectives," *Annals of the American Academy of Political and Social Science* 501 (January 1989): 105–119.
5. Adapted from Marta Tienda and Franklin D. Wilson, "Migration and the Earnings of Hispanic Men," *American Sociological Review* 57 (1992): 661–678.
6. Ibid.
7. Pew Hispanic Center, "Cubans in the United States," August 2006. Retrieved from https://www.pewresearch.org/hispanic/2006/08/25/cubans-in-the-united-states/
8. The U.S. Census 2000, IPUMS (Integrated Public Use Microdata Series).
9. The relationship between sample size and interval width when estimating means also holds true for sample proportions. When the sample size increases, the standard error of the proportion decreases, and therefore, the width of the confidence interval decreases as well.
10. Janet Fanslow and Elizabeth Robinson, "Help Seeking Behaviors and Reasons for Help Seeking Reported by a Representative Sample of Women Victims of Intimate Partner Violence in New Zealand," *Journal of Interpersonal Violence* 25, no. 5 (2010): 929–951.
11. Ibid.
12. Rachel Morgan and Jennifer Truman, *Criminal Victimization, 2017*. NCJ 252472, Bureau of Justice Statistics, December 2018.
13. Jacob Poushter, "Americans, Canadians Differ in Concern About Climate Change," Pew Research Center, March 9, 2016. Retrieved from https://www.pewresearch.org/fact-tank/2016/03/09/americans-canadians-differ-in-concern-about-climate-change/
14. Women's Bureau, U.S. Department of Labor, "Mothers and Families: Labor Force Participation Chart 1," 2016.
15. Data from Enthusiastic Voters Prefer GOP by 20 Points in 2010 Vote, Gallup Organization, April 27, 2010.
16. Jim Norman, "Millennials Like Sanders, Dislike Election Process," Gallup Organization, May 11, 2016. Retrieved from https://news.gallup.com/poll/191465/millennials-sanders-dislike-election-process.aspx
17. Aaron Smith, "Cell Phones, Social Media and Campaign 2014," Pew Research Center, November 3, 2014. Retrieved from https://www.pewinternet.org/2014/11/03/cell-phones-social-media-and-campaign-2014/
18. Data from *The Millennials: Confident. Connected. Open to Change*, Pew Research Center, February 24, 2010. *Note:* Pew operationalizes Millennials differently from the Gallup Organization as presented in Exercise 10. Retrieved from https://www.pewsocialtrends.org/2010/02/24/millennials-confident-connected-open-to-change/

CHAPTER 7

1. Jeff Somer, "Numbers That Sway Markets and Voters," 2012. Retrieved from http://www.nytimes.com/2012/03/04/your-money/rising-gasoline-pricescould-soon-have-economic-effects.html?pagewanted=all&_r=0
2. Steve Hargreaves, "Gas Prices Hit Working Class," 2007. Retrieved from http://money.cnn.com/2007/11/13/news/economy/gas_burden/index.htm
3. American Automobile Association, "Gas Prices," March 25, 2019. Retrieved from https://gasprices.aaa.com/
4. Ronald Wasserstein and Nicole Lazar, "The ASA's Statement on P-Values: Context, Process, and Purpose," *The American Statistician* 70, no. 2 (2016): 129–133.
5. Jan Hoem, "The Reporting of Statistical Significance in Scientific Journals," *Demographic Research* 18 (2008): 438.
6. To compute the sample variance for any particular sample, we must first compute the sample mean. Since the sum of the deviations about the mean must equal 0, only $N - 1$ of the deviation scores are free to vary with each variance estimate.
7. Kayla Fontenot, Jessica Semega, and Melissa Kollar, *Income and Poverty in the United States: 2017*, U.S. Census Bureau, Current Population Reports, P60-263, 2018.
8. Degrees of freedom formula based on Michael Allwood, *The Satterthwaite Formula for Degrees of Freedom in the Two-Sample t-Test* (CollegeBoard

Advanced Placement Program AP Statistics, 2008). Retrieved from https://secure-media.collegeboard.org/apc/ap05_stats_allwood_fin4prod.pdf

9. Lloyd D. Johnson, Patrick M. O'Malley, Jerald G. Bachman, and John E. Schulenberg, *Monitoring the Future National Results on Adolescent Drug Use: Overview of Key Findings, 2008* (Bethesda, MD: National Institute on Drug Abuse, 2009) and NIH News in Health, "Vaping Rises Among Teens," February 2019. Retrieved from https://newsinhealth.nih.gov/2019/02/vaping-rises-among-teens
10. Pew Research Center, *Second-Generation Americans: A Portrait of the Adult Children of Immigrants* (Washington, D.C.: Pew Research Center, February 7, 2013). http://www.pewsocialtrends.org/2013/02/07/secondgeneration-americans/
11. The sample proportions are unbiased estimates of the corresponding population proportions. Therefore, we can use the Z statistic, although our standard error is estimated from the sample proportions.
12. Robert E. Jones and Shirley A. Rainey, "Examining Linkages Between Race, Environmental Concern, Health and Justice in a Highly Polluted Community of Color," *Journal of Black Studies* 36, no. 4 (2006): 473–496.
13. Ibid.
14. Ibid., p. 485.
15. U.S. Census, "Quick Facts United States." Retrieved April 19, 2019, from https://www.census.gov/quickfacts/fact/table/US/SEX255217
16. Pew Research Center, *On Views of Race and Inequality, Blacks and Whites Are Worlds Apart*, June 27, 2016. Retrieved from https://www.pewsocialtrends.org/2016/06/27/on-views-of-race-and-inequality-blacks-and-whites-are-worlds-apart/

CHAPTER 8

1. U.S. Census Bureau, CPS Historical Time Series Tables, Table A-2, 2018.
2. Emily Forrest Cataldi, Christopher T. Bennett, and Xianglei Chen, *First-Generation Students: College Access, Persistence, and Postbachelor's Outcomes* (Washington, D.C.: National Center for Education Statistics, 2018).
3. Because statistical independence is a symmetrical property, the distribution of the independent variable within each category of the dependent variable will also be identical. That is, if first-generation status and gender were statistically independent, we would also expect to see the distribution of first-generation status identical in each gender category.
4. Although this general formula provides a framework for all PRE measures of association, only lambda is illustrated with this formula. Gamma, which is discussed in the next section, is calculated with a different formula. Both are interpreted as PRE measures.
5. Latinx is a gender-neutral term. Data come from a 2018 report from the U.S. Department of Education entitled "Profile of Undergraduate Students: 2011–12." Retrieved from https://nces.ed.gov/pubs2015/2015167.pdf
6. Bharti Varshney, Prashant Kumar, Vivek Sapre, and Sanjeev Varshney, "Demographic Profile of the Internet-Using Population of India," *Management and Labour Studies* 39, no. 4 (2014): 423–427.
7. Ibid.
8. Ibid.
9. Paul Mazerolle, Alex Piquero, and Robert Brame, "Violent Onset Offenders: Do Initial Experiences Shape Criminal Career Dimensions?" *International Criminal Justice Review* 20, no. 2 (2010): 132–146.

CHAPTER 9

1. U.S. Census Bureau, Educational Attainment in the United States: 2018, Table 1.
2. Since the N in our computational example is small ($N = 21$), the assumptions of normality and homogeneity of variance are required. We've selected a small N to demonstrate the calculations for F and have proceeded with Assumptions 3 and 4. If a researcher is not comfortable with making these assumptions for a small sample, she or he can increase the size of N. In general, the F test is known to be robust with respect to moderate violations of these assumptions. A larger N increases the F test's robustness to severe departures from the normality and homogeneity of variance assumptions.
3. Young Kim, Liz Rennick, and Marla Franco, "Latino College Students at Highly Selective Institutions: A Comparison of Their College Experiences and Outcomes to Other Racial/Ethnic Groups," *Journal of Hispanic Higher Education* 13, no. 4 (2014): 245–268.
4. Ibid., p. 252.
5. Ibid., p. 253.

6. Sandra Hofferth, "Childbearing Decision Making and Family Well-Being: A Dynamic, Sequential Model," *American Sociological Review* 48, no. 4 (1983): 533–545.

CHAPTER 10

1. Andrew Perrin and Maeve Duggan, *Americans' Internet Access: 2000–2015*, June 6, 2015. Retrieved from https://www.pewresearch.org/wp-content/uploads/sites/9/2015/06/2015-06-26_internet-usage-across-demographics-discover_FINAL.pdf
2. Ibid.
3. Refer to Paul Allison's *Multiple Regression: A Primer* (Thousand Oaks, CA: Pine Forge Press, 1999) for a complete discussion of multiple regression—statistical methods and techniques that consider the relationship between one dependent variable and one or more independent variables.
4. Michael L. Benson, John Wooldredge, Amy B. Thistlethwaite, and Greer Litton Fox, "The Correlation Between Race and Domestic Violence Is Confounded With Community Context," *Social Problems* 51, no. 3 (2004): 326–342.
5. Michael L. Benson and Greer L. Fox, *Economic Distress, Community Context and Intimate Violence: An Application and Extension of Social Disorganization Theory*, U.S. Department of Justice, Document No.: 193433, 2002.
6. Ibid.
7. Katherine Purswell, Ani Yazedjian, and Michelle Toews, "Students' Intentions and Social Support as Predictors of Self-Reported Academic Behaviors: A Comparison of First- and Continuing-Generation College Students," *Journal of College Student Retention* 10, no. 2 (2008): 191–206.
8. Ibid., pp. 199–200.
9. Ibid., p. 199.

INDEX

$$Z = \frac{p_1 - p_2}{S_{p_1 - p_2}}$$

$$s_{p_1 - p_2} = \sqrt{\frac{p_1(1-p_1)}{N_1} + \frac{p_2(1-p_2)}{N_2}}$$

CHAPTER 8 – THE CHI-SQUARE TEST AND MEASURES OF ASSOCIATION

$$f_e = \frac{\text{(Column Marginal)(Row Marginal)}}{N}$$

$$\chi^2 = \sum \frac{(f_o - f_e)^2}{f_e}$$

$$df = (r - 1)(c - 1)$$

$$\chi^2 = \sum \frac{(|f_o - f_e| - 0.5)^2}{f_e}$$

$$V = \sqrt{\frac{\chi^2}{N(m)}}$$

CHAPTER 9 – ANALYSIS OF VARIANCE

$$SSB = \sum n_k (\bar{Y}_k - \bar{Y})^2$$

$$SSW = \sum (Y_i - \bar{Y}_k)^2$$

$$SSW = \sum Y_i^2 - \sum \frac{(\sum Y_k)^2}{n_k}$$

$$SST = \sum (Y_i - \bar{Y})^2 = SSB + SSW$$

$$df_b = k - 1$$

$$df_w = N - k$$

$$\text{Mean square between} = SSB/df_b$$

$$\text{Mean square within} = SSW/df_w$$

$$F = \frac{\text{Mean square between}}{\text{Mean square within}} = \frac{SSB / df_b}{SSW / df_w}$$

$$\eta^2 = \frac{SSB}{SST}$$